Choosing Environmental Policy

Comparing Instruments and Outcomes In the United States and Europe

EDITED BY

Winston Harrington
Richard D. Morgenstern
Thomas Sterner

RESOURCES FOR THE FUTURE
Washington, DC, USA

Printed in the United States of America

An RFF Press book
Published by Resources for the Future
1616 P Street NW
Washington, DC 20036–1400
USA
www.rffpress.org

Library of Congress Cataloging-in-Publication Data

Choosing environmental policy : comparing instruments and outcomes in the United States and Europe / Winston Harrington, Richard D. Morgenstern, editors.
p. cm.
ISBN 1-891853-87-2 (hardcover : alk. paper) — ISBN 1-891853-88-0 (pbk. : alk. paper)
1. Environmental policy—United States. 2. Environmental policy—Europe. I. Harrington, Winston. II. Morgenstern, Richard D.
HC110.E5C49 2004
363.7'0561--dc22 2004014087

The paper in this book meets the guidelines for permanence and durability of the Committee on Production Guidelines for Book Longevity of the Council on Library Resources. This book was designed and typeset in Bembo and Gill Sans by Peter Lindeman. It was copyedited by Sally Atwater and Bonnie Nevel. The cover was designed by Devin Keithley.

ISBN 1-891853-87-2 (cloth) 1-891853-88-0 ISBN (paper)

About Resources for the Future *and* RFF Press

RESOURCES FOR THE FUTURE (RFF) improves environmental and natural resource policymaking worldwide through independent social science research of the highest caliber. Founded in 1952, RFF pioneered the application of economics as a tool for developing more effective policy about the use and conservation of natural resources. Its scholars continue to employ social science methods to analyze critical issues concerning pollution control, energy policy, land and water use, hazardous waste, climate change, biodiversity, and the environmental challenges of developing countries.

RFF PRESS supports the mission of RFF by publishing book-length works that present a broad range of approaches to the study of natural resources and the environment. Its authors and editors include RFF staff, researchers from the larger academic and policy communities, and journalists. Audiences for publications by RFF Press include all of the participants in the policymaking process—scholars, the media, advocacy groups, NGOs, professionals in business and government, and the public.

Dedication

This book is dedicated to the memory of Allen V. Kneese (1930–2001). Allen was our colleague at RFF and was the man who, as much as anyone else, invented environmental economics, and whose work laid the foundation for the use of economic instruments to improve environmental quality. Allen's pathbreaking work on incentives is one reason it is appropriate to dedicate this book to him. In addition, Allen's work always had a strong comparative and international element; indeed, one of his most influential early works was concerned with water pollution control in Germany.

One of Allen's greatest strengths was his ability to foster collaborative undertakings on the part of researchers scattered around the world. We'd like to think that he'd be particularly pleased by this volume.

Contents

Preface

THE ORIGINAL IDEA for this book arose from a series of discussions among the co-editors in 1999–2000 when Thomas Sterner was in residence at RFF as a Gilbert White Fellow. At that time Winston Harrington and Richard D. Morgenstern had just completed a study that compared the *actual* costs of government environmental, health, and safety regulations to the cost *estimates* prepared prior to implementation of the new rules. The goal was to test the oft-heard proposition that costs are often overestimated. Interestingly, we found this to be true as far as total costs were concerned, but we also found the effectiveness of those regulations to be overestimated almost as much.

Thus, when actual emission reductions were taken into account, the cost of the regulations in our sample showed little bias. Among the other surprising outcomes of that study, at least to us, were the following: First, very few credible estimates were available in the literature on the cost of environmental regulations based on actual data collected after implementation. Second, of those few, a relatively large number were so-called economic incentive (EI) instruments, which rely on financial penalties or rewards to encourage behavior that will improve environmental quality. Third, the unit costs of those EI instruments were vastly overestimated. We were led to a tentative conclusion that EI instruments were more effective at pollution abatement than regulators expected, and that the more traditional approach—so-called command-and-control (CAC), or direct regulation—was less so. In turn, this seemed to suggest that one of the criticisms made of EI instruments when they were first proposed decades ago—namely, that direct regulation might be more costly but would produce emission reductions more effectively and with greater certainty—might be mistaken. While this finding, if validated, might not be the staple of newspaper headlines, it could potentially influence the nature of many policies aimed at improved public health and environmental protection. Indeed, if EI instruments were shown to be both more cost-effective and more effective, the quality of life might be improved for countless numbers of people worldwide.

As economists, the three of us were cheered by the application of EI approaches to a variety of environmental problems over the past twenty years, but as empiricists we were not convinced that the new popularity of these approaches was justified by their actual performance. In addition, we wanted to examine the perception that EI instruments are used to different extents and in different ways on the two sides of the Atlantic. There is quite a literature on policy instruments, but very little of it, to our knowledge, sets out to do systematic comparisons. We felt the time had come to see what could be learned by comparing the actual performances of different policy instruments in different jurisdictions.

Our 12 case studies were selected with this objective in mind. As we got further into the comparisons, however, we found that the actual policies that were put in place were much more complicated than what the terms "economic incentive" or "direct regulation" commonly suggest. Most policies were actually mixtures, many underwent evolution, and in one case the actual policy swung from one approach to the other and then back again. The experience of comparing these approaches reminded some of us of the words of Rick Blaine, bibulous nightclub owner and hero of *Casablanca*. When Captain Renault asked him what brought him to Casablanca, he replied, "My health. I came to Casablanca for the waters."

"The waters? What waters? We're in the desert!"

"I was misinformed."

Our strategy for this project was to find case study pairs featuring the same environmental or resource problem in Europe and the United States, such that an EI instrument was used on one side of the Atlantic and a CAC instrument on the other. Were we misinformed?

Well, we knew it would be difficult to find clean comparisons. Interest among policymakers in EI instruments today is strong, but relatively recent, so few EI policies have a sufficient track record for analysis. We had to take EI instruments where we found them.

While the lack of really clean comparisons was disappointing in some respects, ultimately, we believe, it enriched the study in a number of ways. It not only led to comparisons of different types of instruments, but it invited us to speculate on whether typically "American" or "European" approaches to policymaking even exist. We had to consider why policies seemed to change so much over time. Above all, it reminded us that our contrasting policy instruments were really textbook models of policies, whereas real policies are both more complicated and more interesting.

This book would not have been possible without generous support from the Smith Richardson Foundation and the U.S. Environmental Protection Agency, as well as from the Swedish International Development Cooperation Agency (SIDA). We are particularly indebted to Mark Steinmeyer of Smith Richardson, who encouraged us to pursue this approach in its early stage of development.

Needless to say, there is no implied agreement on the part of our funders with any of our findings, nor is there any implied endorsement of any products or technologies mentioned in the text. The timely and crisp appearance of the volume is due to the unstinting efforts of John Deever, Meg Keller, and Don Reisman of RFF Press, who kept the manuscript moving through the editorial process with patience and unflappability. In its early stages, we benefited from the careful editorial work of Bonnie Newell and Sally Atwater.

We would also like to thank Alex Cristofaro, Scott Farrow, Art Fraas, Barry Korb, Albert McGartland, and Jonathan Wiener, who served as discussants of the initial drafts of the individual papers at a workshop held in Washington in December 2002. Gordon Binder and David Driessen provided useful review comments, as did several anonymous reviewers.

We were privileged to work with an outstanding group of chapter authors. In preparing and revising their manuscripts not only were they knowledgeable and cooperative but graciously timely as well. Special appreciation goes to J. Clarence (Terry) Davies who, in addition to co-authoring the last chapter, served as a critic-at-large in various phases of the project. Finally, we acknowledge the roles played by our spouses, Devra Davis, Marilyn Harrington, and Lena Sterner Persson. We thank them for their patience and understanding: even an edited volume involves far more effort and late-night revisions than originally contemplated.

Winston Harrington, Washington, DC
Richard D. Morgenstern, Washington, DC
Thomas Sterner, Gothenburg

Contributors

HANS TH.A. BRESSERS is scientific director of the Center for Clean Technology and Environmental Policy of the University of Twente (Netherlands), where he is a professor of policy studies and environmental policy. He has served on many government advisory committees on environmental policy and sustainable development. Bressers' recent publications include *Achieving Sustainable Development: The Challenge of Governance across Social Scales,* edited with Walter A. Rosenbaum, and "Understanding the Implementation of Instruments: How to Know What Works, Where, When, and How" in *Governance for Sustainable Development: The Challenge of Adapting Form to Function.*

DALLAS BURTRAW is a senior fellow at RFF whose research interests include the social costs of environmental pollution, benefit–cost analyses of environmental regulation, and the design of incentive-based environmental policies. He recently served as a member of the National Research Council Committee on Air Quality Management in the United States.

J. CLARENCE (TERRY) DAVIES is a senior fellow at RFF. While serving as a consultant to the President's Advisory Council on Executive Organization, he coauthored the reorganization plan that created the U.S. Environmental Protection Agency (EPA). Davies has held positions as assistant professor of public policy at Princeton University; executive vice president of the Conservation Foundation; assistant administrator for policy at the EPA; and executive director of the National Commission on the Environment. His books include *Pollution Control in the United States: Evaluating the System* (with Jan Mazurek).

DAVID A. EVANS is a research associate at RFF and a doctoral candidate in the Department of Economics at the University of Maryland who has studied the

U.S. electricity sector extensively. His research interests include the design and effects of pollution control instruments, with particular attention to the efficacy of emissions trading programs in the U.S.

HENRIK HAMMAR is a researcher at the department of economics, Gothenburg University. His recent research focused on policies combating climate change, with a particular focus on transportation policies on national levels. He recently served as a secretary in a Swedish governmental commission on road taxation.

JAMES K. HAMMITT, professor of economics and decision sciences at the Harvard School of Public Health, is director of the Harvard Center for Risk Analysis. Hammitt's research applies risk analysis, game theory, and mathematical modeling to health and environmental policy on topics such as global climate change, stratospheric ozone depletion, food safety, and the characterization of social preferences over health and environmental risks.

WINSTON HARRINGTON is a senior fellow at RFF whose research interests include problems of estimating the costs of environmental policy. He has worked extensively on the economics of enforcing environmental regulations, the health benefits derived from improved air quality, the costs of waterborne disease outbreaks, endangered species policy, and the economics of outdoor recreation. His books include *Economics and Episodic Disease: The Benefits of Preventing a* Giardiasis *Outbreak* (with Alan J. Krupnick and Walter O. Spofford, Jr.).

KRIS R. D. LULOFS is a senior research associate at the Center for Clean Technology and Environmental Policy of the University of Twente. A senior member of the Netherlands Institute of Government, the Dutch research school for public administration and political science, he focuses on assessment of sustainability related policy strategies. His recent publications cover assessments of voluntary approaches, negotiated agreements, and economic incentives.

MIRANDA LOH is a doctoral candidate at the Harvard School of Public Health. Her research focuses on the assessment of human exposure to volatile organic compounds in nonresidential microenvironments using both field and modeling methods. Her previous work with the World Resources Institute focused on air pollution and health policy issues.

ÅSA LÖFGREN is an assistant professor of environmental economics at Gothenburg University. Her main research focus is on environmental policies. She also works and teaches extensively on applications broadly related to the energy sector.

KATRIN MILLOCK is a researcher at EUREQua (CNRS–University of Paris I) who specializes in environmental and resource economics and agricultural policy. Her recent research projects focus on technology adoption, *ex post* evaluation of environmental taxes in France, and the economic aspects of environmental negotiated agreements for industry.

RICHARD D. MORGENSTERN is a senior fellow at RFF whose research focuses on the design of environmental policies, including economic incentive measures, in relation to regulatory issues and climate change policies. As a senior economic counselor to the undersecretary for global affairs at the U.S. Department of State, he participated in negotiations for the Kyoto Protocol. At the U.S. Environmental Protection Agency, he acted as deputy administrator and as assistant administrator for policy, planning, and evaluation; he was director of its Office of Policy Analysis. Morgenstern is editor of *Economic Analyses at EPA: Assessing Regulatory Impact.*

RICHARD G. NEWELL is a senior fellow at RFF whose research focuses on the economic analysis of technological change and incentive-based policy, with applications to climate change, energy technologies, and air pollution.

KAREN PALMER is a senior fellow at RFF whose recent research has focused on policies to promote renewable energy and multi-pollutant control in the electricity sector. Her book publications include *Alternating Currents: Electricity Markets and Public Policy* (with Timothy J. Brennan and Salvador A. Martinez) and *A Shock to the System: Restructuring America's Electricity Industry* (with Brennan et al.).

KRISTIAN ROGERS is a recent graduate of Kenyon College in Ohio. While working on this book he was an intern at RFF, and he now works as a research assistant at the Federal Reserve Board.

THOMAS STERNER, professor of environmental economics at the University of Gothenburg, Sweden, is a university fellow at RFF whose main research focus is the design of environmental policies. His most recent book, *Policy Instruments for Environmental and Natural Resource Management,* presented theory and compared application of environmental policies in industrialized and developing nations.

FRANK WÄTZOLD is a senior researcher in the Department of Economics at UFZ Centre for Environmental Research, Leipzig, Germany. His research focuses on the economic analysis of environmental policy instruments and biodiversity economics. He recently served as an adviser to the EU Commission (DG Research and DG Enterprise). His publications include a book on the implementation of environmental policy in Germany.

Overview

Comparing Instrument Choices

Winston Harrington, Richard D. Morgenstern, and Thomas Sterner[1]

ENVIRONMENTAL POLICY debates have evolved considerably over the past four decades, moving from the political fringe, through a Manichean phase pitting "the public interest" against "capitalist greed," and finally into the mainstream, where environmental policies could be considered on their merits, rather than as symbols. If in fact we are all environmentalists now, the central issues today are what works, what doesn't, and what it costs.

The chapters in this volume focus on the means of national-level pollution abatement policy: the actual performance of environmental regulations, measured after the fact; and the issue of "instrument choice"—that is, the mechanism used to achieve the environmental objective. Performance can be measured by many standards, but the most common are effectiveness (is the policy meeting its goals?) and efficiency (are the goals being achieved in a cost-effective manner?). A further element is whether the policy is consistent with broadly held values, such as equity or fairness, nonintrusiveness, and public participation. The first and still most common type of policy instrument is some form of direct regulation of the actions of firms and households. This is often referred to as command and control (CAC)—rather unfortunately, since this language not only conjures up an image of Stalinist ethics and lack of efficiency but also obfuscates some interesting distinctions within the broad category of physical regulation. Examples for pollution abatement include both emissions discharge limits that leave the firm a great deal of discretion concerning the choice of technical or other solutions, and mandates to use certain specified kinds of technology. Within the same broad category we also have regulations that are particularly designed for flexibility in the spatial or temporal dimensions: zoning and regulations that particularly apply to certain seasons or times of day, for example.

The main alternative category of instruments is often referred to as economic incentives (EI). Rather than commands or requirements, they provide penalties or rewards to encourage behavior that will improve environmental quality. Again this category is not fully watertight. Sometimes regulations include economic penalties, such as fines, or implicit rewards, such as recognition and appreciation by regulators, authorities, or consumers. Although these categories are far from perfect, we believe they do capture two strands in policymaking that can generally be separated and are sufficiently clear to be compared.

Inherently, these topics are abstract. They might not immediately excite the public, which is more interested in the direct environmental outcome and economic consequences. Yet, however abstruse and distant the subject may seem, the choice of policy instrument directly influences the effectiveness, economy, and nearly all other aspects of pollution abatement policy. The polluters will always raise the question, "How much will it cost me (or my company)?" Different instruments imply different total costs as well as different distributions of this cost among polluters and between polluters and other segments of society. These factors in turn determine how much effort the firms will put into complying with or fighting against the legislation—and thus of course the environmental and economic outcomes.

The individual chapters contain case studies that allow comparison of the actual performance of policy instruments to deal with major environmental problems. Two approaches, one from the United States and one from countries of Western Europe, are examined for each of six problems. To the extent possible the cases were chosen so that a regulatory policy on one side of the Atlantic is paired with an incentive policy on the other.[2] Another major factor in the selection of cases has been a desire to reflect on some of the most prominent environmental issues. Today global warming is a salient issue, but since we wanted to include *ex post* evaluation, we have focused on some of the top environmental priorities of the past decades—water pollution, the precursors to acid rain (sulfur and nitrogen oxides), and health hazards such as leaded gasoline and hazardous solvents.

Only recently has it been possible to find enough EI policies to carry out such a project. Until about 15 years ago the environmental policies in effect were heavily dominated by regulatory approaches. This applies particularly to the United States, where a great volume of new federal regulation to promote environmental quality was enacted during the 1970s, none of which could be characterized as economic incentives. Since then, however, there has been a remarkable surge of interest in EI approaches in environmental policy (U.S. EPA 2001). Now, whenever new environmental policies are proposed, it is almost inevitable that economic incentive instruments will be proposed and will receive a respectful hearing. Researchers and policy analysts are also looking carefully to find ways of incorporating elements of economic incentives in existing policies.

The reasons for this newfound popularity of EI policies are unclear. Perhaps it is due to the growth in awareness of economic incentive approaches among policymakers and policy analysts between 1970 and 1990. In the 1970s these approaches were quite unfamiliar to those outside the economics profession. An important factor is probably the emergence of tradable emissions permits in the late 1970s. Before then, the main EI alternative to the regulatory policies being implemented was the effluent fee. As we discuss further below, effluent fees could only encourage pollution sources to reduce pollution; they could not offer assurances that the sources would actually do so. Furthermore, effluent fees placed an additional burden on a firm: not only did it have to pay for abatement but (at least with simple fee structures) it also had to pay a form of "rent" for the unabated pollution. Effectively this placed property rights with society rather than with the polluter. It appears that this approach was relatively acceptable in some European countries, where environmental taxes have been the typical EI instrument. In the United States, however, where the concept of prior appropriation of rights to the environment by beneficial users is stronger, environmental taxes have generally been resisted very effectively. By the 1980s the policy community was generally aware of a "quantity-based" EI alternative—tradable emissions permits—that seemed to provide the same overall assurances of achieving environmental goals that were thought possible via direct regulation.

Another possible cause is the widespread disappointment with outcomes of the regulations adopted in the 1970s. The U.S. experience between 1970 and 1990 repeatedly raised questions about the presumed effectiveness of direct regulation of pollution. Even though Congress had passed air and water pollution statutes requiring stringent regulations on pollutant sources and tight timetables for implementation, it proved very difficult for the Environmental Protection Agency (EPA) to implement such programs. The regulations that were promulgated were also administratively complex and cumbersome, and the attempt to impose sometimes-rigid regulations on firms in very different situations spawned a raft of legal challenges. In other words, much of the enthusiasm for EI could be attributed to disenchantment with CAC.

Whatever the cause of this turnabout, it is clear that systematic comparison of the actual performance of regulatory versus incentive policy interventions did not play a major role. That is our goal here: the explicit comparison of direct regulation and incentive-based policies and policy outcomes in real-world applications. To provide structure for this comparison, we have compiled a list of assertions or arguments, mostly made for or against these instruments during the 1970s, when Western countries were for the first time forming comprehensive policies for controlling environmental pollution. We compare after-the-fact, *ex post* outcomes with before-the-fact, *ex ante* expectations along a number of dimensions. That is, where possible we compare the actual costs and emission reductions achieved by the particular policy with the goals and expectations of the policymakers.

In other areas of public policy—education and social welfare, for example—it is common to compare the outcomes of distinct policies enacted in different states. Unfortunately, that approach is less useful for evaluating environmental policy, since (1) the federal role is so large and (2) *ex post* analysis of environmental policies is even less common at the state level than at the federal level. An alternative is to compare environmental policies implemented in the United States with those used in other countries that face similar environmental problems. Specifically, this book seeks to assess the effectiveness and efficiency of a range of U.S. policies compared with a different set of policies addressing the same problems in Europe. The focus is on assessing *actual performance* via a series of paired case studies, rather than the more typical hypothetical comparisons.

Environmentalism has passed its first period of enthusiasm and novelty. In the future, we face some serious and expensive challenges that will require not only knowledge and dedication but also professional efficiency. Perhaps it is time to adopt the paradigm from medicine in which all practices are scientifically evaluated before universal adoption. Environmental management is at least as complex as medical science, and we believe that in the future, different policies that are intended to improve the environment through various instruments, property rights structures, or information requirements should be evaluated systematically as part of an international learning process. Naturally, policies cannot automatically be transferred from one area of application, one time period, or one country to another. However, a systematic approach to evaluating policy instruments and other institutional designs is bound to enhance the efficiency of future decisionmaking.

Method

To put our task in the proper context, we need to enlarge upon two subjects: the abundance and scarcity, respectively, of *ex ante* and *ex post* evaluations of environmental policies, and the distinction between CAC and EI instruments. The scarcity of *ex post* studies can be explained both by the difficulty and expense of conducting such studies and by the absence of incentives in the existing policy process for doing so. Without such studies, in turn, it is no wonder that there is little empirical basis on which judgments about the effectiveness and efficiency of environmental policies can be made.

Most environmental policy proposals are subjected to *ex ante* analysis and debate before they are implemented. In the United States these *ex ante* analyses typically cost about $1 million each (Morgenstern and Landy 1997). EPA's recently proposed regulations to control fine particles in the air involved 10 years of study and analysis, at a cost to the federal government in excess of $100 million. Unfortunately, there is little *ex post* analysis of environmental policies. This has hurt policymaking in a number of ways. Ineffective or inefficient policies remain in place long after they should have been reformed. Recent brownfield initiatives, for example, represent in part a (delayed) response to the overly stringent cleanup requirements of the Superfund program. Most importantly, debates over new policies are hampered by a lack of understanding about how well previous initiatives fared.

Since the baseline conditions are inherently somewhat speculative, it is not possible to know with great precision what would have happened if a different regulatory instrument had been chosen or if no regulation had been implemented at all. Thus, even when *ex post* analyses are performed, the appropriate benchmark for "success" is unobservable—and therefore often controversial. An alternative or rather supplementary way of gathering evidence on the success of environmental initiatives is the one employed in this volume. The cases were selected to illustrate the difference between regulatory and incentive approaches to pollution abatement. In other words, we use *ex post* analysis of actual policies carried out in different nations to illuminate one of the most important and contentious issues of contemporary environmental policy analysis: the practical advantages and limitations of economic incentive policies. Upon more detailed study of the various cases, it sometimes turned out that the distinction was not so clear-cut as originally anticipated. In some cases, as policies evolved, purely regulatory approaches may have added flexibility mechanisms, providing subjects of the regulation with more choices than they had originally. In others, certain flexibility mechanisms that were planned were never used. Also, particularly within Europe, policymaking is not very homogeneous across countries. Thus the approaches followed may in some cases vary considerably depending on the nations studied. In a few chapters, this is illustrated by comparisons of several European countries with the United States.

Ex ante *and* ex post *Analyses*

Whenever experts are polled about how to improve environmental management, they invariably point to the need for careful *ex post* analyses of existing programs. How, it is asked, can we design effective programs for the future if we do not have a complete understanding of what has and has not worked in the past? Are there alternative approaches for addressing the same environmental problem that have demonstrated superior performance?

Commonly used criteria for judging environmental policies are effectiveness and efficiency. A policy that is ineffective in reaching its goals is clearly deficient. A policy that meets its goals at excessive cost is wasting societal resources that might better be used for other purposes. A third criterion for judging environmental priorities is whether the policy is consistent with broadly held societal values, such as equity or fairness, unintrusiveness, and public participation.

The majority of economic analyses of environmental policies in the United States and abroad are performed in the course of developing the policies themselves. In the United States, *ex ante* analyses are required of all major federal regulations. These studies, called regulatory impact analyses (RIAs) or economic analyses (EAs), involve extensive quantitative and qualitative examination of the costs, effectiveness, and often the benefits of the regulations. In most cases the RIAs influence the design of the regulations, and they usually serve as a reference point for discussions among competing interests inside and outside government. Similar, albeit scaled-down versions of the RIAs are often performed for smaller, so-called "nonmajor" regulations. Of all the components of RIAs, the monetization of costs and benefits of regulation is by far the most difficult to estimate.

To an economist, the cost of a good or service is the maximum value of the opportunities forgone in obtaining that good or service.[3] Regulatory analyses generally account for the most obvious categories of costs, such as private sector capital and operating expenditures undertaken to comply with the regulation. Hard-to-measure costs, such as government administrative expenses, transition costs, and what are referred to in economics jargon as general equilibrium effects are rarely considered in regulatory cost estimates. For one thing, often it makes sense to calculate these costs only with respect to regulation in the aggregate, rather than for specific regulations. The cost of administration of environmental statutes is usually omitted because of a joint cost allocation problem; besides, the government's costs are thought to be small relative to those of the private sector. For individual regulations focused on a single sector—and thus not involving many spillover impacts among sectors—one can say on *a priori* grounds that general equilibrium effects are likely to be *de minimis*.[4] As for the other costs, the principal reason they are excluded is the lack of credible information or analytical resources to apply whatever data or models do exist. Thus, additional management resources or disrupted production are plausibly important elements of regulatory costs, but they are not included in most regulatory analyses.

Although no specific models have been developed to explain possible bias in *ex ante* estimates by regulatory bodies, a less formal literature addresses the procedures used in agency rulemaking and reveals problem areas that can lead to overestimates or underestimates. Some argue that inadequacies in EPA methods bias the cost estimates upward. Others believe the costs are more likely to be underestimated because of errors of omission during the rulemaking process.

In contrast to *ex ante* estimates of costs and effectiveness, *ex post* studies of the effectiveness of regulation are uncommon, and of costs, downright rare. Despite the general call for more information on program performance in the Government Performance and Results Act of 1993, rulemaking agencies have neither a legislative mandate nor a bureaucratic incentive to perform such analyses.[5] In fact, the conduct of *ex post* studies may detract from an agency's mission by using limited resources and by generating outcomes that may prove embarrassing. Although the General Accounting Office and a few other organizations in the federal government have an explicit responsibility for program evaluation, these agencies rarely undertake probing, in-depth studies. Not surprisingly, most detailed *ex post* studies have been carried out by independent researchers.

Despite the dearth of *ex post* information on program performance, two recent studies have examined the performance of environmental policies according to several criteria. In a comprehensive assessment, Davies and Mazurek (1998) considered the successes and failures of the U.S. environmental regulatory system over the past three decades. Overall, the study found the U.S. regulatory regime "excessively intrusive" and called into question both the effectiveness and the efficiency of the entire system.

A paper by Harrington et al. (2000) surveyed the literature to find two dozen U.S. and foreign regulations for which detailed *ex post* and *ex ante* comparisons had been completed. Like the Davies and Mazurek research, this study also found that regulatory performance generally did not measure up to expectations of effectiveness. Interestingly, and contrary to conventional wisdom, the handful of economic incentive regulations included in the study consistently met or bettered performance goals. In terms of costs, the authors found that estimates of costs per ton of emissions reduced (so-called unit costs) were generally in line with expectations, although total costs were often overstated, reflecting the lower-than-expected emissions reductions achieved (see box, opposite).

Command-and-Control versus Economic Incentives

Although there may be subtle distinctions within each category, there is still a general difference between regulatory and incentive instruments in the amount of discretion granted to pollution sources in determining their pollutant discharges. In a regulatory policy the discretion belongs mainly to the regulator. For

Difficulties of Making *ex ante* and *ex post* Cost Comparisons

Suppose a cost estimate for a pollution abatement regulation is to be prepared, based on an industry of 100 plants, with preregulatory emissions averaging 100 units per day. Assume that the regulation calls for emissions to be reduced to 25 units per day, at a cost of $200,000 per plant. After implementation, a survey is conducted to estimate the real cost of the regulation. To simplify the discussion, we assume the baseline is identical to the *ex ante* estimate. Some of the possible outcomes are shown in the table below.

	Ex ante *estimate*	*Alternative* ex post *outcomes*			
	(baseline)	*1*	*2*	*3*	*4*
Number of plants	100	100	150	100	100
Preregulation emissions	100	100	100	50	100
Postregulation emissions	25	25	25	25	50
Cost per plant	$200,000	$100,000	$200,000	$200,000	$200,000
Aggregate cost	$20,000,000	$10,000,000	$30,000,000	$20,000,000	$20,000,000
Emissions reductions	7,500	7,500	11,250	2,500	5,000
Cost per emissions unit	$2,666	$1,333	$2,666	$8,000	$4,000

The first of these cases is an example of misestimation of per plant costs. The next three are examples of various ways in which the "quantity" of regulatory output—that is, emission reduction—is different from the prediction.

Case 1: The cost per plant is overestimated by a factor of 2, while all other quantities are estimated correctly, so costs per emissions unit as well as costs per plant are overestimated. This is probably the situation that most observers have in mind when they assert that costs are overestimated.

Case 2: Costs are estimated correctly on a per plant basis, but an underestimate of the number of plants means that the total costs exceed the estimate. This type of uncertainty would include the case in which the total number of plants was known but the number of plants with a given characteristic or technology was not, and it might apply, for example, to landfill sites subject to corrective action requirements.

Case 3: Costs per plant are estimated accurately, but the preregulatory emissions are much less than originally thought. This could be considered a case of accurate estimation because the costs per plant are estimated accurately and the environmental goal is met. Or it could be considered underestimation because the cost-effectiveness (cost per unit emissions reduction) is underestimated.

Case 4: Costs per plant are estimated accurately, but the postregulatory emissions are not. With CAC regulation in which the emissions are set by the regulator, this will not happen. However, it could occur if the regulation is not enforced, if it calls for the installation of a specific technology rather than the achievement of an emissions target, or if performance is defined as emissions per unit of output.

Complexity increases with the heterogeneity among firms. Half the firms, for example, might abate at a cost of virtually $0 while some 10% find it so expensive that they fail to comply or are driven to bankruptcy. Or, existing firms might exit the market while others enter, making the number of firms (and thus the production levels) nonconstant.

Source: Harrington, W., R.D. Morgenstern, and P. Nelson. 2000. "On the Accuracy of Regulatory Cost Estimates," *Journal of Policy Analysis and Management*, 19(2): 297–322.

each regulated source the regulation normally specifies the limits for each pollutant and sometimes even the technology used to achieve those limits.

An economic incentive policy, in contrast, provides incentives to abate but does not specify the abatement methods or even the quantity that is permissible to discharge. The discretion belongs to the regulated source. Greater discretion gives sources more flexibility, which means, in turn, that the source can minimize the cost of compliance. We are mainly concerned with two EI instruments: effluent fees, in which a charge is levied on each unit of pollutant discharge, and tradable emissions permits, in which a fixed number of discharge permits representing an allowable amount of emissions are distributed among the pollutant sources and allowed to be traded. After Weitzman (1974), these are often called "price" and "quantity" instruments, respectively. They are the textbook examples of economic incentives, and every environmental economics textbook contains an extensive discussion of their properties. Economists are fond of these textbook instruments because they offer the prospect of harnessing the genius and power of self-regulated markets and the self-interested behavior of individuals to solve nonmarket problems. Given a set of disparate firms discharging pollutants into the environment, the textbook instruments offer a practical way to achieve a given reduction in emissions at least cost. In other words, they are cost-effective.[6]

A regulator or an environmentalist, however, might not put much value on the cost-effectiveness attribute and instead worry about a loss of control. Instead of being able to specify both the aggregate emissions level and the source-by-source distribution of those emissions, with a tradable permit scheme regulators can only specify the aggregate emissions, not their spatial distribution. With effluent fees, they cannot even specify the aggregate emissions level. Environmentalists' skepticism runs even deeper. They tend to view unbridled economic growth as the cause of environmental problems to begin with and fear that reliance on economic instruments will only make things worse. Who can believe that real-world companies will behave like the hypothetical firms in the grossly oversimplified models of economists? For example, they might argue, what is to prevent a firm subject to an effluent fee from paying the fee and then continuing to pollute? There are all manner of real-world complications—the prevalence of oligopolies or other forms of restricted competition, difficulties in monitoring of emissions, spatial variation, among others—that make the application of economics to environmental problems very difficult. Neglecting these complexities, one might say, and applying all-too-simplified textbook models may easily be hazardous.

In 1978 Steven Kelman conducted 63 interviews among environmentalists, lobbyists in trade associations, and congressional staffers to elicit their attitudes toward effluent charges. Of the 61 who had heard of effluent charges, 35 were opposed under any circumstances, 14 supported their use, and 10 advocated their use in experiments but opposed full implementation (Kelman 1981). The most popular reason for opposition involved ethics—principally the beliefs that effluent charges were a "license to pollute" and would put prices on nature and human health. Not surprisingly these ethical attitudes were cited by most of the environmentalists (74%), but they were also mentioned by 25% of the industry representatives. In addition, participants, especially environmentalists, doubted whether firms would respond to effluent fees by cutting pollution rates, or (alternatively) whether it would be politically feasible to raise the rates to the level required to elicit a response from the firms.[7] The latter has in fact turned out to be a realistic concern.

Given those doubts and the crisis atmosphere surrounding the first Earth Day in 1970, it is probably not too surprising that Congress enacted an environmental program that relied entirely on direct regulation of sources of air and water pollution. The new statutes established ambitious goals and very tight timetables for their achievement. According to the 1970 Clean Air Act, the national air quality goals were to be attained by 1977. The 1972 Water Pollution Control Act Amendments set an even more ambitious goal of zero discharge of pollution by 1985, with an "interim goal" to make the waters "fishable and swimmable" by 1983. Stringent regulations, which generally consisted of limitations specific to pollutants and types of sources, were to be written and enforced by EPA or state environmental agencies. The situation in Europe was much the same. Environmental policies began to be adopted in the Nordic countries in the early 1970s, at about the same time or slightly before the United States, and the rest of Western Europe followed shortly thereafter. The period 1985–1995 can be considered as a second wave in which the level of ambition rose strongly and more integral policies were developed.

Not surprisingly, implementation of the new environmental program in the United States began to run into difficulties almost immediately. EPA was overwhelmed. Having neither the information required to write sensible regulations nor the personnel to conduct analyses, the agency had to rely on industry for the one and outside consulting firms for the other. EPA had to follow demanding and time-consuming administrative procedures in issuing the regulations. The affected industries were not slow to challenge the new rules, both during the rulemaking process and then after promulgation, in the courts. Although the courts generally upheld EPA's authority to write industrywide regulations for air and water pollution, often they also delayed or prevented enforcement of those rules in particular situations (Melnick 1983).

These difficulties led, ironically, to the appearance of the first marketlike mechanism in U.S. environmental policy. By 1976 it was clear that the Clean Air Act's national ambient air quality standards were not going to be met in some urban locations.[8] In these "nonattainment areas," as specified in the 1970 act, no increase in emissions by stationary sources was to be permitted. Taken literally, this provision meant that no new source could locate in the area, nor could any existing source expand. In other words, no growth was permitted. With economic and political disaster impending, the notions of emissions "offsets" and "bubbles" emerged. These were devices for encouraging existing sources to reduce emissions to make room for new ones (Oates 2000).

Meanwhile, economists continued to investigate the properties of economic incentives and their possible use in a variety of environmental contexts, notwithstanding the relatively modest interest in economic incentives on the part of policymakers. They closely followed the experiments under way with bubbles and offsets (and later, banking, a mechanism that allowed emissions to be bartered across time). With the exception of the program to phase down use of lead in gasoline instituted in the mid-1980s, the performance of these instruments was widely regarded as disappointing, as few transactions were observed (Hahn and Hester 1989; Tietenberg 1990). The lack of trades has generally been attributed to the informal nature of the permits, the difficulty of quantifying emissions reductions, and the high transaction costs involved. This experience was instructive later during the design of the acid rain trading program.

The continued study of EI instruments has generated several theoretical results on the differences one might expect in the performance of EI and CAC instruments. These results were not always favorable to EI approaches—far from it; beginning with Rose-Ackerman (1973) there has been a strain of skepticism in the profession about use of EI instruments for environmental pollution. Furthermore, a survey of economists in the United States and Europe showed that support for economic incentives was strongest in the academy and got progressively weaker the closer one got to the trenches where environmental policy was actually made (Frey et al. 1985). Policymakers, businessmen, and environmentalists have also occasionally ventured opinions on the subject, in some cases to explain why EI instruments were not used.

The Hypotheses

The past three decades has seen a good deal of speculation and dispute over the differences between the two types of instruments in practice, leading to the development of a fairly long list of assertions or hypotheses about these differences. Unfortunately, if you ask 10 knowledgeable people, the 10 lists you get will differ—probably dramatically so. That's because some of the characteristics of these policies vary depending on policy details, or may be true only under some circumstances. Different observers may have different policies or circumstances in mind. On other characteristics, the advocates and skeptics tend to agree—yet disagree on the importance of the criterion. We do our best to be clear about what our assumptions are. In the last chapter we will revisit the following hypotheses in light of the case studies.

1. *Static efficiency*. Incentive instruments are more efficient than regulatory instruments.

An "efficient" economic outcome is one in which everyone in the society is as well off as he or she can possibly be, given the initial allocation of wealth. The only way to make one person better off is to make someone else worse off. Economists like markets because under the standard assumptions of perfect competition (most importantly, price-taking firms and perfect information), the market outcome is efficient. That is, on the demand side, goods are purchased by those who have the highest willingness to pay for them. On the supply side, firms produce goods at the least cost. Finally, prices adjust so that goods are produced in exactly the quantities that they are demanded.

In the real world of goods and services, of course, departures from perfect competition are not uncommon, but despite those departures, the market enjoys nearly universal support as the best way to organize an economy. This is especially true since the 1991 collapse of socialism, the market's only remaining rival.

In applying market tools to environmental problems, only one of the three mechanisms above can be brought to bear: a price can be levied on pollutant discharges that will limit the use of environmental services by the firm.[9] Unlike ordinary goods, however, there is no corresponding mechanism on the demand side. The environment is a "public good"; unlike other goods, individuals can't choose the level of environmental quality they wish to experience. Consequently, there can be no autonomous adjustment in prices to maximize everyone's satisfaction. Instead the price—implicit or explicit—of environmental quality has to be set by the regulator, and ultimately by the political process.

Just as with the market as a whole, real-world considerations affect claims of efficiency for economic incentives. For example, if there are only a few sellers of emissions permits, it is possible that a cost-effective allocation of emissions will not be found. There are also complications that don't arise in the general case of markets. If environmental impacts differ substantially across time and space, for example, then the cost-effective outcome is no longer the efficient outcome.

Finally, the superior cost-effectiveness of EIs depends on the existence of big differences in abatement cost among the polluters (Sterner 2002). In the presence of these big differences, a well-informed regulator may be able to capture considerable efficiency gains through sensible and case-sensitive regulation. If negotiations with industry include a reasonable degree of trust and openness, the regulator may be able to get individual abatement requirements roughly "right" and thus at least limit the inefficiencies inherent in a more rudimentary regulation. On the other hand, even a savvy regulator must issue regulations before the subject firms have to comply, and hence she must rely on *ex ante* estimates of regulatory cost in setting the regulation. When an economic incentive is employed, the firm makes its decisions on the basis of actual costs rather than cost estimates made prior to setting the regulation.

2. Information requirements. Generally, EI instruments require less information than CAC instruments to achieve emissions reductions cost-effectively.

We refer to the information that must be available at the time of implementation, not the information after the policy is set and the regulated parties react to it. Information requirements are crucial elements of environmental policies—elements that affect some of our other hypotheses—so it behooves us to examine them with care. Even in the simplest case of the uniformly mixed pollutant, the information required to implement effluent standards or an effluent fee depends on the goals of the regulator.

Regardless of type, policies can require more or less information depending on the policy objective. To meet a cost-effectiveness or net benefit objective with reasonable accuracy, an EI needs reasonable aggregate information on marginal damage and marginal abatement costs. A direct regulation needs the information required by the EI policy, plus the individual sources' marginal abatement costs. The marginal abatement cost is not needed to implement effluent fees because the rational polluter will abate up to the point where it equals the fee; thus an effluent fee is always cost-effective and only requires the regulator to set the fee rate.

However, if the objective of the regulator is to meet an emissions reduction target, it seems that a regulatory instrument can require less information than emissions fees. To use a fee instrument to meet a quantity objective, one needs an estimate of the marginal abatement cost function, and after implementation one must be prepared to modify the fee as actual source responses are observed. This is not an issue for tradable permits, since the quantity of permits is set, not their price.

3. Dynamic efficiency. The real advantages of EI instruments are realized only over time, because they provide a continual incentive to reduce emissions and permit a maximum of flexibility in the means of achieving those reductions, thus encouraging more efficient production and abatement technology.

The effects of direct regulation on technology are potentially complex and depend on the details. For example, new source performance standards, which are a common feature of regulatory policies in the United States, were intended to encourage dissemination of advanced abatement technology as old plants were retired and replaced by new facilities with current technology. However, the very requirement to install new technology conceivably discourages research in new abatement methods by pollutant dischargers, since discovering ways to reduce emissions can become the basis of even more stringent standards. This phenomenon has been called the regulatory ratchet. (It would not discourage innovation by the pollution abatement industry, of course.) New source performance standards have the stated objective of promoting new technology, which they may do, but they could also have the pernicious effect of postponing retirements of older, dirtier plants and increasing barriers to entry by outside firms.

Finally, there may be dynamic effects through the ways in which firms and industrial associations interact in the long run with policymakers. This will no doubt depend on how costs are shared among polluters and between polluters and society—that is, the nature of the evolution of environmental policymaking will depend on the instruments chosen.[10]

4. Effectiveness. Regulatory instruments achieve their objectives faster and with greater certainty than incentives.

As discussed above, in the early 1970s environmental regulation looked like a straightforward application of the government's police power. Disinterested experts would develop emissions standards for each point source industry based on the technology criteria established by Congress. This approach might not find the least costly abatement opportunities, but at least it would establish clear rules and identify specific ways of complying with those rules, thus expediting compliance.

Concerns about effectiveness were probably the main reason for the early reluctance to adopt EI instruments. Emissions fees bore especially heavy criticism in this respect, deriving from the uncertainty about the emissions reductions that would result from a particular fee. The earliest emissions offset policies also raised a concern—the possibility of "paper trades" or fanciful estimates of emissions credits. Several instances of bogus trades gave validity to this fear (Liroff 1986). In response, restrictions on trades were tightened, which adversely affected the efficiency of the program and led to the establishment of cap-and-trade programs of the sort we are most concerned with here.

With marketable permits, there is a cap on aggregate emissions, so presumably effectiveness is high. However, it is still likely that if we compare direct regulation with a cap-and-trade policy having the same aggregate emissions rate, we will find lower overall emissions with the regulatory policy. Under direct regulation, plants routinely overcomply with emissions permits. If these permits can be traded, then this emissions gap suddenly has value and will very likely be traded to a source that will use it (Oates et al. 1989 describe a similar mechanism for ambient standard setting).

5. Regulatee burden. Regulated firms are more likely to oppose EI regulations than CAC instruments because they fear that they will face higher costs, despite the greater efficiency of EI instruments.

The assertion of greater cost-effectiveness of EI instruments refers to *social* costs, the sum of costs to all members of society. When it comes to the *private* costs imposed on regulated firms, the burden of incentives may be greater than that of direct regulation. This depends on the type of EI. Under regulation, a polluting firm pays the cost of pollution abatement. Under an emissions fee policy, the firm pays the cost of abatement plus a fee for the remaining pollution discharged. The firm is better off only if the abatement cost is lower by an amount at least as great as the fee payments. Buchanan and Tullock (1975) point out that this could account for much of the opposition of the business community to effluent fees during the 1970s.

Under some circumstances it will be possible to use the fee revenues to overcome such opposition by revenue recycling—redistributing the fee revenues to pollution sources. To preserve the incentive effects, the redistribution has to be made on some basis other than the amount of the firms' emissions.[11] With tradable permits, firms' opposition to the costs they will incur can be overcome by distributing permits *gratis* to emitters rather than auctioning them off. However, such reimbursements subsidize the use of environmental resources in production and in the long run encourage overproduction of output, as discussed above.

6. Administrative burden. Regulatory policies have higher administrative costs.

Administrative burden is closely related to information costs and hence to policy objectives. Beyond information, administrative costs are determined by the amount of interaction between the regulator and the regulated sources. There are several reasons to think these costs might be greater under regulation than, for example, under an emissions fee. Establishing a regulation requires setting specific requirements for each regulated source, whereas with an effluent fee, only one rate (or at most a small number) needs to be set and is applicable to all sources. The multiplicity of individual standards, and the possibility of changing them, might encourage more lobbying and negotiation by affected sources as well. One important exception to this rule consists of regulations that ban certain activities altogether. In this case, the permitted rate is the same for all sources—zero—and if violations can be easily observed, a ban may be simple to administer.

After implementation, abatement policies must be monitored and enforced. We discuss monitoring separately below. As for enforcement, regulatory policies very likely have higher costs because violations quickly pass from the administrative system to the legal system. Fee collections, on the other hand, are another case of tax collections, for which the authorities usually have an administrative system established and which lead to legal difficulties only in exceptional cases. An additional advantage of fees concerns the incentives they offer to regulated sources to contest the policy. By their very nature, fee collections for increased emissions tend to rise gradually, whereas with direct regulation there is a bright line that separates compliance from violation. In principle, that means that there is a step discontinuity in the penalty function. The potentially high incremental cost at the point of violation gives sources an incentive to defend themselves legally rather than accept sanctions.

Of all the hypotheses we examine, this is one of the most informal and *ad hoc*. However, we think most observers who have spent time dealing with bureaucracies have an intuitive idea of what administrative costs are, and therefore it makes sense to ask whether some policy instruments impose more than others.

7. *Hotspots and spikes.* The performance of all pollution abatement instruments is seriously compromised for pollutants with highly differentiated spatial or temporal effects, but more so for incentive than for regulatory instruments.

As noted, one of the regulators' chief concerns with EI instruments is the limited source-specific control that can be exercised over discharges at individual facilities. In a CAC system it is easier to require more stringent emissions reductions at those plants where the emissions cause greater damage. If there were a few sources associated with high damages, a CAC instrument that targeted those sources directly would very likely be superior to an EI instrument (Rose-Ackerman 1973). Likewise, during unusual weather conditions that make ordinary emissions discharges hazardous, a short-term regulatory instrument is likely to be more effective than an economic instrument.

There have been attempts to design EI instruments to address this problem, such as spatially or temporally varying emissions fees or so-called ambient permit markets (i.e., separate permit markets for each receptor, with each source required to hold a portfolio of permits for each receptor). Some of these schemes have been analyzed by Montgomery (1972), Krupnick (1986), and McGartland and Oates (1985). Another possible remedy is a constant emissions fee, with some of the revenues used to subsidize more extensive abatement at certain sources. With the exception of the RECLAIM program for controlling NO_x emissions in Southern California, however, these proposals have not been implemented, probably because of their unwieldiness. Local congestion and transport fees in London and other major cities may prove to be interesting examples.

8. *Monitoring requirements.* The monitoring requirements of EI policies are more demanding than those of CAC policies because they require credible and quantitative emissions estimates, whereas regulatory policies at most require evidence of excess emissions or the absence of abatement technology.

Certainly, it is easier to detect compliance with a CAC standard that requires use of a designated technology than for any EI instrument. Many of the monitoring methods that have been used to determine compliance in CAC regimes cannot be used for EI regimes because they don't measure mass emissions, which are typically required for determining compliance with EI instruments. Examples include opacity tests, property line measurements, and inferences drawn from equipment malfunctions. However, with the long-term decline of the cost of monitoring in the past two decades, sources of pollution are increasingly required to have continuous emissions monitoring or frequent emissions sampling, so the significance of this issue may lessen over time.[12] One should also note that having firms pay real cash for emissions raises the profile of emissions figures, making monitoring and emissions data more visible to managers, which has both advantages and disadvantages.

9. Tax interaction effects. Adverse tax interaction effects exist for both types of instruments. but are likely to be larger with EI than with CAC policies achieving the same emissions reductions.

In recent years a good deal of attention has been paid to the efficiency of new policies in the presence of preexisting tax policies that distort the market decisions of households and firms. Studies have shown that environmental instruments can either aggravate or counteract this tax distortion, depending on the characteristics of the preexisting tax, the particular regulatory instrument, and for EI instruments, the disposition of the proceeds from the fee or the auction of permits.

Though this hypothesis is not empirically testable directly, researchers have examined these tax interactions in computable general equilibrium models and found that the importance of preexisting tax distortions depends strongly on the details of particular policies. Because an environmental policy has to have major effects in the broader economy before the tax interaction effects become noticeable, it is of special interest in the debate over global climate change policy, where the positive revenue-recycling effects of a tax (or auctioned permits) would be considerable.

10. Effects on altruism. Economic incentives encourage the notion that the environment is "just another commodity" and reduce the willingness of firms and citizens to provide environmental public goods voluntarily. Regulatory policies are consistent with a norm that requires every discharger to "do his best" and thus provide a better basis for a change in social and personal attitudes about one's responsibility to the environment (Kelman 1981).

In a CAC context, altruism is easy enough to define: voluntary limitation of emissions discharges to rates lower than the unconstrained level or than what the regulation allows. It can also be readily observed, for one can usually observe both the emissions standard and the actual emissions rate. Although there may be several other reasons why plants overcomply (such as indivisibilities in abatement equipment or concern about excess emissions during process upsets), whether this is truly "altruism" is less important than the fact that emissions are less than expected.

In an EI context, the definition of altruism is also straightforward: lower emissions than what is economically justified based on the emissions fee or the permit price. This outcome almost certainly represents voluntary emissions reductions, because in an EI regime the other justifications for emissions reductions are not present. However, it is more difficult to observe, because determining whether the emissions are "economically justified" requires the observer to know the marginal abatement costs—and marginal abatement costs are usually estimated by equating them to the observed price. In other words, in an EI regime the only way to conclude that the firm is behaving altruistically is to assert that marginal abatement costs are higher than the emissions fee or permit price, but it is not clear what the basis of such an assertion would be.

Perhaps that is another reason for skepticism about the presence of altruism in EI regimes. Presumably a firm engages in altruistic behavior to gain other, non-monetary benefits, such as a reputation for public-spiritedness. If the good behavior cannot be conclusively observed, how can the firm earn this reputation? It would be better off choosing another venue for altruism.

11. Adaptability. Compared with CAC instruments, EI instruments can be changed more quickly and easily in response to changing environmental or economic conditions.

Changing any policy to which regulated sources and others have adapted is likely to provoke considerable resistance, even if the change makes the policy less stringent, but it is likely to be particularly difficult for direct regulation. Behind this hypothesis is the observation that EI instruments are defined by a small number of parameters. To change the stringency of an emissions fee system, just raise or lower the emissions fee or the number of tradable permits. For taxes that are determined by parliament, the barriers to change may be much higher than for fees that are set administratively by the bureaucracy. With permits, one would have to be careful not to confiscate permits held by firms or destroy their value by issuing new permits.

Changing a CAC policy, on the other hand, could require readjusting many regulations because point-source standards tend to be tailored to individual sources or categories of sources. This is more difficult administratively—and probably also politically, because the multiplicity of separate regulatory actions creates plenty of opportunities to fight the change.

12. Cost revelation. With EI instruments, it is easier to observe the cost of environmental regulation.

Theory tells us that for a firm subject to an emissions fee, the marginal cost is the same as the fee rate; in a tradable permit regime, the marginal cost is the market price of the permits. Under a regulatory instrument, a firm must abate to a prescribed quantity; there are no fees or permit prices from which marginal costs can be observed.

International Trade Effects. Inasmuch as our method involves international comparisons, it is natural to consider international trade implications. Trade and the environment have become a tense political issue in both the United States and Europe because of concern over the effect of environmental regulations on industrial competitiveness. In particular, if free trade occurs among countries with different environmental standards, nations with more lax standards will tend to develop a comparative advantage in environmentally sensitive goods, which will result in "havens" for the world's dirty industries. A more recent focus has been on a corollary proposition: if free trade occurs among countries with different environmental standards, environmentally sensitive industries in nations with more stringent standards will force the weakening of standards—a race to the bottom—to ensure their survival (Anderson and Blackhurst 1992; Bhagwati and Hudec 1996).

The race to the bottom concerns the stringency of regulations, but might there not also be concerns about the instrument used as well? Indeed, several of the hypotheses above have implications for international trade, including the following:

(a) The complexity and often opaqueness of construction and operating permit requirements, which typically accompany direct regulations, tend to favor domestic industry over foreign-owned firms.

(b) By favoring end-of-pipe treatment and more stringent requirements for new plants, regulatory instruments provide innovation incentives for domestic abatement technology producers, giving them an advantage in world markets.

(c) By imposing greater regulatory burdens on regulated firms, incentives leave those plants more vulnerable to import competition.

A Concluding Comment on the Hypotheses

It should be evident that there can be many variations on the basic design of both regulatory and incentive instruments. In particular, policy details can be altered to accommodate many of the criticisms of these instruments that are implicit in these hypotheses. Often, however, improvements in some areas cause or add to problems in others. For example, regulatory policies can individualize abatement requirements for each source. This can greatly increase cost-effectiveness of the regulations or reduce opposition to the regulations (or both), but it also greatly increases the information required. Similarly, the revenue-generating potential of an EI policy could remedy identified defects or obstacles—if, say, fees were returned to firms to reduce impacts or used to correct a hotspot by subsidizing additional abatement at certain sources—but that would sacrifice economic efficiency. In the case studies, it will be evident that policies often contain features that attempt to deal with some of the supposed defects identified by their critics.

The Cases

Below we describe briefly the policies examined.

SO_2 emissions from utility boilers: permit market (United States) versus sulfur emissions standards (Germany).

Sulfur dioxide (SO_2) is a by-product of burning fossil fuels (primarily coal) containing small amounts of sulfur. In the atmosphere, SO_2 is converted into acid rain, which threatens forests and lakes in the northeastern United States and in Europe, and into sulfate particles, which have been implicated in elevated mortality in U.S. cities. The U.S. SO_2 trading program was set up to achieve a 10-million-ton aggregate reduction in SO_2 emissions by 2010—approximately a 50 percent reduction. To achieve this, EPA in 1994 established a market in emissions allowances for electric utilities. By 2000, actual emissions reductions exceeded the scheduled reductions, at considerably lower cost than predicted. The German case applies to large combustion boilers and dates to the early 1980s. In 1983, in response to accumulating evidence of forest dieback attributable to acid deposition, the former West German government imposed strict emissions limitations on large sources of SO_2, mostly utility boilers and district heating plants. Between 1983 and 1993, annual emissions of SO_2 declined from 2.9 million to 0.7 million tons.

Industrial water pollution: effluent fees (the Netherlands) versus effluent guidelines and permits (United States).

The concern here is with all water pollution generated by industrial point sources.[13] In 1970, the Netherlands implemented an effluent charge for industrial wastewater containing organic pollutants. By 1990 effluent discharges had dropped by almost 75% even though industrial production increased over the period. In the United States, the federal Water Pollution Control Act Amendments of 1972 required every industrial facility discharging wastewater into the nation's waterways to have a permit specifying the amounts of various pollutants that it could discharge. To assist the states and EPA regional offices in the preparation of these permits, EPA was directed to prepare a very detailed set of technology-based pollutant discharge standards in about 60 major polluting industries. Separate guidelines were defined for new versus existing plants and direct dischargers versus dischargers into publicly owned treatment facilities. Further, the standards became more stringent over time. To get meaningful standards (and probably for reasons of equity), the 60 industries were often subcategorized by process, location or product. In all, hundreds of specific regulations were written.

NO_x emissions from utility boilers: emissions taxes (Sweden and France) versus performance standards and pollution trading (United States).

Oxides of nitrogen (NO_x) are both a component of acid rain and main constituents in the complex series of reactions that produce photochemical smog. Our U.S. case focuses on NO_x control programs that affect coal-fired utility boilers. Among the regulatory programs requiring NO_x emissions reductions from coal boilers are the new source performance standards and the acid rain program of the Clean Air Act amendments of 1970 and 1990, respectively. There is also a discussion of emissions trading programs employed in the Northeast. The European case provides an additional comparison between NO_x taxes on utility boilers in Sweden and France. In 1992 the Swedish government imposed a revenue-neutral fee on emissions of NO_x from utility and industrial boilers. Taxes are paid into a fund according to the quantity of emissions from each plant and are refunded to sources according to their electricity outputs. In France a NO_x emissions tax was imposed with no attempt at revenue neutrality and at a much lower rate than in Sweden.

Chlorofluorocarbons (CFCs) permit market (United States) versus mandatory phaseouts (other industrial countries).

In 1974, it was discovered that CFCs were destroying the stratospheric ozone layer, which prevents excessive ultraviolet radiation from the sun from reaching the earth's surface. Disappearance of stratospheric ozone would have far-reaching and possibly catastrophic consequences for human health, many natural ecosystems, and global climate. The United States used a tradable permit policy to phase out production of CFCs; European nations used specific regulations to phase out both production and use.

Leaded gasoline: marketable permits for leaded fuel production (United States) versus mandatory lead phaseouts plus differential taxes to prevent misfueling (most European countries).

The initial impetus to remove lead from gasoline came from auto emissions standards requiring the use of catalytic converters. Lead additives would quickly poison the catalyst and render the converters ineffective. At the same time, strong evidence emerged that leaded gasoline was the primary source of the alarmingly high levels of lead in the blood of children. The United States acted almost a decade ahead of Europe. Beginning in 1974 and coincident with the introduction of catalyst-equipped vehicles, manufacturers were required to offer unleaded gasoline for sale. The U.S. policy began as a regulation and switched to economic incentives in 1981. In 1995, leaded gasoline was banned for all vehicles. Most European countries relied on what we would call a "sunset" policy, a date after which sale of leaded fuel would be illegal, coupled with a differential tax to ensure that the price of leaded fuel exceeded that of unleaded, to discourage misfueling.

Chlorinated solvents: source regulation (United States) versus three distinct policy approaches in Europe.

Chlorinated solvents threaten the environment in a variety of ways. Most of these compounds are toxic and strongly suspected of being carcinogenic. Others threaten the stratospheric ozone layer. The principal reason for regulation was to reduce local ambient and worker exposures. Here we offer comparisons both between the United States and Europe and within Europe, since the Swedish, German, and Norwegian policies varied considerably. Sweden imposed a ban on a narrow set of compounds, notably trichloroethylene (TCE); Germany created a system of detailed technology-based regulations; and Norway imposed a tax on a variety of chlorinated solvents. In the United States, solvents are regulated under several statutes with a strong but not exclusive reliance on regulatory instruments. Some incentive-based policies are also included in the mix.

A Note on the Process

Although each of the cases was prepared independently, the editors established a common framework for the individual analyses. The papers were commissioned in spring 2002 and then presented in draft form at an authors' workshop held at Resources for the Future in December 2002. Current or former U.S. government experts were enlisted as reviewers of the individual papers. Their comments, along with those of other authors offered at the workshop and individually, were incorporated into the revised papers, which appear as chapters in this volume. Each of the case studies can and should be read as a stand-alone assessment of the performance of a particular regulation, but the selection of cases—similar environmental problems addressed via different policy approaches in different nations—presents an opportunity to examine the broader implications of the full set of cases. The last chapter draws some cross-regulatory, cross-instrument, and other lessons from the case studies.

Notes

1. The authors would like to thank J. Clarence Davies, David Evans, and David Driessen for helpful comments on an earlier draft.

2. As illustrated by the case studies, it does not seem that Europe is more in favor of EI, as defined here, than the United States, or vice versa. Nor do we believe that either continent can be considered more ambitious when it comes to environmental policy in general, though each is more ambitious in some areas. Wiener (2003) comes to similar conclusions about how the United States and Europe use the precautionary principle in environmental regulations concerning risk.

3. More precisely, the cost of a regulation is equal to "the change in consumer and producer surpluses associated with the regulation and with any price and/or income changes that may result"(Cropper and Oates 1992, 721).

4. A recent paper by Garber and Hammitt (1998) suggests that the public valuation of a firm's stock may be reduced by the uncertainty associated with pollution liability under, for example, Superfund.

5. Over the past decade, Congress has shown greater interest in *ex post* information. In addition to the Government Performance and Results Act, the Clean Air Act Amendments of 1990 required EPA to develop a retrospective assessment of the overall benefits and costs of the first 20 years of the act. The Small Business Regulatory Enforcement Fairness Act of 1996 also contains requirements for retrospective studies.

6. Cost-effective emissions reduction ordinarily requires that the marginal cost of emissions reduction—i.e., the cost of the last unit of pollution reduction—be the same for all sources. If that were not true, then it would be possible to achieve the same reductions at lower costs by increasing emissions reductions at the plant with the lower marginal costs, and reducing them by the same amount at the plant with the higher marginal costs.

7. At the time of this survey, the concept of tradable emissions permits was familiar to only a few environmental economists. If the economists' work had been more widely known, perhaps the answers would have been different, at least with respect to the effectiveness of permits in reducing emissions. However, it is unlikely that the ethical concerns would have been allayed. Former Senator Eugene McCarthy, for example, once compared the trading of emission permits to the medieval practice of selling papal indulgences (McCarthy 2000).

8. It was primarily the standards for ozone and particulates that were not met (in many large urban areas). For other standards (lead, NO_X and others) attainment was nearly universal.

9. Both a price instrument (an emissions fee) and a quantity instrument (a system of marketable permits) put a price on environmental services.

10. Furthermore, the dynamics of how innovations are made and spread may also be affected.

11. See, for instance, Chapter 5 on the policy for NO_X emissions in Sweden.

12. Note that this need not apply to all industries, since there may be an opposite trend toward increasing product and process complexity.

13. Excluded are pollution that is generated by households and discharged into municipal wastewater treatment systems, and nonpoint source pollution, which results primarily from agricultural runoff.

References

Anderson, K., and R. Blackhurst. 1992. *The Greening of World Trade Issues.* Ann Arbor: University of Michigan Press.

Avery, D.T. 1999. We Are All Environmentalists Now. *American Outlook* Summer: 35–37.

Beatty, J. 2000. We Are All Environmentalists Now. Right? *Atlantic Unbound* Roundtable Discussion, September.

Bhagwati, J., and R. Hudec. 1996. *Fair Trade and Harmonization: Prerequisites for Free Trade?* Cambridge, MA, and London: MIT Press.

Buchanan, J., and G. Tullock. 1975. Polluters' Profits and Political Response: Direct Controls versus Taxes. *American Economic Review* 65(1).

Cropper, M., and W. E. Oates. 1992. Environmental Economics: A Survey. *Journal of Economic Literature* 30(2):675–740.

Davies, J.C., and J. Mazurek. 1998. *Pollution Control in the United States.* Washington, DC: Resources for the Future.

Frey, B., F. Schneider, and W.W. Pommerehne. 1985. Economists' Opinions on Environmental Policy Instruments: Analysis of a Survey. *Journal of Environmental Economics and Management* 12(1).

Garber, S., and J. K. Hammitt. 1998. Risk Premiums for Environmental Liability: Does Superfund Increase the Cost of Capital? *Journal of Environmental Economics and Management* 36(3):267–94.

Hahn, R., and G. Hester. 1989. Where Did All the Markets Go? An Analysis of EPA's Emissions Trading Program. *Yale Journal on Regulation* 6(1): 109–53.

Harrington, W., R.D. Morgenstern, and P. Nelson. 2000. On the Accuracy of Regulatory Cost Estimates. *Journal of Policy Analysis and Management* 19(2): 297–322.

Kellogg, M. 1994. After Environmentalism. *Regulation* 17(1).

Kelman, S. 1981. *What Price Incentives? Economists and the Environment.* Dover, MA: Auburn House.

Kingsnorth, P. 2000. Drowning in a Green-Wide Sea. *The Ecologist* (July).

Krupnick, J. 1986. Cost of Alternative Policies for the Control of Nitrogen Dioxide in Baltimore. *Journal of Environmental Economics and Management* 13(2): 189–97.

Liroff, R.A. 1986. *Reforming Air Pollution Regulation: The Toil and Trouble of EPA's Bubble.* Washington, DC: Conservation Foundation.

McCarthy, E. 2000. How Do You Keep Them Busy Once They've Left the White House? *Progressive Populist* 6(12):www.populist.com/00.12.html (accessed July 12, 2004).

McGartland, A., and W.E. Oates. 1985. Marketable Permits for the Prevention of Environmental Deterioration. *Journal of Environmental Economics and Management* 12: 207–28.

Melnick, R.S. 1983. *Regulation and the Courts: The Case of the Clean Air Act.* Washington, DC: Brookings Institution.

Montgomery, W.D. 1972. Markets in Licenses and Efficient Pollution Control Programs. *Journal of Economic Theory* 5: 395–418.

Morgenstern, R.D., and M.K. Landy. 1997. Economic Analysis: Benefits, Costs, Implications. In *Economic Analyses at EPA: Assessing Regulatory Impact,* edited by R.D. Morgenstern. Washington, DC: Resources for the Future.

Oates, W.E. 2000. From Research to Policy: The Case of Environmental Economics. *University of Illinois Law Review* 2000(1): 135–54.

Oates, W.E., P.R. Portney, and A.M. McGartland. 1989. The Net Benefits of Incentive-Based Regulation: A Case Study of Environmental Standard-Setting. *American Economic Review* 79: 1233–42.

Rose-Ackerman, S. 1973. Effluent Charges: A Critique. *Canadian Journal of Economics* 6(4).

Sterner, T. 2002. *Policy Instruments for Environmental and Natural Resource Management.* Washington, DC: Resources for the Future.

Tietenberg, T.H. 1990. Economic Instruments for Environmental Regulation. *Oxford Review of Economic Policy* 6: 17–33.

U.S. Environmental Protection Agency (EPA). 2001. The United States Experience with Economic Incentives for Protecting the Environment. Washington, DC.

Weitzman, M. 1974. Prices versus Quantities. *Review of Economic Studies* 41(4):477–491.

Wiener, J.B. 2003. Whose Precaution After All? *Duke Journal of Comparative and International Law* 13:207–62.

CHAPTER 1

SO_2 *Emissions in Germany*

Regulations to Fight *Waldsterben*

Frank Wätzold

ONE OF THE MOST SERIOUS environmental problems in recent German history has been the decline of forest vegetation caused by air pollution, a process known as *Waldsterben*, "forest death." Arousing great public attention, it coincided with (and may even have partly caused) the emergence of the Green Party in Germany. *Waldsterben* created enormous pressure on politicians and industry to reduce the emissions believed responsible for this environmental damage—sulfur dioxide (SO_2). Because large combustion plants in the electricity sector were by far the largest source of SO_2 emissions, it was obvious that these emissions had to be reduced significantly if the environmental situation was to be alleviated. Consequently, stringent regulations entitled the Großfeuerungsanlagen-Verordnung (GFA-VO, Ordinance on Large Combustion Plants) were compiled and took effect on July 1, 1983.

Following the enactment of GFA-VO, the electricity sector embarked upon a tremendous (and expensive) reduction program that led to a sharp decline in SO_2 emissions. Electricity generators in the biggest German federal state, North Rhine–Westphalia, voluntarily agreed to cut emissions of SO_2 and nitrogen oxides (NO_x) even faster than required. Though highly ambitious, the reductions envisaged by GFA-VO and the voluntary agreement were actually exceeded.

Many economists seem to have a common opinion about the properties of command-and-control instruments like the GFA-VO: In a nutshell, they believe that such instruments are usually reasonably good at meeting the desired emissions reductions but poor at achieving the least-cost allocation of abatement activities (static efficiency) and stimulating environmentally friendly technological progress (dynamic efficiency). Against this assessment of command and control instruments the obvious success of GFA-VO in reducing emissions—but also the high costs necessary to achieve those reductions—prompt a number of questions:

1. What were the reasons for the tremendous reduction in SO_2 emissions, and how important was it that the German SO_2 policy relied on a command-and-control approach?
2. Were the high costs of GFA-VO attributable to the failure to achieve the efficient allocation of abatement efforts?
3. Or were the high costs due to negative effects of the command-and-control policy and its effect on technological progress?

The aim of this case study is to answer these questions—to explain why GFA-VO was successful in emission reductions and to evaluate both its static and its dynamic efficiency properties. The focus of the analysis is on SO_2 emissions from existing large combustion plants (LCPs) and on the electricity generating industry, which accounts for the majority of LCPs.[1]

The case study is structured as follows. First, some background information about *Waldsterben* and the electricity sector is provided. The next sections describe the political evolution of and substance of GFA-VO and the voluntary agreement in North Rhine–Westphalia. Then, GFA-VO and the voluntary agreement in terms of emission reductions and static as well as dynamic efficiency are evaluated. The final section discusses the German SO_2 policy against the background of what economists usually consider the properties of command-and-control policies.

The Environmental Problem

Waldsterben

Before the 1970s, the general concern about SO_2 emissions was their effect on the environment close to the emissions location. The obvious solution was to construct tall chimney stacks that distributed the emissions over a larger area. In the 1970s it became evident that such a policy led to the deterioration of air quality and vegetation in formerly pollution-free zones. The first signs of *Waldsterben* appeared, and at the end of the 1970s and early 1980s it spread rapidly. By 1984, 50% of the German forests were affected, with 33% considered slightly damaged and 17% severely damaged (UBA 1994). The signs of *Waldsterben* were visible not only to experts but also to laypersons in many mountainous regions in Germany.

German scientists believed that high SO_2 emissions were one of the main reasons for *Waldsterben*.[2] Table 1-1 categorizes SO_2 emissions in West Germany in 1980 by source.

Contributing close to 60% of all SO_2 emissions in 1980, the electricity sector was by far the largest emitter. It was clear that substantial emissions reductions could be achieved only if the sector made significant abatement efforts. To understand the reaction of the affected industry on the demands for cutting SO_2 emissions, a closer look at this sector is needed.

Table 1-1. *Sources of SO_2 Emissions in West Germany, 1980*

	kt	Share (%)
Industrial processes	110	3.5
Road traffic	67	2.1
Other transport	20	0.6
Households	196	6.2
Small consumers	142	4.5
Industrial combustion	750	23.7
Electricity supply (including district heating)	1,879	59.4
Total	3,164	100.0

Source: UBA 1997, *135–39*. Kt = kilotons.

German Electricity Sector

Germany was and still is one of the largest electricity producers (and consumers) in the European Union. In 1986 West Germany generated 408.3 billion kWh (and consumed 386.0 billion kWh) (Statistisches Bundesamt 1991). The German electricity sector is fragmented and has a rather complex structure. Eight large companies own and operate the national high-voltage grid and the majority of generating capacity. But there are also nearly 1,000 regional and local companies that primarily distribute electricity. Although the public sector owns or holds majority shares in many of the regional and local companies, the eight large companies are dominated by private shareholders.

Prior to the recent liberalization of the European electricity markets, German electricity suppliers enjoyed regional monopolies. The sector was exempt from competition and antitrust laws. Electricity prices were fixed by electricity suppliers but had to be approved by public authorities. To raise prices, suppliers had to demonstrate a corresponding rise in production costs. The electricity suppliers began to lose this comfortable position with the 1997 European regulation on a single electricity market.

In 1997, the power stations produced 486,768 GWh (up from 355,048 GWh in 1987), which was 88.6% (84.9%) of the total electricity generated in Germany. In 1995, 79% of the electricity generated by the roughly 1,000 suppliers was produced by the largest companies, which then numbered nine.[3] The approximately 80 regional suppliers provided about 10%, and the remaining 11% came from the small local utilities. For electricity distribution, however, the regional and local companies have shares of 36% and 31%, respectively. Only 33% of the electricity sold to households, companies, and public institutions came from the nine large suppliers, which often provide other suppliers with electricity instead of selling it directly to consumers.

Coal and uranium have been the main energy sources in Germany. In 1987, the sector produced 20.7% of its electricity from lignite, 29.5% from hard coal, and 36.5% from nuclear energy; 5.5% was produced from gas and 2.1% from oil. Hydroelectric power stations accounted for 5.1%. Other sources (including renewable energy sources, such as waste, wind, and solar energy) contributed less than 1% to electricity generation. Since 1987, the contributions of the energy sources have not changed significantly (Bültmann and Wätzold 2000).

Germany has large coal reserves. Because German hard coal is much more expensive than coal available on the world market, the German government has traditionally intervened to ensure that indigenous hard coal is used for electricity generation (Ikwue and Skea 1996). Between 1964 and the early 1970s, electricity suppliers were encouraged through tax benefits and subsidies to build power stations that burned hard coal. As of 1974, the use of hard coal was supported by a long-term contract between electricity suppliers and the coal industry stipulating that the electricity sector would buy 33 million to 47.5 million tons of German hard coal each year and pay a price sufficient to cover the costs of the mining companies. Electricity suppliers that ran hard coal generators got subsidies that were financed by a levy on electricity prices *(Kohlepfennig)*. The contract expired in 1995, and the amount of hard coal the electricity sector buys and the price it pays are no longer regulated. Now the mining companies sell hard coal at world market prices but are compensated for the difference between the price they get and their production costs. The subsidies come from the federal budget and are paid for only a limited amount of coal.

National Political Response to *Waldsterben*: Strict Legislation

Rising concern about *Waldsterben* led to legislation that sought a fast and drastic reduction of SO_2 emissions from the German electricity sector. The next section describes the policy process that preceded enactment of GFA-VO as well as the lead actors. The main content of the GFA-VO is then summarized.

Political Evolution of GFA-VO[4]

Leading Actors and Their Motives. The Bundesministerium des Inneren (BMI, Ministry of the Interior) was in charge of pollution control at the end of the 1970s. It reacted to the problem of *Waldsterben* and the public discussion about it by resolving to significantly reduce SO_2 (and NO_x) emissions. In pressing for tighter emissions limits, BMI understood that private and industrial electricity consumers would pay the pollution abatement equipment in the end, through higher electricity prices. Despite its insistence on strict emissions standards, BMI consulted with industry on feasible technological options to reduce SO_2 emissions. Unlike electricity suppliers, BMI was convinced that reliable desulfurization techniques existed that could enable an emissions limit of 400 mg SO_2/m^3.

The industry groups involved in the discussion about GFA-VO were the electricity sector, the coal industry, and industrial associations such as the Bundesverband der deutschen Industrie. The electricity industry generally welcomed definite regulations on SO_2 (and NO_x) emissions at the national level. The legislation on which air pollution control policy was based at that time (TA Luft, Technical Instructions on Air Quality Control) demanded only that

SO_2 emissions be reduced "as much as possible." This led to varying regulations in the German states. Nevertheless, the electricity suppliers were opposed to GFA-VO because they considered the emissions limits and deadlines too strict. They stressed that the paucity of experience with denitrification and desulfurization in Germany meant they would have to rely on Japanese experience and therefore could not guarantee compliance with the strict emissions limit (400 mg/m^3 SO_2). Additionally, electricity suppliers said the deadlines were too short to allow testing of desulfurization techniques in pilot plants first, and any shortcomings and optimization methods could not be discovered before the techniques had to be applied to the entire fleet of power stations (Bertram and Karger 1988). The high costs incurred in installing desulfurization techniques were less important for the electricity suppliers because they held regional monopolies and thus could rather easily transfer the costs to their customers via higher electricity prices. It was precisely this mechanism that led industry, especially energy-intensive sectors, to resist GFA-VO. Industry was afraid its international competitiveness would be jeopardized if only German companies had to pay higher electricity prices. Moreover, the coal industry was concerned that strict emissions limits for SO_2 would force electricity suppliers to stop using high-sulfur German coal.

In the governmental discussions that preceded adoption of the ordinance, the Bundeswirtschaftsministerium (Ministry of Economics) was a strong supporter of business interests. The agency especially tried to prevent the strict SO_2 emissions limit.

The suppliers of abatement equipment were asked to provide information about the technologically feasible emissions limits. The suppliers had an economic incentive to offer equipment that could reduce emissions limits as low as possible: the lower the emissions limits, the more complex and expensive the equipment. But the suppliers also had to guarantee that LCP operators could meet the low limits. To be on the safe side, most incorporated a safety margin in the emissions limits they gave policymakers and guaranteed to LCP operators: the limits they said were achievable were slightly higher than the actual limits.

The decision about GFA-VO was influenced by leading politicians, especially at the level of the federal government. Until October 1982, the federal government was a coalition between the Social Democrats (SPD) and the Liberals (FDP). It was succeeded by a coalition between the Liberals and the Conservatives (CDU/CSU), which also won the March 1983 general election. The growing concern of the population about *Waldsterben* put enormous pressure on politicians, not least because it was an issue in the 1983 election. Furthermore, the Green Party scored its first electoral successes at this time. It was especially important for the Social Democrats to stop the rise of the Greens, since many of their voters sympathized with environmental ideas.

Policy Process. At the end of 1977, growing concern about atmospheric SO_2 led BMI to start working out a concept for an ordinance dealing with emissions from power stations. In May 1978 a preliminary draft was presented and discussed with representatives of industry. BMI suggested an emissions limit of

400mg/m^3 SO_2, which was strongly opposed by representatives of the electricity and coal sectors. Although this limit was technologically feasible, it was feared that it could not be met with high-sulfur German coal (used in 25% of the German coal combustion plants) and would require companies to switch to imported coal. Because the department in charge of the ordinance was involved in the highly controversial revision of other air pollution laws, work on the ordinance came to a standstill.

As *Waldsterben* became increasingly conspicuous, pressure to resume work on the ordinance mounted. In December 1980, the Umweltbundesamt (Federal Environmental Agency) presented a second draft, which included an emissions limit of 650 mg/m^3 SO_2 for all combustion plants with a capacity of more than 1 MW_{th}. This draft was discussed with representatives of industry and technical experts at a meeting on September 28–30, 1981. It was decided to commission a working group comprising representatives of BMI and the German states as well as technical experts, such as suppliers of abatement equipment, to elaborate the draft.

In May 1982 the working group presented a proposal for an ordinance, which was discussed by the Cabinet on September 1, 1982. The proposal included an emissions value of 400mg/m^3 SO_2. The limit had been reduced through the personal intervention of the minister of the Interior, Gerhard Baum (FDP). This reduction was opposed by the minister of Economics, Otto Graf Lambsdorff (FDP), because it implied higher costs for industry. However, growing public concern about *Waldsterben* led the leader of the Liberals and the foreign minister, Hans-Dietrich Genscher, to push this limit through within FDP. Chancellor Helmut Schmidt (SPD) also supported a stricter emissions standard for the same reason.

The proposal was then sent to the German states and relevant organizations for feedback. The official hearing was on November 29, 1982, and a revised proposal was presented in January 1983. However, this revision was influenced less by the hearing than by a conference of the federal ministers of the Environment (November 11–12, 1982), who favored a stricter ordinance. In a third revised proposal some ideas of industry were finally taken into account—for example, special regulations for plants with heat recovery technology. However, industry was not able to achieve substantial modifications. This third version of the proposal was accepted by the Cabinet on February 23, 1983, and sent to the Bundesrat (Upper House of Parliament) for approval. The Bundesrat agreed to the proposal with minor amendments, most of which provided for even stricter regulations. After the federal government accepted the modifications proposed by the Bundesrat, GFA-VO came into force on July 1, 1983.

The striking characteristic of the political evolution of the GFA-VO is that it constituted a continuous development toward a stricter ordinance. Industry lost nearly all the battles as *Waldsterben* became manifest in many forests and the public pressured politicians to act. The fears of the electricity sector—that desulfurization technologies were not advanced enough to ensure an SO_2 emissions limit of 400 mg/m^3 and that the schedule was too tight to allow the techniques to be tested in pilot plants—were only partly considered.

Table 1-2. *Features of GFA-VO (Verordnung über Großfeuerungsanlagen–13. BImSchV)*

Type of fuel	*Capacity*	*Remaining operation time*	*Emissions limits*
Solid or liquid fuel	Rated thermal input of >300 MW_{th} and a remaining operating time of	≤10,000 hours	As agreed in the permits
		> 10,000 and ≤30,000 hours	2,500 mg/m^3
		> 30,000 hours	As for new plants (400 mg/m^3)
	Rated thermal input of ≤ 300 MW_{th} and a remaining operating time of	≤10,000 hours	As agreed in the permits
Gaseous fuel	No limits set	> 10,000 hours	2,500 mg/m^3

Notes:
Scope of application (§1).
1. GFA-VO only covers combustion plants with a rated thermal input of more than 50 MW (100 MW if gaseous fuel is used).
2. SO_2 emission limits for existing plants (§20).
3. The German ordinance sets emissions limits for individual plants, differentiated by fuel, capacity. and remaining operating time.

The above limits for SO_2 emissions applied until April 1, 1993. Thereafter existing plants also had to meet the requirements imposed on new plants, generally 400 mg/m^3 and 650 mg/m^3 SO_2 for plants using coal with a high or fluctuating percentage of sulfur. A plant is regarded as in compliance when all daily averages and a certain percentage of the half-hourly averages for a year meet the limits. In addition to the emissions limits, a desulfurization rate of 85% is stipulated. Exemption: Authorities may grant exemptions from requirements of GFA-VO if their fulfillment is impossible or entails excessive costs (§33). Authorities are also entitled to establish requirements that go beyond the provisions of the ordinance (§34). Deadlines: Generally, plants are given five years to comply with the emissions limits for SO_2. If the operators seek to achieve the emissions limits by fuel switching, they have to comply with the SO_2 emission limits within two years (§36 (2)).

Provisions of GFA-VO

As a result, the outcome of the policy process, GFA-VO, can be regarded as a fairly strict piece of legislation. Its main features are summarized in Table 1-2.

Beyond Compliance: The Voluntary Agreement

Implementation of GFA-VO was accompanied in North Rhine–Westphalia (NRW), the German federal state with the most large combustion plants, by a voluntary agreement that the NRW government negotiated with the North Rhine–Westphalian electricity industry in 1984. It was called Emissionsminderungsplan für Großfeuerungsanlagen der öffentlichen Energieversorgung in NRW (EMP, Emissions Reduction Plan for Large Combustion Plants of the Electricity Supply Industry in NRW) and intended to reduce SO_2 and NO_x emissions as quickly and effectively as possible. The next two sections describe the motives of the NRW government and the NRW electricity supply industry, as well as the negotiation process, and a short description of the content of EMP follows.

Negotiating the Agreement

The NRW government sought to exceed the requirements of GFA-VO because it wanted to respond to high public concern about *Waldsterben*. The pressure on public authorities to improve the situation was unrelenting, and as the German state with the highest number of large combustion plants, NRW was determined to set a good example and demonstrate that the emissions limits and deadlines were realistic and attainable.

The electricity suppliers participated in EMP to show their willingness to help solve environmental problems, chiefly *Waldsterben*. But they had more than their reputations in mind. They benefited from EMP because the plan's tight timetable required permitting authorities to carry out the authorization procedures swiftly and thus gave electricity suppliers more certainty in their planning. Just as for complying with the federal ordinance, it was relatively easy for the electricity suppliers to promise ambitious emissions reductions because their regional monopolies enabled them to transfer the costs to their customers.

By mid-1984, the electricity suppliers had to declare which of their plants they planned to shut down or reduce in capacity (rated thermal input). These plants had to be closed once their remaining operating time had expired—by January 1993 at the latest. From the declarations it followed that in West Germany, power stations that accounted for about 88% of the total capacity on a fossil fuel basis would continue to operate (Bertram and Karger 1988). In NRW 80% of the stations that burned fossil fuel would remain in operation (EMP 1984). At this point the NRW government initiated talks with the large North Rhine–Westphalian electricity suppliers and the Vereinigung Deutscher Elektrizitätswerke (VDEW, Association of German Electricity Suppliers).

The initial talks were held between the NRW president and members of the management boards of the five large North Rhine–Westphalian electricity suppliers (Elektromark, RWE, STEAG, VEW, and VKR). Although a basic consensus to formulate an emissions reduction plan was reached relatively quickly, more meetings were necessary to agree on its content. It was agreed to set up an emissions reduction schedule with annual emissions ceilings for individual combustion units. Although the major features were developed during the high-level meetings between the NRW president and the management board members, the actual ceiling values were negotiated at working-level meetings, which were also attended by the local electricity suppliers.

The government of NRW did not specify the emissions ceilings but demanded that the EMP requirements go beyond those of GFA-VO. The ceilings were worked out in close cooperation with the electricity suppliers (Figure 1-1). The government asked LCP operators to reveal their actual 1983 emissions and indicate to what extent and how fast they could reduce them. The operators gave emissions reductions based on planned plant closures and the abatement measures they envisaged to meet the emissions limits set in the GFA-VO. As they took into account the need for safety margins,[5] in some cases promised addi-

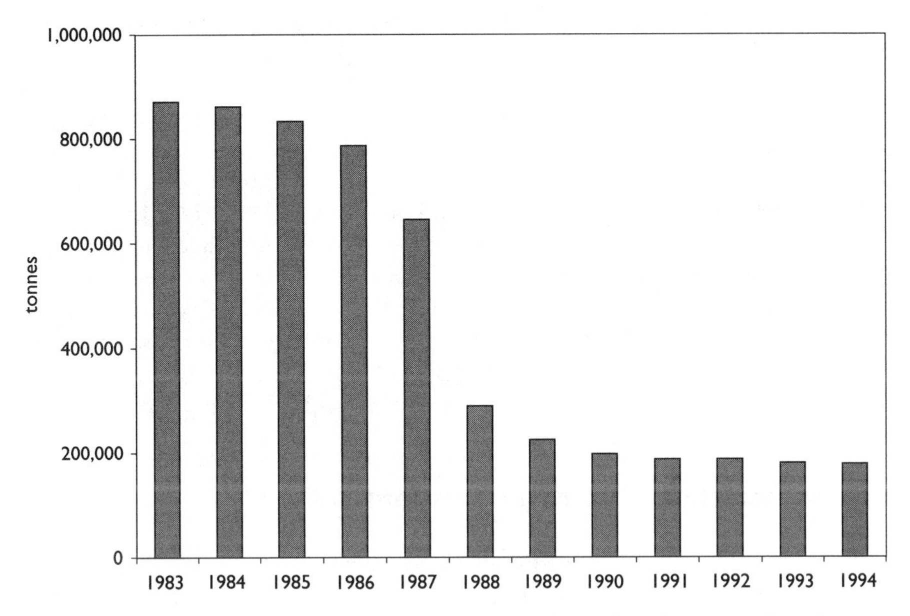

Figure 1-1: *Annual Ceilings for SO_2 Emission from the Electricity Supply Industry in NRW*

Source: EMP 1984, *12*.

tional measures, and agreed to reduce emissions as soon as possible, they went beyond the ordinance. The emissions targets and timetables proposed by the companies were here and there tightened on demand of the NRW government and then integrated in EMP. The negotiations were concluded in November 1984.

Despite some reservations—for example, regarding future fuel quality and the timely execution of authorization procedures—in the end both parties regarded EMP as a realistic action plan (EMP 1984). To resolve the companies' doubts about authorization, government officials promised that the procedures would be carried out as swiftly as possible.

Provisions of EMP

EMP did not have the status of a contract under public law and therefore could not be enforced. It was a special form of a voluntary agreement expressing serious declarations of intent on the part of the LCP operators.

The agreement applied to all LCPs run by electricity suppliers in NRW; in total they accounted for a capacity of 82,540 MW_{th}. The plan did not set new SO_2 emissions limits but was based on those in GFA-VO. By setting annual emissions ceilings, it provided a timetable for the gradual realization of SO_2 emissions reductions that would go beyond the ordinance. Emissions ceilings were set for each individual power station. In total, the electricity suppliers faced the following SO_2 emission targets.

According to GFA-VO, the emissions limits had to be met by June 1988. Although EMP provided for additional reductions in this year, a remarkable decline in emissions was reached even beforehand. This can mainly be explained by a special agreement between the NRW government and one electricity supplier, RWE, which stipulated that RWE decreases its SO_2 emissions by about 110,000 tons between 1984 and 1987 by applying a technique called Trocken-Additiv-Verfahren (TAV, dry additive technique) (EMP 1984). TAV involves the addition of lime before or during the combustion process—a simple technique that allows SO_2 emissions to be quickly reduced, but only to a limited extent. Although GFA-VO could not be met by means of TAV, the technique was useful as an interim measure until desulfurization plants were installed. The company then stopped using TAV and switched to flue gas desulfurization. EMP itself did not prescribe or promote the application of any particular desulfurization techniques.

Ex post Evaluation of German Emissions Policies[6]

The purpose of this section is to evaluate the implementation of GFA-VO and EMP: did large combustion plants in the electricity sector in fact reduce SO_2 emissions, and did they achieve static and dynamic efficiency? This section starts with a brief description of monitoring and enforcement activities, which usually have a strong influence on an environmental policy's effectiveness. Next, actual emissions with GFA-VO and EMP requirements are compared. Then, the overall abatement costs are estimated and the static efficiency of GFA-VO is evaluated. Finally, the repercussions of GFA-VO in terms of dynamic efficiency are analyzed.

Emission Reductions

Monitoring and Enforcement.[7] GFA-VO requires LCP operators to install equipment to continuously measure, record, and process SO_2 emissions (§25(4–5)). Once the instruments have been installed, the whole system needs to be approved by the supervisory authority. GFA-VO requires LCP operators to have their measuring equipment checked (once a year) and calibrated (every three or five years). To prevent manipulation, the entire measuring system is sealed. At the end of each year, a report on annual emissions values is submitted to the supervisory authority (§27(1)). Additionally, supervisory authorities are authorized to carry out on-site controls whenever they think this necessary (§52(2) BImSchG).

The equipment used to record and process emissions values have undergone a remarkable evolution. In the beginning the data were recorded on punched ticker tape. Since the end of the 1980s and early 1990s, electricity suppliers have used computers that not only record but also process the data and automatically calculate half-hourly and daily mean values. In 1998 NRW started connecting companies to the Emissions-Fernüberwachung (EFÜ) system, a system for the telemetric transfer of emissions data.

The technology of recording and processing equipment has made monitoring much easier. At first supervisory authorities had to analyze hundreds of meters of punched ticker tape. When computers were introduced, the annual emissions report was just a single piece of paper categorizing the emissions values (half-hourly and daily means) in different classes indicating whether the values were inside or exceeded the limits. This enabled noncompliance to be detected at a glance. With EFÜ, the emissions values are automatically transmitted to the permitting authority once a day. The authority automatically has a message on its computer screen saying whether the emissions limits have been exceeded during the past 24 hours. EFÜ makes it impossible to exceed the emissions limits without being detected.

If emission limits are exceeded, the companies are obliged to inform the supervisory authority and explain the reasons. By comparing the data in the annual emissions report with the cases of noncompliance the companies have reported over the year, permitting authorities can see whether they were always informed whenever the emission limits were exceeded.

According to GFA-VO, combustion units may be run without desulfurization plants for 72 successive hours and a total of 240 hours a year. In only two cases in NRW have electricity suppliers failed to fix their flue gas desulfurization (FGD) systems within the time stipulated. In the first case the FGD system did not work at all and had to be replaced. In the second case the coating of the absorber building caught fire, necessitating a new coating. In both cases the electricity suppliers applied for an exemption under §33 GFA-VO, seeking special permission to run the combustion units without FGD until the FGD systems had been repaired. In both cases the supervisory authorities refused to grant an exemption and closed down the combustion units for one year and three to four months, respectively.

Apart from those two cases, North Rhine–Westphalian supervisory authorities have never had to use any means of coercion. If the issue is not settled informally, it has been sufficient to threaten the companies with coercion. However, the authorities have demonstrated that they do not hesitate to apply tougher sanctions if necessary.

Because problems with FGD systems initially occurred relatively often, the companies actually needed the chance to exceed emissions limit values for 72 or 240 hours. However, in recent years companies have only rarely made use of the 72/240-hour rule. On average emissions limits are exceeded for not more than 100 to 150 hours a year. Usually the FGD systems are fixed in 10 to 15 hours. In fault-free operation, the emissions values of the electricity suppliers are well below the GFA-VO limits.

Actual Emissions under GFA-VO and EMP. The GFA-VO sets emissions limit values expressed in mg/m^3 waste gases, which vary depending on the fuel, size, and remaining operating time of the plant. As a general rule it can be said that nearly all large combustion plants had to comply with an emissions limit of $400 mg/m^3$ SO_2 as of April 1, 1993. Table 1-3 shows the development of the average SO_2 emissions of West German combustion plants in the electricity supply industry for selected years between 1980 and 1995.

Because the GFA-VO sets emissions limits for individual plants on a daily and half-hourly basis, the yearly average emissions can only indicate—but not

Table 1-3. *Average SO_2 Emissions of Combustion Plants in the Electricity Sector in West Germany*

	1980	*1982*	*1985*	*1988*	*1989*	*1990*	*1992*	*1995*
Emissions of SO_2 (mg/m^3)	2,154	2,160	1,847	582	270	290	250	154
Percentage of GFA-VO limit	538.5	540.0	461.8	145.5	67.5	72.5	62.5	38.5
Percentage reduction from 1980	–	–0.3	14.3	73.0	87.5	86.5	88.4	92.9

Source: VDEW 1999.

Table 1-4. *SO_2 Emissions from Combustion Plants in West Germany*

	1980	*1985*	*1990*	*1993*
SO_2 emissions (1,000 t/annum)	1,879	1,506	295	236

Source: Bültmann and Wätzold 2000.

Table 1-5. *Actual SO_2 Emissions from Large Combustion Plants in the Electricity Sector and EMP Emissions Targets*

	1983	*1984*	*1985*	*1986*	*1987*	*1988*	*1989*	*1990*
Actual SO_2 emissions (1,000 t/annum)	870	845	750	745	640	205	90	95
EMP emissions targets (1,000 t/annum)	870	860	833	786	645	289	224	197
Percentage of EMP limit	100.0	98.3	90.0	94.8	99.2	70.9	40.1	48.2

Source: MURL 1992, *12*.

prove—compliance with GFA-VO. However, the individual combustion plants actually reached emissions that were well below the legal limits. Thus, massive overcompliance with the requirements of GFA-VO can be observed.

The emissions targets of GVA-VO focus on the concentration of SO_2 in flue gas, not on the overall SO_2 emissions from combustion plants. Thus, it is possible that even though the concentration of SO_2 in flue gas is reduced, these reductions are completely or partly offset by an overall increase in the amount of flue gas resulting from an increase in electricity production. However, Table 1-4 shows that similar to the sharp decrease in the concentration of SO_2 in flue gas, the German policy also led to a sharp drop in the overall SO_2 emissions from combustion plants.

There are indications that recently[8] some LCP operators have increased their emissions to approach the emissions limit of 400mg/m^3 SO_2 in GFA-VO. By reducing the lime input in the FGD systems, operators can reduce costs (albeit only to a very limited extent). This can be explained as a response to the recent liberalization of the European electricity market and the loss of the energy suppliers' comfortable position as regional monopolists (Bültmann and Wätzold 2000).

The EMP emission targets and the emissions levels finally achieved are given in Table 1-5.

Table 1-5 shows that the targets of the emissions reduction plan were met. Only in 1987 were they nearly exceeded, but in 1990 overall emissions were roughly 50% lower than agreed between the NRW government and electricity suppliers. Thus overcompliance exists with not only GFA-VO but also EMP.

The question whether the level of SO_2 abatement was adequate to halt *Waldsterben* or whether a lower level would have been sufficient is difficult to answer. SO_2 emissions are only one of many factors believed to cause *Waldsterben,* and scientists are unable to realistically quantify the effects of various levels of SO_2 emissions (Schütt 1988). Forest health has not significantly improved since the 1980s, but neither has it further deteriorated—and that may be attributed to the significant reduction in SO_2 emissions (UBA 2000).

Static Efficiency

Overview of Abatement Costs. The considerable emissions reductions achieved by GFA-VO were not free. The German association of electricity suppliers conducted a survey in which members were asked to indicate the investments in FGD they had made to comply with GFA-VO. The survey covered 70% of the FGD systems installed at power stations of the electricity supply industry and concluded that for the whole of West Germany, DM14.3 billion (= €7.3 billion) was spent (Jung 1988a). This corresponds to specific investment costs (arithmetic mean) of DM150/KW_{th} (= €77/KW_{th}) (Jung 1988b). In the survey, electricity suppliers were also asked to provide data on the running costs of their FGD systems. On the basis of their answers, Jung estimated that the specific operating costs amount to DM0.019/kWh (= €0.01/kWh). The figure includes costs for resource and energy use, personnel, maintenance, capital servicing, and miscellaneous costs (Jung 1988a). The investments in desulphurisation plants in the North Rhine-Westphalian electricity supply industry amounted to over DM8 billion (= €4.1 billion) (MURL 1992).

These high costs were not solely a result of the strict emissions limits. They also arose partly from the pressure to reduce emission even more quickly, through EMP. The nearly simultaneous installation of desulfurization equipment in LCPs all over Germany led to a surge in demand for this equipment with a resulting increase in prices.[9] Furthermore, because Germany had little experience with the necessary technology, no learning effects were achieved. For example, shortcomings that should have come to light before the systems were introduced in the entire fleet of power stations remained undiscovered. As a result, these shortcomings had to be remedied in all power stations.

Efficient Allocation of Abatement Activities. The information on costs given in the previous section clearly shows that the abatement efforts initiated by GFA-VO were extremely costly. This leads to the question of whether the same amount of emissions reductions could have been achieved at lower cost if the abatement activities had been allocated in a different, more efficient, way. Efficiency is attained when an environmental policy ensures that the least-cost implementation of abatement activities is achieved to reach a given goal, such as a certain level of emissions (e.g., Baumol and Oates 1988). This implies that the allocation of abatement activities among polluters is such that a reallocation would not produce any cost savings.

Therefore, to diagnose a failure to attain static efficiency, we have to ask whether it would have been technologically feasible to reallocate abatement activities without changing the overall level of emissions. This implies that some LCPs (those for which emissions reduction is more costly) could have reduced their SO_2 emissions less at the expense of other LCPs (those for which emissions reduction is less costly) that would otherwise have reduced their emissions even further.

According to Bültmann and Wätzold (2000), the choice of a so-called bubble solution would have been less expensive than the allocation of abatement activities actually applied. However, the cost savings would have been rather small because the potential for reallocating abatement activities would have been limited.[10] Once FGD systems are installed, emissions can be fine-tuned by increasing or decreasing lime input. Since the lime input develops proportionally with the emissions, no cost savings can be achieved by reducing the lime input at one plant and adding a comparable amount at another. Therefore, options for cost savings exist only for the installation of FGD systems. Cost savings could have been realized by installing smaller or no FGD systems at all in some plants and compensating for the resulting increase in emissions by raising abatement at other plants.[11] But the increase in emissions (unpurified flue gas contains about 2,300–2,400 mg SO_2/m^3) would be rather high, and the options for further reductions are limited, given the ambitious target of drastic reductions of SO_2 emissions. This means it would have been impossible to run large power stations without FGD because the increase in emissions could not have been compensated for. Realistically, cost savings could have been realized only by higher abatement at large power stations and saving comprehensive FGD systems in one or more small plants.[12]

Dynamic Efficiency

Although GFA-VO did not stimulate the development of new abatement technologies, it was very successful in speeding up the diffusion of FGD systems in Germany on a large scale. When the standards for SO_2 emissions were being set, BMI argued for the limits of what seemed technologically feasible at that time. The ministry succeeded in setting a strict standard even though it required technologies that at the time had not been well tested in Germany. In insisting that advanced technologies could be implemented, BMI received support from firms that would supply the abatement equipment and believed they could guarantee that the technology would meet the standards.

Discussion of the Case Study

The most striking feature of the German SO_2 policy is its tremendous success in reducing emissions. The requirements of GFA-VO and EMP were not merely fulfilled; instead, a high degree of overcompliance was observed. How can this be explained?[13]

An effective monitoring and enforcement system was the *precondition* for this success. EFÜ makes it impossible to exceed emissions limits without being noticed by the supervisory authority, and the authorities showed a willingness to apply severe sanctions for noncompliance.

The potential for overcompliance existed because of some technical and learning aspects: the emission values of LCPs fluctuate, for example, because of the varying sulfur content of coal. To make sure emissions limits are always met, operators need a safety margin so that on average, they remain below the limits. Suppliers of abatement equipment add another safety margin, because they have to guarantee that their equipment will meet certain values. While the emission limits are set on a daily or hourly basis, our emission data represent annual averages of emissions (it is easier to meet an emission limit on the yearly than the hourly average). Because there was little experience of FGD systems, the effectiveness of the system was underestimated. Additionally, learning effects in the operation of the FGD systems also enabled LCP operators to reach even lower emissions values.

The potential for overcompliance was realized for two reasons. First was the enormous public pressure on politicians, public authorities, and LCP operators to reduce SO_2 emissions dramatically to stop the increasingly visible *Waldsterben*. The perceived urgency of dealing with this environmental problem affected not only the political evolution of the GFA-VO but also its implementation, including the voluntary EMP agreement in NRW. Second, the regional monopolies enabled electricity suppliers to shift their abatement costs to their customers. Although the initial outlays were high, marginal abatement costs for overcompliance were not: once the FGD systems were installed, the emissions values could be fine-tuned by adding lime—a rather inexpensive way of reducing emissions.

Thus, overcompliance stemmed from a combination of factors and had little to do with the choice of a command-and-control instrument per se. This finding is supported by the facts that other German command-and-control policies were clearly less successful, and there are strong indications that overcompliance was reduced once the liberalization of the energy market took place.

Traditional economists' view is that command-and-control instruments are less efficient than economic instruments (e.g., taxes and tradable permits) because they generally do not allow a differentiation of abatement activities according to cost criteria. However, our analysis suggests that the German command-and-control approach was most likely the best choice in terms of efficiency. The reason is that the policy aim was to reduce SO_2 emissions as much and as soon as possible. That ambitious objective could be achieved only if all the sources reduced their emissions to the extent that was technologically feasible, leaving little room for differentiation.

This observation suggests that the efficiency of instruments partly depends on the extent of the emissions reductions intended. A (possibly simplified) generalization may be that economic instruments are to be preferred in terms of cost minimization when the aim is to reduce emissions to a medium extent; otherwise command-and-control instruments are equal or even preferred. For small emissions reductions, overall abatement costs may be low, and thus cost differentiation cannot create considerable savings; for dramatic emission reductions, the potential for cost differentiation hardly exists because sources with high abatement costs also have to reduce emissions to achieve the target. Only with reductions of a medium size are gains through the differentiation of abatement activities substantial enough to outweigh any costs of economic instruments that may arise through establishing a permit market or a tax collection scheme.

The traditional economists' view also considers command and control instruments inferior to economic instruments in terms of dynamic efficiency. Yet GFA-VO succeeded in achieving a rapid diffusion of what was then very advanced SO_2 abatement technologies in Germany.

A command-and-control approach is said to discourage innovation by polluters, since discovering new ways of reducing emissions can become the basis of a more stringent standard. This view does not capture the situation of the case study, however, where independent suppliers of abatement technologies existed. These suppliers had an interest in a very stringent standard because this would increase their sales. Indeed, they provided the regulators with information about the best available technology and (with a margin of safety) its capabilities.

Another reason why economists prefer economic instruments over command-and-control instruments is that they provide a continuous incentive for emissions reductions and thus promote new pollution-saving technologies. It is difficult to relate this argument to the German SO_2 policy because we cannot know whether a comparable economic instrument would have done a better job of generating new and better technologies than the chosen command and control approach. The implementation of an SO_2 emissions trading scheme in the United States has not led to the development of technologies that could have been used to achieve the ambitious German reduction aims at lower costs.[14] However, one should be careful in making a simple comparison, since the goals of the U.S. emissions trading program were much less ambitious.

To conclude, the case study on SO_2 policies in Germany contradicts some of the traditional economists' views on environmental policy instruments' ability to achieve desired emissions reductions and their static and dynamic efficiency. In this way, the study reminds us of the dangers of prejudiced generalization when assessing environmental policy instruments and of the need for a detailed analysis of each policy.

Notes

1. Most of the empirical information for this case study stems from a report by Alexandra Bültmann and Frank Wätzold for the European Commission (Bültmann and Wätzold 2000).

2. For a comprehensive discussion of the factors that caused *Waldsterben,* see Schütt (1988) and Forschungsbeirat Waldsterben/Luftverunreinigungen (1989).

3. Two companies (Badenwerk Holding AG and Energie-Versorgung Schwaben Holding AG) merged in 1997 to form Energie Baden-Württemberg AG, making the current number eight.

4. Large parts of this chapter are based on Dose (1997).

5. To ensure that not even peak emissions exceed the limits imposed by GFA-VO, companies needed to include a safety margin in the construction of the abatement equipment and therefore on average achieve emission values that are well below the legal limits.

6. The policy evaluation is carried out from an *ex post* perspective. To our knowledge no explicit *ex ante* study has been conducted.

7. The information on monitoring and enforcement activities is based on research in NRW. However, the overall picture—intense monitoring and tough enforcement measures—is the same all over Germany.

8. This behavior is not yet reflected in the existing data.

9. Bültmann and Wätzold (2000, *184*) quote estimates that nowadays the installation of an FDG system would be one-third cheaper than in the mid-1980s.

10. It should be noted that static as well as dynamic efficiency aspects related to a shift away from hard coal to other sources of energy are not analyzed. This shift is considered not feasible for political reasons.

11. The installation of smaller FGD systems implies that only parts of the flue gas are desulfurized. Those parts that are not purified are not cooled down in the desulfurization process. Therefore, by mixing the two flue gas streams before they are emitted, the required temperature of 72°C can be met without costly reheating. This means that cost savings rather result from saving reheating than from smaller FGD systems.

12. This would not have caused "hotspot" problems because pollution around power stations had been kept low through the construction of tall chimney stacks, which distributed the emissions over a large area.

13. Compliance with environmental law is not typical for all areas of German environmental policy. There are several studies that find an "implementation gap" in German environmental law; see e.g. Mayntz et al. (1978), Lübbe-Wolff (1993), and for an overview of German implementation research, Bültmann and Wätzold (2002).

14. For the effects of the SO_2 emissions trading scheme on the development of new technologies, see Chapter 2.

References

Baumol, W.J., and W.E. Oates. 1988. *The Theory of Environmental Policy.* Cambridge: Cambridge University Press.

Bertram, J., and R. Karger. 1988. Entscheidungen gefordert. Planung, Bau und Inbetriebnahme von REA-Anlagen in Steinkohlekraftwerken. *Sonderdruck aus Energie* 40(7): 98–109.

Bültmann, A., and F. Wätzold. 2000. The Implementation of National and European Environmental Legislation in Germany: Three Case Studies. UFZ-report No. 20/2000. Leipzig.

———. 2002. Der Vollzug von Umweltrecht in Deutschland—Ökonomische Analyse und Fallstudien. Marburg.

Dose, N. 1997. Die verhandelnde Verwaltung. Eine empirische Untersuchung über den Vollzug des Immissionsschutzrechtes. Baden-Baden.

Emissionsminderungsplan für Großfeuerungs (EMP). 1984. Anlagen der öffentlichen Energieversorgung in Nordrhein-Westfalen, November.

Forschungsbeirat Waldschäden/Luftverunreinigungen (Hrsg.) 1989. Dritter Bericht. Waldschäden durch Luftverunreinigungen. Karlsruhe.

Ikwue, A, and J. Skea. 1996. The Energy Sector Response to European Combustion Emission Rdegulations. In *Environmental Policy in Europe*, edited by F. Lévêque. Cheltenham, UK, and Brookfield, MA.

Jung, J. 1988a. Die Kosten der SO_2- und NO_x-Minderung in der deutschen Elektrizitätswirtschaft. *Elektrizitätswirtschaft* 87(5): 267–70.

———. 1988b. Investitionsaufwand für die SO_2- und NO_x-Minderung in der deutschen Elektrizitätswirtschaft. *VGB Kraftwerkstechnik* 68(2): 154–57.
Lübbe-Wolff, G. 1993. Vollzugsprobleme der Umweltverwaltung. *Natur und Recht* 5: 217–29.
Mayntz, R., et al. 1978. Vollzugsprobleme der Umweltpolitik. Empirische Untersuchung der Implementation von Gesetzen im Bereich der Luftreinhaltung und des Gewässerschutzes. Wiesbaden.
Ministerium für Umwelt, Raumordnung und Landwirtschaft (MURL) des Landes Nordrhein-Westfalen. 1992. Bilanz und Erfolg des Emissionsminderungsplanes für Großfeuerungsanlagen der öffentlichen Energieversorgung in NRW (EMP). Düsseldorf.
Schütt, P., unter H. Mitarbeit von Blaschke, et al. 1988. Der Wald stirbt an Streß. München.
Statistisches Bundesamt. 1991. *Statistisches Jahrbuch 1991*. Wiesbaden.
Umweltbundesamt (UBA). 1994. Daten zur Umwelt 1992/93. Berlin.
———. 1997. Daten zur Umwelt 1997. Der Zustand der Umwelt in Deutschland. Berlin.
———. 2000. Daten zur Umwelt 2000. Der Zustand der Umwelt in Deutschland. Berlin.
Vereinigung Deutscher Elektrizitätswerke (VDEW). 1999. Data provided on April 6, 1999.

Relevant Regulations

Gesetz zum Schutz vor schädlichen Umwelteinwirkungen durch Luftverunreinigungen, Geräusche, Erschütterungen und ähnliche Vorgänge (Bundes-Immissionsschutzgesetz – BImSchG) of 15 March 1974, version of 14 May 1990 (BGBl. I p. 880) – (BGBl. III 2129-8).

Dreizehnte Verordnung zur Durchführung des Bundes-Immissionsschutzgesetzes (Verordnung über Großfeuerungsanlagen – 13. BImSchV) of 22 June 1983 (BGBl. I 1983 p. 719, BGBl. III 2129-8-1-13, BGBl. I 1990, p. 2106).

Erste Allgemeine Verwaltungsvorschrift zum Bundes-Immissionsschutzgesetz (Technische Anleitung zur Reinhaltung der Luft – TA Luft) of 27 February 1986, GMBl. p. 95.

CHAPTER 2

SO_2 Cap-and-Trade Program in the United States

A "Living Legend" of Market Effectiveness

Dallas Burtraw and Karen Palmer

TITLE IV OF THE 1990 Clean Air Act Amendments regulates emissions of sulfur dioxide (SO_2) from electricity-generating facilities under an emissions trading program that is designed to encourage the electricity industry to minimize the cost of reducing emissions. The industry is allocated a fixed number of total allowances, and firms are required to surrender one allowance for each ton of sulfur dioxide emitted by their plants.[1] Firms may transfer allowances among facilities or to other firms, or bank them for use in future years.

A less widely acknowledged innovation of Title IV is the annual cap on average aggregate emissions by electricity generators, set at about one-half the amount emitted in 1980. The cap accommodates an allowance bank, such that in any year aggregate industry emissions must be equal to or less than the number of allowances allocated for the year plus the surplus accrued from previous years.

The Title IV program combines a solid environmental goal, in the form of a cap on the annual allocation of emissions allowances, with the flexibility to trade or bank allowances. Rather than forcing firms to emit SO_2 at a uniform rate or to install specific control technology, emissions allowance trading enables power plants operating at high marginal pollution abatement costs to purchase SO_2 emissions allowances from plants operating at lower marginal abatement costs. Such trading, by equalizing marginal abatement costs among generating units (boilers), should limit SO_2 emissions at a lower cost than traditional command-and-control approaches.

In this chapter, we provide an overview of the origin, design, and performance of Title IV of the Clean Air Act Amendments of 1990, otherwise known as the SO_2 trading program or acid rain program. We analyze the program's specific features and its adaptability as a model for addressing other pollution problems, such as control of nitrogen oxides (NO_x) or carbon dioxide (CO_2) emissions.[2] We provide further background on the regulation of SO_2 emissions from coal-fired power plants. Subsequently, we describe Title IV in more detail and survey the *ex ante* and *ex post* estimates of the costs of the program. Finally, we offer conclusions, drawing on a set of hypotheses that are common to the case studies addressed in this project, and looking to the future of SO_2 regulation.

Background

Sulfur dioxide is a ubiquitous threat to the environment and public health. Along with NO_x, SO_2 contributes to wet sulfate deposition (acid rain) and the degradation of ecosystems.[3] Potential secondary impacts of acid rain include increases in the amounts of aluminum and methylmercury in lakes and fish.

As a gas emitted directly from power plants, SO_2 is regulated as a criteria air pollutant affecting human health.[4] Increasingly, however, the largest threat of SO_2 from the standpoint of public health is its role as a precursor to secondary particulates, which are a constituent of particulate matter (PM), another criteria air pollutant. Sulfur dioxide and PM are associated with human morbidity and mortality.

In areas that have not attained national ambient air quality standards (NAAQS), states are required under the Clean Air Act to develop implementation plans to demonstrate reasonable progress toward the standards. These plans typically include regulation of existing sources. In the 1970s, many local areas failed to meet the NAAQS for SO_2, and this nonattainment status was the basis for establishing emissions rates to protect human health based on ambient air quality within the vicinity of a power plant. As part of the remedy for the contribution of SO_2 emissions to local air quality problems in the 1970s, utility companies constructed 429 tall stacks, many more than 500 feet tall, on coal-fired boilers (Regens and Rycroft 1988). This solution helped alleviate local SO_2 pollution by emitting the pollutants high into the atmosphere. As a consequence, the vast majority of urban areas in the 1980s attained the national ambient air quality standards for SO_2, and this gas is no longer widely perceived as a local health problem.

The smokestack remedy to local air quality problems, however, contributed to the deterioration of air quality at a regional level. Emitted high in the atmosphere, SO_2 emissions from coal plants travel hundreds of miles and convert to sulfates that, as particulates, degrade air quality and damage human health and visibility. Furthermore, deposition of sulfuric compounds in soils and waterways in regions distant from the source of emissions contributes to acidification of forests and lakes.

Although acid rain and secondary particulates are national problems, they are most severe in the eastern United States, because most high-sulfur coal is found in the Appalachians and the Midwest. Western coal, such as that found in the Powder River Basin of Wyoming and Montana, is mostly low-sulfur coal. Because a significant part of the delivered cost of coal is the cost of its transportation, coal-fired power plants have tended to burn coal from nearby sources. These types of power plants, located predominantly east of the Mississippi River, generally have burned high-sulfur coal. The pollution problem is reinforced by weather patterns that largely blow airborne emissions from west to east, and by political forces, because eastern states have encouraged utilities to burn high-sulfur coal to protect in-state coal mining jobs.

New Source Performance Standards

The 1970 amendments to the Clean Air Act implemented performance standards for *new sources,* as well as for those that undertake major modification, based on emissions per unit of heat input. Collectively, these standards are known as new source performance standards (NSPS). The first generation of NSPS was an emissions rate standard of 1.2 pounds of SO_2 per million Btus of heat input at a facility. This standard eventually became a touchstone for the allocation of allowances under Phase II of the SO_2 trading program.

The NSPS were amended in 1977 and took effect in 1978.[5] Since then, new coal-fired power plants have faced an emissions rate–based standard that requires a 90% reduction in SO_2 emissions from previously uncontrolled levels, or a 70% reduction if the facility uses low-sulfur coal. Regardless of coal sulfur content, this approach effectively required new plants to install scrubbers (flue gas desulfurization equipment)—the only available technology that could achieve such reductions—and essentially eliminated the opportunity for compliance through other means, such as process changes or demand reduction. Although the emissions limitation was nominally a performance standard, it basically dictated technological choices in a typical command-and-control fashion. Operators of new facilities could not switch to low-sulfur coal to achieve comparable emissions reductions, and the only technology available to meet the standard was scrubbing.

The NSPS are seen as a way to improve environmental performance over time. However, the power plants that existed before 1978 are not regulated by the NSPS and seem to have an almost indefinite life, which many argue has been lengthened by their cost advantage relative to new sources that face NSPS (Ellerman 1998; Nelson et al. 1993). As a consequence, NSPS have been an ineffective tool for addressing SO_2 emitted from existing coal-fired plants. Hence, prior to the 1990 Clean Air Act Amendments, federal clean air policy treated coal-fired facilities unevenly. The 1990 Clean Air Act Amendments bridged this gap by regulating SO_2 emissions from all new and existing power plants in a uniform and cost-effective manner.[6]

Political Preconditions for the "Grand Experiment"

In the 1980s, more than 70 pieces of legislation were introduced to address the issue of SO_2 emissions and associated environmental problems. One prominent proposal in 1983 sought to roll back emissions by 10 million tons from 1980 levels—about the same amount as eventually required under Title IV—by requiring the installation of scrubbers at the 50 dirtiest plants, which represented 89% of the nation's pre-NSPS coal-fired capacity.[7] The estimated annual cost of this proposal ranged from about \$7.9 billion to \$11.5 billion per year (1995\$).[8] An important part of this proposal would have been a fund to redistribute 90% of the capital cost for compliance to electricity consumers in utility systems or regions away from where emissions reductions would have been achieved.[9]

The debate over the cost of controlling acid rain, and the search for an alternative to forced scrubbing, culminated in Title IV of the Clean Air Act Amendments of 1990. The "cap with trading" design enabled a compromise between environmental interests, which sought a 10–12 million ton reduction in annual SO_2 emissions, and industry groups, which argued that such reductions would be prohibitively expensive.[10] The result has been called the "grand experiment" (Stavins 1998).

Program Design

Under Title IV, the annual allocation of allowances for SO_2 emissions from electric utility power plants is capped ultimately at 8.95 million tons, approximately 10 million tons less than the amount emitted by utility facilities in 1980.[11] Reductions to achieve the 8.95-million-ton cap took place in two phases. Phase I began in 1995 and affected 263 generating units at the dirtiest 110 coal- and oil-fired electricity-generating facilities, plus 174 voluntary units. Phase II started in 2000 and covered all other electricity-generating facilities greater than 25 megawatts of capacity, plus smaller ones using fuel with a sulfur content greater than 0.05%.

The law assigns allowances to each affected power plant unit based on its heat input during a historical base period (1985–1987), multiplied by an emissions rate calculated such that aggregated emissions equal the target emissions cap. One SO_2 allowance entitles its holder to emit one ton of SO_2. Other industrial sources are excluded from the mandatory program, but they may voluntarily subscribe after establishing a historical emissions profile.

To save costs, operators of affected facilities can trade emissions allowances between their own facilities or with other firms. Individual facilities can therefore implement abatement measures that depart from an engineering prescription of the cleanest technological possibility for that facility. If a plant reduces its emissions below its endowed level of emissions allowances, it can switch them to another of its units, bank them for future use, or sell them. If a plant emits at a level greater than its endowed level, it must compensate another plant or firm to reduce emissions commensurately. This provision allows for programmatic cost savings by creating incentives for the plants that face the lowest costs of SO_2 reduction to make more of the reductions, thereby minimizing overall compliance costs.

Ex ante Analysis of the SO_2 Trading Program

One of the earliest studies of the cost under an allowance trading system was Elman et al. (1990), who estimated the marginal cost of compliance and used this as the value of an emissions allowance. Under a perfect trading market, this study predicts marginal costs (presumed to equal allowance prices under perfect trading) to be $742 to $1,032 (1995$). Costs for imperfect trading (i.e., trading that does not equate marginal cost over the entire market, perhaps because of state constraints on the use of local coal) are estimated to average $1,935, ranging up to $2,580 or even $5,160 at several utilities. However, a more relevant study by the consulting firm ICF (1990) conducted for the U.S. Environmental Protection Agency (EPA) was available prior to enactment of the legislation. This study captured more accurately the ultimate design of the regulation and projected marginal costs of $579 to $760 (1995$) for full compliance under the program. This and a number of other studies are summarized in Table 2-1 on page 50.[12]

In the 1980s, the federal government spent more than $500 million studying scientific aspects of the acid rain problem through an interagency effort called the National Acid Precipitation Assessment Program (NAPAP). That research informed the congressional debate but did not directly lead to the compromise that ultimately emerged. By 1990, few economic studies estimated the benefits of reducing SO_2. In one study, NAPAP considered a change in acidic deposition in just New York and parts of New England; for coldwater recreational fishing alone, it estimated willingness-to-pay for improvements comparable to those expected to result from Title IV to be $4.2 million to $14.7 million.[13] The only economist we know of who ventured an opinion of the overall relative benefits and costs of Title IV in 1990 wrote that expected benefits and costs appeared to be about equal (Portney 1990).

The First Years of the SO_2 Trading Program

Two types of measures are most important in assessing the performance of the SO_2 trading program. One is its performance in achieving the goal of reducing emissions and the related environmental and health benefits. The second is the cost of achieving these emissions reductions and the extent to which potential gains from trading allowances have been realized. Related issues include the efficiency of allowance banking, the effects of the program on incentives for technical change, and the administrative and compliance costs of the program.

Emissions Reductions

The SO_2 provisions of Title IV have led to dramatic declines in emissions of SO_2 from power plants over the past 10 years (U.S. EPA 2002). During Phase I of the program, SO_2 emissions fell dramatically relative to previous levels, and also fell (although somewhat less dramatically) relative to levels that likely would have been obtained in the absence of Title IV (Ellerman et al. 2000). Total emissions in 1995, the first year of the program, were 11.87 million tons—25% below 1990 levels and more than 30% below 1980 levels (see Figure 2-1).

Phase I resulted in substantial overcompliance by Phase I units, with total emissions from these units falling well below capped levels throughout the Phase I period. These units surpassed the emissions reduction goals of the program because companies were allowed to bank unused emissions allowances for use in future years. The unused allowances created by this overcompliance were added to a bank that totaled nearly 11.6 million allowances by the end of Phase I.

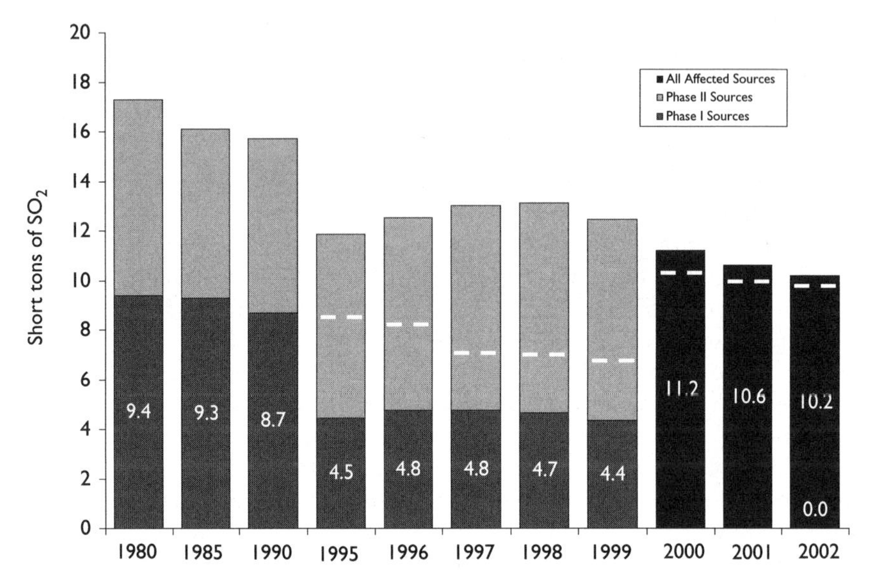

Figure 2-1: *Annual Ceilings for SO_2 Emission from the Electricity Supply Industry in NRW*

Source: EMP 1984, *12*.

Although emissions from the Phase I units remained relatively flat over the ensuing four years, emissions during the unconstrained Phase II period rose, causing total emissions to climb somewhat between 1995 and 1999. However, during the first year of Phase II, total SO_2 emissions declined further and continued to decline in 2001, reaching a level of 10.63 million tons—almost 40% below 1980 levels.

The ability to bank allowances for future use proved crucial to the success of the program. Once firms had built up a bank of unused allowances, they had a vested interest in maintaining the value of those banked credits, and thus in furthering the program itself.

During the two years after Phase II began, in 2000, emissions exceeded the annual allowance allocations by roughly 1 million tons each year as utilities begin to draw down the bank. Emissions are expected to continue to be above the annual cap through the remainder of this decade as they gradually decline to roughly 9 million tons per year. Nonetheless, despite drawdown of the allowance bank over the decade, emissions during the decade are expected to be substantially below the levels predicted in the absence of Title IV. (U.S. EPA 2001).

Environmental Effectiveness

The environmental improvements associated with the decline in emissions also have been substantial. According to EPA, acid rain in the eastern United States fell by as much as 25% during Phase I (U.S. EPA 2001).

However, even though the SO_2 program's principal political motivation was the desire to reduce acid rain, the lion's share of the benefits are expected to come from improvements to human health. Air quality has also improved with the decline of ambient concentrations of sulfate particles, especially in those areas where sulfate concentrations historically have been high. The human health benefits of these sulfate reductions are believed to be substantial. EPA estimates that by the year 2010, the total annual health benefits associated with SO_2 emissions reductions under the program will be more than $50 billion per year (U.S. EPA 2001). In an independent assessment, Burtraw et al. (1998) used an integrated assessment model developed on behalf of NAPAP to study the benefits and costs of Title IV regulations. They simulated both the environmental consequences and associated health benefits arising from the program's expected reductions in emissions of SO_2 and NO_x and the costs of reducing emissions. The study finds that the human health benefits from particulate reductions alone are expected to be 7 times the costs of controlling emissions during Phase I. As emissions fall even further during Phase II, the benefits are expected to increase to at least 10 times higher than the costs, and expectations are even higher under more aggressive assumptions about future growth in electricity demand and power plant lifetimes. This is a startling change compared with the only *ex ante* estimate, which in 1990 suggested benefits and costs were about equal. The difference stems from a greater understanding today of the expected benefits associated with particulate reductions and the unanticipated low realized costs of emissions reductions under the program.

Effects of Trading on the Environment. Some observers feared the SO_2 trading program might cause large regional shifts in emissions and concentrate pollution in "hotspots." Originally, the SO_2 trading program was designed for two trading regions, one in the East and one in the West, to make sure that emissions were adequately reduced where problems with acid rain were the most severe—in the East. Ultimately, the two-region model was abandoned and replaced by a single national SO_2 market with a single national cap, largely because the single-market approach was expected to result in greater cost savings from allowance trading (Hausker 1992). Some argue that unfettered trading in a single national market is a mistake because it fails to adequately protect sensitive areas in the Northeast, particularly in New York State (Solomon and Lee 2000).

However, despite those fears, the evidence points in the other direction. Burtraw and Mansur (1999) modeled the effects of allowance trading and banking under the SO_2 program within a simulation model, and they find that trading should be expected to lead to inconsequential, though positive, changes in the health-related benefits of the SO_2 program (holding total emissions constant). Advocates have expressed strong concern that trading could lead to an increase in pollutant concentrations in heavily populated areas along the east coast of the United States. However, Burtraw and Mansur find that the entire eastern seaboard can expect benefits because of the expected pattern of trading, in addition to the aggregate decrease in emissions (Figure 2-2). Furthermore, they expect trading to contribute to decreases in acid deposition in sensitive regions.

Swift (2000) argues that banking also has not generated hotspots. He shows that, largely as a result of allowance banking, emissions in virtually all states fell below the allocations from 1995 to 1998. Swift shows that during this four-year

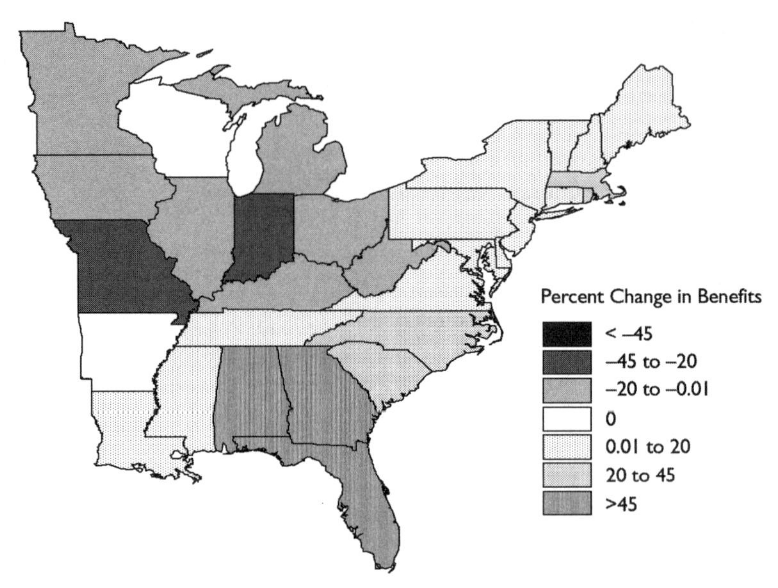

Figure 2-2. *Percent Change in Title IV Baseline Benefits Attributable to Trading for 1995*

period, more than 80% of the allowances used to offset emissions came from the same state as the emitting source. Consistent with Burtraw and Mansur, he suggests that allowance trading may have cooled potential hotspots because the Phase I plants with the largest emissions are the ones that have cleaned up the most. EPA's own analysis of the potential geographic shifts in emissions resulting from trading during each year of Phase I shows that sellers and ultimate buyers of SO_2 allowances tend to be located within 200 miles of each other (U.S. EPA n.d.). Finally, a geographic shift is unlikely because trades of SO_2 at a national level cannot lead to violations of local ambient air quality standards for SO_2. Sources must comply with local standards as well as with the national aggregate cap-and-trade program.

Efficiency of SO_2 Allowance Markets

Allowance prices have fluctuated since trading under the program began. The price of an allowance started at close to \$150 per ton and fell to about \$70 by early 1996. During 1999, prices rose above \$200 per ton but fell again in 2000 to a point closer to \$150. Allowance prices since the beginning of Phase II have fluctuated between roughly \$70 and \$210 per ton, with an average price during the period of close to \$170 per ton. Figure 2-3 illustrates allowance prices over time.

One frequently cited measure of the success of the SO_2 allowance market has been the observation that allowance prices are substantially lower than predicted by EPA and others at the time the program was adopted (Table 2-1). Both the former administrator of EPA and the former chair of the Council of Economic Advisers have suggested that *ex post* realized allowance prices are much lower than *ex ante* estimates of marginal abatement costs, and that this fact demon-

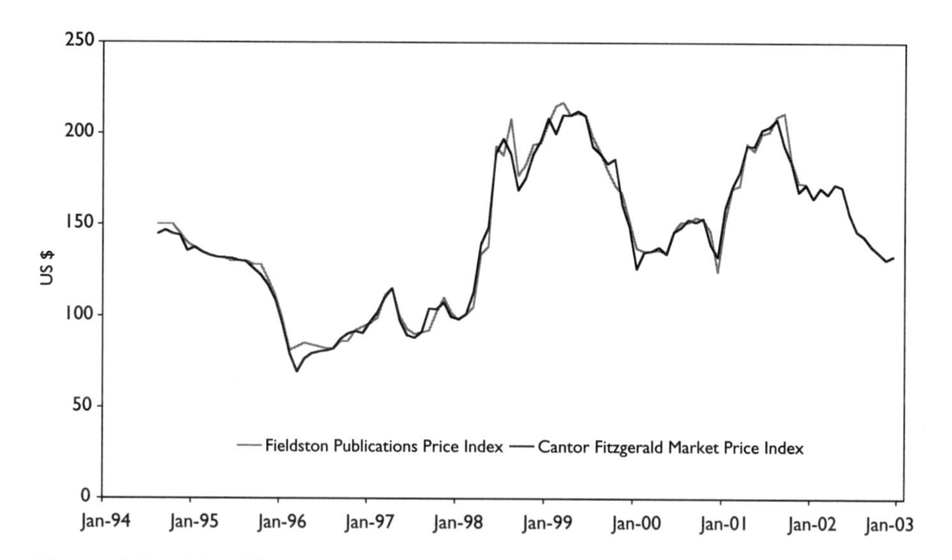

Figure 2-3. *SO_2 Allowance Prices, 1994–2001*

Source: U.S. EPA, http://www.epa.gov/airmarkt/trading/so2market/prices.html (accessed February 3, 2003).

strates that the allowance trading market is responsible for reducing the cost of curbing SO_2 emissions.[14]

Allowance prices have been lower than expected *ex ante;* however, if used as a proxy for program costs, these differences are exaggerated (Smith et al. 1998). This can be important because the SO_2 program has been cited as a model for the regulation of other pollutants.[15] Furthermore, there is no need to exaggerate the savings of the SO_2 program. A proper accounting of the total costs of the program compared with a well-defined baseline indicates that savings have been substantial.

Economic Measure of Cost Savings

Measuring the cost savings attributable to allowance trading requires comparing total costs under trading with a counterfactual baseline description of what would have happened in the absence of the program. This approach to measuring the size of the gains from trade under the SO_2 trading program is explored in an econometric model by Carlson et al. (2000). Most previous studies, including those listed in Table 2-1, rely on engineering-based models of compliance options and their costs, but Carlson et al. use a simulation model based on marginal abatement cost functions derived from an econometrically estimated long-run total cost function for electricity generation for a sample of more than 800 generating units from 1985 to 1994.[16] From an economic perspective, this approach is superior because it takes into account substitution among inputs in response to changes in relative input prices.

Carlson et al. estimate that the potential cost savings attributable to formal emissions trading, compared with the counterfactual of a uniform emissions rate

Table 2-1. *Estimates of Long-run (2010) Annual and Marginal Cost (1995$) of Sulfur Dioxide Emissions*

Study	Annual cost (billions)	Marginal cost per ton SO_2	Average cost per ton SO_2
Carlson et al. (2000)	$1.1	$291	$174
Ellerman et al. (2000)	1.4	350	137
Burtraw et al. (1998)	0.9		239
White (1997) [EPRI]		436	
ICF (1995) [EPA]	2.3	532	252
White et al. (1995) [EPRI]	1.4–2.9	543	286–334
GAO (1994)	2.2–3.3		230–374
Van Horn Consulting et al. (1993) [EPRI]	2.4–3.3	520	314–405
ICF (1990) [EPA]	2.3–5.9	579–760	348–499

Note: EPRI = Electric Power Research Institute.

standard, were $250 million (1995$) during Phase I of the program. They estimate the savings to be $784 million per year during Phase II, or about 43% of total compliance costs under a uniform emissions rate standard. When compared with a forced scrubbing scenario, cost savings are estimated to be almost $1.6 billion (1995$) per year.

The Carlson et al. estimate is compared with several other estimates in Table 2-1. The most rigorous of these is Ellerman et al. (2000), which is based on an extensive survey of the industry, with extrapolation to estimate long-run compliance costs.[17] Ellerman et al. estimate the cost savings from emissions trading, inclusive of savings attributable to banking, to be about 55% of total compliance costs under a command-and-control approach. Earlier studies (Van Horn Consulting et al. 1993; GAO 1994) suggest that an efficient allowance market would result in twice the cost savings estimated by Carlson et al. In all these studies, the command-and-control approach that is modeled is "enlightened": it is a performance standard (emissions rate or emissions tonnage standard) applied to each facility, calibrated to achieve the same level of total emissions.[18] This approach implicitly encompasses many of the beneficial incentives of the SO_2 trading program compared with a technology-forcing approach by providing individual facilities with flexibility in achieving the standard. Other command-and-control approaches that were seriously considered in the United States, such as forced scrubbing at larger facilities, could have cost substantially more. Forced scrubbing is also the approach embodied in the NSPS for SO_2 emitted by power plants.[19]

A striking feature of the studies summarized in Table 2-1 is that as a group they have successively estimated a sequence of declining projections of annual and marginal costs of compliance. There are several contributing reasons for this trend. One is that the trading program ignited a search for ways to reduce emissions at lower cost (as theory suggests is likely to occur with this type of regulation), and the fruitful results of this enterprise are measured by later studies. A second reason for the declining projections in the successive studies is that advantageous trends in fuel markets contributed to a decline in emissions rates, making it easier for utilities to attain the goals of the program and thereby reducing program costs (Ellerman and Montero 1998; Fieldston Company 1996; Burtraw 1996). Carlson et al. (2000) find that declining fuel prices lowered mar-

Table 2-2. *Contribution of Price and Technological Change to Annual Compliance Costs (1995$)*

Scenario	*Command-and-control (millions)*	*Efficient trading (millions)*	*Marginal abatement cost ($/ton)*	*Potential gains from trade (millions)*
Benchmark estimate (1995 prices and 1995 technology)	$2,230	$1,510	$436	$720
Benchmark with (1989 prices and 1989 technology)	2,670	1,900	560	770
ICF (1990)	—	2,300–5,900	579–760	—
Preferred estimate	1,820	1,040	291	780

ginal control costs by about $200 per ton over the decade preceding 1990. In Table 2-1, the right-hand column (reporting average cost per ton estimated by the various studies) reflects this decline.[20] A third reason for lower cost estimates in later studies points to the role of unanticipated exogenous technical change that would have occurred in the absence of the program. Carlson et al. (2000) show that technical improvements, including improvements in overall generating efficiency, lowered the typical unit's marginal abatement cost function by almost $50 per ton of SO_2 over the decade preceding 1995.

The Carlson et al. model can help sort out these different factors' contributions to changing marginal abatement costs. Table 2-2 presents several estimates using this model with varied assumptions about fuel prices and technological changes. The columns present the annual cost of a command-and-control approach (uniform emissions rate standard), the annual cost of efficient trading, its associated marginal abatement cost, and finally the estimated gains from trade that are available from efficient trading.

The first row in Table 2-2 reports numbers for a benchmark scenario in which relative fuel prices remain stable at 1995 levels and technology, including the utilization rate of scrubbers, is characterized at 1995 levels.[21] This benchmark predicts long-run marginal abatement costs will be $436 (1995$). The second row presents an estimate with prices held to their 1989 level (implying a higher price for low-sulfur coal relative to high-sulfur coal than obtained in 1995) and the time trend for technological change (factor productivity) also held at 1989 levels.[22] From this vantage point, marginal abatement costs rise to $560, a 28% increase. Notably, this is not far from the estimate offered by ICF (1990), calculated with comparable information and reported in the third row.

The last row in Table 2-2 presents the Carlson et al. (2000) preferred estimate. Compared with the benchmark, this scenario adopts 1995 prices and 2010 technology. It assumes that utilization rates and performance of in-place scrubbers continue to improve. It assumes a slower retirement rate of coal-fired facilities, with one-half of all retired facilities replaced by gas. It also assumes that using continuous emissions monitoring (CEM) systems[23] in place of the historical measure of emissions will raise emissions estimates and necessitate a greater level of control.[24] Explicitly missing from consideration in all of these studies is the influence of other potential regulatory actions, such as further control of particulates or NO_x emissions or actions to address global warming.

Economic Performance in the Early Years

There is ample evidence of cost savings in Phase I of the trading program. Ellerman et al. (2000) estimate savings of $350 million per year—about half of their measured cost of compliance. However, there is also evidence that the market did not perform perfectly in the early years.

Economic theory suggests that the marginal cost of compliance activities should be the same at all facilities (except as may be constrained by local ambient air quality restrictions). Contrary to theory, Carlson et al. (2000) find that, in the first two years of Phase I, actual compliance costs exceeded the least-cost solution by $280 million in 1995 and by $339 million in 1996 (1995$). To be fair, we note that a command-and-control counterfactual also would be unlikely to achieve emissions reductions in the least-cost manner. Nonetheless, evidence suggests that the allowance market did not achieve the least-cost solution during the first two years, even though allowance prices and marginal abatement costs were approximately equal. Carlson et al. appear to confirm what many others suggest—that unfamiliarity with the new program led many to pursue a policy of "autarchy" (no trade) and self-sufficiency in compliance (Bohi 1994; Bohi and Burtraw 1997; Ellerman 2000; Hart 2000; Swift 2001).

Other studies also find that utilities failed to take full advantage of the allowance market at the beginning of Phase I. Swinton (2002), for example, analyzes the efficiency of SO_2 control in Florida during Phase I. Using a panel of data on Florida power plants from 1990 through 1998, he estimates an output distance function and then uses this function to find the shadow prices of SO_2 emissions reductions. He finds that power plants in Florida did not use the allowance market to its fullest potential: several plants controlled emissions when purchasing allowances would have been a more economic option.

Analysts have pointed to state public utility regulations and other state laws as influences that have tended to undermine the efficiency of the SO_2 market.[25] If that is the case, the effect can be significant. Prior to implementation and during the early years of the program, a great deal of uncertainty existed in many states about how regulators were going to treat allowance transactions in setting regulated rates, and this uncertainty damped utilities' enthusiasm for using the allowance market (Burtraw 1996; Bohi 1994). Rose (1997) suggests that public utility commission (PUC) activities have discouraged the use of the market in favor of strategies such as fuel switching. Rose et al. (1993) and Lile and Burtraw (1998) document a number of PUC actions that promote use of local high-sulfur coal coupled with scrubbing.

Empirical studies have used econometric techniques to examine the extent to which PUC regulations have affected the performance of the SO_2 market. Arimura (2002) suggests that such an analysis should focus on compliance decisions at the generating-unit level; he endeavored to identify whether units with relatively low marginal abatement costs were reducing emissions while those with higher costs were buying permits. Using data from the Phase I time period, he estimates a probit model of the choice between fuel switching and allowance purchases coupled with continued use of high-sulfur coal as compliance strategies. [26] He finds that, holding abatement costs and other characteristics fixed, generating units facing PUC regulations are more likely to rely on fuel switching for compliance rather than on the allowance market. [27] He also finds that in states with high-sulfur coal, where efforts were made to protect local coal producers, there was greater use of allowance purchases than fuel switching or blending for compliance.

Arimura then uses the model to simulate the compliance strategies that would be adopted in the absence of PUC regulation and without protection of high-sulfur coal markets. He finds that PUC regulation has a stronger effect on compliance behavior than does the protection policy. He also finds that the allowance price would have been higher in the absence of PUC regulation. This finding helps explain why allowance prices during Phase I were lower than anticipated.

Sotkiewicz (2002) used utility data for 1996 and exercised a simulation production cost model to evaluate facility performance. He too finds that utility compliance costs due to state-level PUC regulation were above least cost—ranging from 4.5% to 139% above. PUC regulations governing cost recovery for investment in scrubbers led to the majority of cost increases.

In sum, the question about performance in the early years of the SO_2 trading program is one of perspective. Emissions trading was a new institution that might be expected to pass through a period of transition before reaching a mature and efficient market. The issue of performance in the early years may be relevant to the design of trading programs in the future. However, all the literature appears to express uniformly the expectation for, or to provide evidence of, improved performance in the allowance market over time. The volume of trading doubled each year over the first three years of the program, suggesting a process of learning on the part of firms (Kruger and Dean 1997). Also, increasing competition in the electricity sector pressured firms to reduce costs and take better advantage of trading.

Cost Savings over Time

Analysts who offer a direct comparison of current short-run allowance prices with estimates of long-run marginal costs of emissions reductions tend to exaggerate the cost savings from allowance trading. Economic theory suggests that short-run and long-run measures should indeed be related, but they are not directly comparable. The two measures should be related by the opportunity cost of holding emissions allowances, which is the interest rate.

In the short term, firms should be expected to reduce emissions to the point where the marginal cost of doing so equals the discounted present value of marginal costs in the long run. Over time, one would expect the bank to be drawn down in a smooth manner such that the allowance price in one year is related to that in future years by the rate of interest (Rubin 1996; Cronshaw and Kruse 1996). This relationship from economic theory provides one way to check on the performance of the market and the likely accuracy of estimated costs. Applying a discount rate of 8%, the present discounted value of a long-run marginal cost estimate of $291 in 2010 (Carlson et al. 2000) is about $157 in 2002. Allowance prices hovered between $130 and $170 for most of 2002, suggesting that the Carlson et al. model is roughly consistent with current experience, and that intertemporal arbitrage is working to an important degree.

Ellerman and Montero (2002) also argue that banking has been an efficient means of compliance. Likewise, Ellerman et al. (2000) estimate that savings from banking totaled $1,339 million over 13 years in their central case, with the Phase II bank drawn down over 8 years. This is 7% of total savings over the 13-year period, and just slightly less than the savings from spatial trading of allowances during Phase I (9%). It is substantially less than savings from spatial trading in Phase II (84%). However, the authors emphasize that although banking is a relatively minor source of total savings, it is a valuable feature of a successful trading program because it lets firms "avoid the much larger losses associated with meeting fixed targets in an uncertain world" (Ellerman et al. 2000, *285*).

Consequently, despite evidence of differences in marginal costs among firms (suggesting some lost opportunity to reduce costs in the short run), firms in the aggregate appear to be behaving efficiently over time. The trend in allowance prices over time suggests intertemporal planning consistent with economic theory.

General Equilibrium Costs

A full accounting of costs must take into account the interaction of the program with the full economy. An important body of literature has formed regarding the interaction of regulatory programs with the preexisting tax system, such as the tax on labor income. A tax imposes a difference between the before-tax wage (or the value of the marginal product of labor to firms) and the after-tax wage (or the opportunity cost of labor from the worker's perspective). Any additional regulation that raises product prices potentially imposes a hidden cost on the economy by further lowering the real wages of workers. This can be viewed as a "virtual tax" magnifying the significance of previous taxes, with losses in productivity as a consequence.[28]

Economic instruments are likely to impose a greater cost through the tax interaction effect than prescriptive approaches because the economic instruments have a greater effect on product prices, and this tends to offset some of the reduction in compliance costs. When economic instruments are used, firms must not only comply with environmental standards but also internalize the opportunity cost of the remaining emissions. In the SO_2 program, this occurs through the cost of emissions allowances.

Goulder et al. (1997) investigated the magnitude of the tax-interaction effect in the context of the SO_2 program using both analytical and numerical general equilibrium models. They find that this effect will cost the economy about $1.06 billion per year (1995$) in Phase II of the program, adding 70% to their estimated program compliance costs. Their estimate would pertain in the long run if the entire electricity sector sets prices in the market rather than basing them on cost of service. If prices were based on cost of service, then the regulatory burden would be much lower because allowances under Title IV were distributed at

Table 2-3. *General Equilibrium Cost of SO_2 Allowance Trading as Percentage of Partial Equilibrium Least-Cost Compliance*

Percentage values normalized around first cell	*Least-cost compliance (%)*	*Command-and-control performance Standard (%)*
Partial equilibrium measure	100	135
General equilibrium measure		
With revenue	129	n/a
Without revenue	171 (Title IV)	178

zero original cost. The hidden cost of the tax interaction effect would be reduced substantially, but not entirely, if instead the government auctioned the permits and used the revenues from the auction to reduce preexisting distortionary taxes. However, under grandfathering, the revenue is not available for this purpose.

If the entire industry were deregulated, the cost of the tax interaction effect could be substantial. Table 2-3 illustrates the relative potential cost savings from allowance trading and the hidden costs of grandfathered emissions allowances, compared with the costs under a command-and-control approach. The values in this table are expressed in percentage terms, normalized around the values in the cells of the first row. The value in the first cell in the first row represents the least-cost estimate of compliance in 2010, or the partial equilibrium cost, estimated by Carlson et al. (2000). The second cell in the first row shows that compliance (partial equilibrium) costs under the command-and-control scenario that was modeled in that study are 35% higher than costs under the least-cost approach.

The remaining rows reflect estimates of costs in a general equilibrium context. The first column summarizes the Goulder et al. (1997) finding that the general equilibrium costs of a market-based policy (emissions tax or auctioned permit system) are about 129% of the partial equilibrium measure of costs in the least-cost solution. The bottom row indicates that the costs of a permit system that fails to raise revenues represents about 171% of the least-cost partial equilibrium estimate.

The last cell in the bottom row of the table yields an estimate of the relative cost of command-and-control policies in a general equilibrium setting. We find that the type of policies modeled in the context of the SO_2 program—a uniform emissions standard applied to all sources—would result in general equilibrium costs that were 178% of those measured in the least-cost solution in a partial equilibrium framework.[29] In other words, the general equilibrium cost of the tradable permit program (171) is only slightly less than the general equilibrium cost of a command-and-control program (178). The example suggests that the failure to raise revenue and to use that revenue to offset distorting taxes squanders much of the savings in compliance costs that could be achieved by a flexible tradable permit system. As the electricity industry moves away from cost-of-service (regulated) prices to market-based (deregulated) prices for electricity, this failure will have greater relevance in the context of the SO_2 program.

Compliance Methods and Technological Change

Despite regulatory issues and less than fully efficient performance by the allowance market in the first years of Phase I, ample evidence suggests that the flexibility associated with allowance trading has contributed to dramatic cost savings compared with traditional regulation. It has also led to efficiency improvements that have reduced the cost of controlling SO_2 emissions. Burtraw (1996) argues that Title IV created competition between intermediate industries. Previously independent factor markets supplying services to utilities (such as coal mining, rail transport, and scrubber manufacturing firms) were thrown into competition with each other by the program's flexible implementation as they raced to supply the electricity sector with low-cost compliance strategies (Heller and Kaplan 1996). This unleashed competitive pressure to find ways to reduce costs in all these markets. The result has been a dramatic decline in the delivered cost of low-sulfur coal and improvements in the performance of scrubbing.

A review of changes in the receipts for coal distinguished by sulfur content reveals that between 1990 and 1994, sales of low-sulfur coal (defined as less than 0.6 pounds of sulfur per million Btus) increased by 28% while prices fell by 9%. Meanwhile, sales of high-sulfur coal (defined as greater than 1.67 pounds of sulfur per million Btus) fell by 18%, although prices fell by only 6% (U.S. EIA 1991; 1995a; 1995b; Resources Data International 1995).

Two trends explain the accelerated decline in the price of low-sulfur coal. The most important has been the reduction in cost of rail transportation of low-sulfur western coal, driven by investment and innovation in the rail industry following railroad industry deregulation. Rail transportation constitutes about 50% of the total cost for low-sulfur coal from the West delivered to the East, since western coal is considerably cheaper to mine. In the 1990s, coal transportation prices in the East averaged 20–26 mills (1 mill = 0.1 cent) per ton-mile. However, competition among the railroads for western coal caused prices in Phase I to drop to an average of 10–14 mills per ton-mile.

A second explanation for the decline in coal prices is that the higher capital and other costs expected for using low-sulfur coal failed to materialize because of the flexibility offered by Title IV (Burtraw 2000). With traditional emissions rate limits, or even with an emissions cap at individual facilities, facilities would be faced with the choice of either installing scrubbers or switching to low-sulfur coal. The associated capital investments in handling facilities would also be included in that choice. However, another compliance strategy that has taken hold in the wake of Title IV is the blending of subbituminous low-sulfur with bituminous high-sulfur coal. Prior to Title IV, blending of different fuels was not thought to be feasible (Torrens et al. 1992), but experimentation in response to the allowance market demonstrated that the detrimental effects of blending low-sulfur coal with other coals were smaller than originally thought.

The costs of controlling emissions at facilities that burn high sulfur coal also have been lower than expected. The capital cost of scrubbers per kilowatt of installed capacity remained fairly constant through Phase I (Keohane 2002).[30] This appears to be because induced innovation for scrubbing technology peaked before Title IV as a result of NSPS requirements to install scrubbers. The earlier requirements appear to have lowered the cost of operating scrubbers before 1990 (Popp 2001). However, patents granted during the 1990s, when the SO_2 trading program provided incentives to improve performance, indicate an improvement in the removal efficiency and reliability of scrubbers that thereby lowered their operating cost (Popp 2001).

Furthermore, the SO_2 program enabled facilities to reduce the capacity needed to achieve roughly equivalent reductions. Prior to Title IV, scrubber systems usually included a spare module to maintain low emissions rates in the event that any other module became inoperative. One estimate indicates that a spare module would increase capital costs by one-third (U.S. EIA 1994). The increased efficiency and reliability of scrubbers not only reduced maintenance costs and increased utilization rates but also reduced the need for spare modules. In addition, the allowances reduced the need for spare modules because allowances could be used for compliance during maintenance periods or unplanned outages.

Keohane (2000) and Taylor (2001) both find that abatement costs per ton of removal have fallen substantially, especially in retrofitted scrubbers installed for compliance in the SO_2 program. In addition, utilization of scrubbed units significantly increased (Ellerman et al. 2000; Carlson et al. 2000). Increased utilization is important to reducing the average cost of scrubbing because it spreads capital costs over a greater number of tons reduced. Before the SO_2 program, scrubbers did not exhibit reliability rates sufficient to achieve the current level of utilization.

Nonetheless, about half as many scrubbers were installed under Phase I as were originally anticipated. This occurred even though the allowance program itself encouraged scrubbing by allocating 3.5 million "bonus" allowances to firms that installed scrubbers as the means of compliance—for the explicit purpose of protecting jobs in regions with high-sulfur coal.

Voluntary Participation

Another source of additional allowances comes from the inclusion of units that voluntarily subscribed to the program under the industrial opt-in provisions and the utility substitution or compensation provisions. These provisions were available to facilities not covered by the first phase of the program. Volunteer facilities had an incentive to participate if they could reduce their emissions at a marginal cost that was less than the allowance price. The volunteers received allowances equal to a forecast of their emissions for industrial facilities, or allowances similar to allocations for other utilities' Phase I units for electricity generators. Participants could sell unused allowances if they reduced their emissions.

Montero (1999) finds that the voluntary provisions led to significant adverse selection. Changes in coal prices and in the utilization of power plants caused the true counterfactual baseline for many participants to differ from the formulas that were applied under the opt-in program. This provided an opportunity for some facilities to harvest allowances for emissions reductions that would have occurred anyway. The program rules encouraged this adverse selection by allowing volunteers to wait until the end of the year before declaring whether to participate, and by allowing them to determine their participation on a year-by-year basis. Although the voluntary provisions amount to only a small part of the overall program, they illustrate a flaw in the program design that could be important in other situations.

Administration and Compliance

Implementation of the SO_2 trading program has been characterized by low administrative costs and a high level of compliance. One reason is that industry shared an interest in the successful implementation of the program and there was relatively little litigation compared to many regulatory programs. For example, if EPA failed to implement regulations in a timely manner, then unit-specific emissions limitations stated in the law would apply to each and every source, without an opportunity for trading (McLean 1997)—a more costly scenario for industry. Also, there is little area for dispute in a cap-and-trade program because the law establishes the standard and the basic allocations. Few lawsuits were filed to slow the implementation of trading, and those that were filed concerned rules for substitution units and other allocation issues (Swift 2001).

The program enjoyed 100% compliance in all years of Phase I, and since then, the program has been marred only by an unimportant compliance failure of two generating units, which were short 11 allowances to cover their emissions for 2001. The failure was an administrative mistake that could have occurred only because compliance activities had become routine. It did not result in excess emissions. This program's excellent performance record compares with the 80% compliance typical of other federal air programs (Swift 2001).

Two technical features contribute to simple administration and successful compliance of the SO_2 trading program. One is the CEM system, which contributes to the confidence that emissions reductions are being achieved.[31] The other is the high and certain penalty levied for noncompliance.[32]

Conclusions

The SO_2 trading program has become an international model for cap-and-trade programs, and several features set its performance apart from traditional prescriptive approaches to regulation. Although state utility regulation and other factors inhibited utilities from taking full advantage of the allowance market, particularly in the early years of the program, ample evidence indicates that allowance trading has achieved cost savings. This is especially true if one evaluates the program relative to realistic possibilities for command-and-control approaches. Trading has allowed utilities to take advantage of trends in fuel markets, especially the expanded availability of low-sulfur coal, in ways that more rigid technology-forcing approaches to regulation would have precluded. The program has also resulted in innovation through changes in organizational technology, in the organization of markets, and in experimentation at individual boilers—much of which arguably would not have occurred under a more prescriptive approach to regulation. In addition, the program is viewed as administratively transparent: penalties are certain and compliance has been virtually perfect. The mechanism of allowance trading for SO_2 has become popular with most industry and environmental interest groups, even when the levels of the cap were in dispute.

In addition to being cost-effective, the SO_2 program has resulted in greater-than-expected benefits. Aggregate benefits are expected to be an order of magnitude greater than aggregate costs once the program is fully implemented. Moreover, despite its lack of geographic resolution, trading under the program has not affected sensitive ecosystems and human health; in fact, it may have benefited them.

One limitation of the trading program is its inability to adapt to new scientific and economic information. Even as new information becomes available about the relative benefits and costs of SO_2 reductions, the cap cannot be changed short of an act of Congress.[33] The emissions cap leaves regulators with their "feet stuck in concrete" and unable to adjust to new information (Zuckerman and Weiner 1998; Swift 2001). A more prescriptive approach, such as the NO_x provisions of Title IV, shares this attribute.

An alternative to an immovable cap would be a cap that adjusted in response to new information. Others suggest similar trigger mechanisms on emissions caps to provide economic relief if costs are greater than expected (Pizer 2002), but such an approach might better be coupled with a mechanism that provides further environmental improvement when costs are less than expected. A safety valve—relaxing the cap when allowance prices hit a specified level and lowering it when allowance prices fall below a floor—would act like a tax system in this regard by incorporating new information about costs.

The perceived success of the SO_2 program in reducing compliance costs, coupled with new information about the benefits of reducing air pollution, has contributed to a new wave of legislative proposals aimed at further reducing SO_2 emissions from power plants.[34] It reflects a remarkable consensus in the policy community that all these proposals would expand the use of a cap-and-trade approach to achieve emissions reductions. The proposals also extend this approach to the regulation of NO_x at a national level. Most would also apply a cap-and-trade approach to achieving reductions in mercury and carbon dioxide.[35] Trading of mercury allowances is controversial, however, because mercury is a hazardous air pollutant, and inclusion of CO_2 is controversial because mandatory CO_2 reductions are absent from the Bush administration's Clear Skies Initiative.

The potential long-run success of efforts in the United States to control SO_2 emissions in an economically efficient manner is illustrated in Figure 2-4, which portrays expected emissions from the electricity sector in the year 2020 under various scenarios. The first is a business-as-usual scenario, reflecting a forecast of emissions levels that might have been obtained in the absence of Title IV and other aspects of the Clean Air Act, such as eventual enforcement of new particulate standards that are likely to require reductions in SO_2 emissions. Also reported is the forecast of about 9 million tons per year of emissions, corresponding for an average year to the annual allocation of SO_2 emissions allocations after the bank achieves an equilibrium under Title IV. The next three bars in the figure represent the three multipollutant proposals. The cap on top of the bar for the Bush administration's Clear Skies Initiative (CSI) indicates that annual emissions will

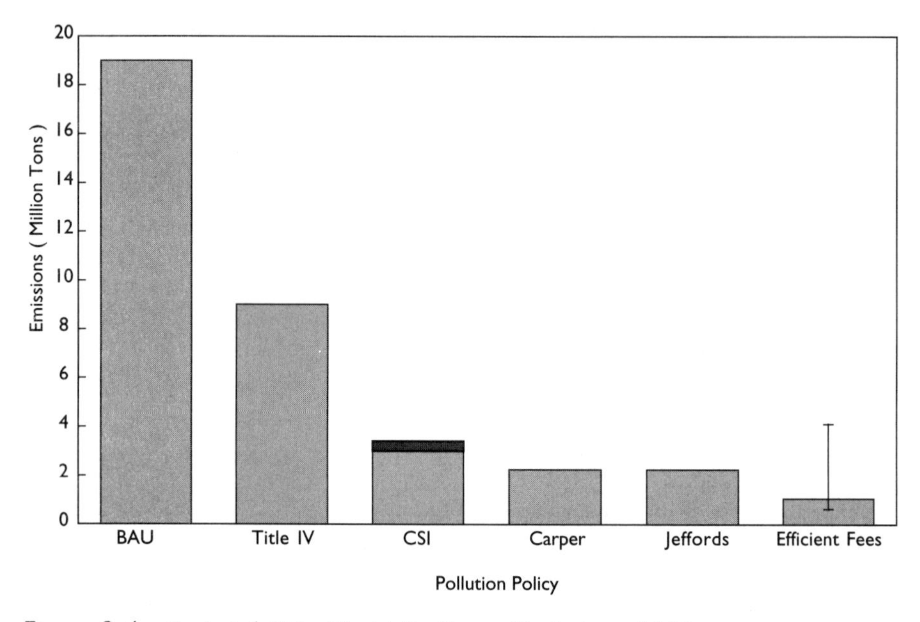

Figure 2-4. *Projected SO_2 Electricity Sector Emissions, 2020.*

be slightly above annual distribution of allowances, because the bank is not expected to be in equilibrium yet under that policy. Finally, at the right is an estimate of the efficient level of SO_2 emissions under a cost-effective regulatory policy, such as emissions fees or a cap-and-trade program coupled with an allowance auction. This estimate is the result of an integrated assessment linking the Tracking and Analysis Framework model of atmospheric transport and health benefits (excluding ozone) from reductions in SO_2 emissions with another model of electricity markets.[36] The bar indicates the 90% confidence interval around the mortality, morbidity, and valuation estimates in the benefits model (Banzhaf et al. 2002).

Figure 2-4 illustrates that if any of the multipollutant proposals on the table were to pass, emissions would fall within the range identified as efficient by cost-benefit analysis. Furthermore, under all these proposals, emissions reductions would be achieved using a cap-and-trade approach that promises to achieve the reductions in a cost-effective manner. This is important because the cost of achieving emissions reductions feeds back into the level of reductions that can be justified on the basis of cost-benefit analysis. If the program were more expensive, as would be expected under a traditional regulatory program or possibly from a cap-and-trade program that was not well designed, then cost-benefit analysis would suggest a higher level of emissions as the efficient target.

The SO_2 program has accomplished a remarkable reduction in emissions. But equally remarkable from a policy perspective are the fundamental changes that the program has wrought in the nature of environmental policy. The program was viewed by many in 1990 as ideologically motivated at best and a fraud at worst (Seligman 1994). Today, from across the political spectrum the cap-and-trade approach is the centerpiece of proposals for U.S. environmental policies.

Notes

1. Allowances are allocated to individual facilities in proportion to their fuel consumption during the 1985–1987 period multiplied by an emissions factor. About 2.8% of the annual allowance allocations are withheld by the U.S. Environmental Protection Agency (EPA) and distributed to buyers through an annual auction run by the Chicago Board of Trade. The revenues are returned to the utilities that were the original owners of the allowances.

2. See also Ellerman (2003).

3. Of the two contributors to acid precipitation, SO_2 is considered the more significant because most affected regions still have significant capacity to buffer excess nitrogen. In the future, NO_x emissions may rise in significance, depending on the region's soil characteristics. See discussion in U.S. EPA 1995.

4. Criteria air pollutants are pollutants for which EPA has designated national ambient air quality standards that all localities are expected to meet.

5. The change in the NSPS that followed the Clean Air Act Amendments of 1977 was promulgated in 1979 and made effective retroactively to September 18, 1978.

6. It is important to note that the NSPS continue to be in force for new plants.

7. Under the Sikorski-Waxman bill in the 98th Congress (H.R. 3400), scrubbing would have been applied to about half of the affected capacity and would have accounted for 70% of the SO_2 reduction. The bill also would have required switching from high- to low-sulfur coal and other improvements at other facilities (Edison Electric Institute 1983).

8. The lower estimate is according to government studies (U.S. OTA 1983), and the higher estimate is according to an industry study (TBS 1983).

9. Another bill (H.R. 4567) in 1986 was aimed at similar environmental gains but promoted cost reductions by applying a target average emissions rate for each utility company. Notably, an industry study suggested the costs would be higher as a result: although intrafirm trading would reduce scrubbing by 14% (to 74 gigawatts of capacity), it would increase reliance on low-sulfur coal, and the resulting premium on low-sulfur coal would raise costs for units already using low-sulfur coal (TBS 1986). This prediction is contradicted by the turn of events under Title IV, when the cost of low-sulfur coal fell with its expanded use. Taking account of changes in fuel and other input prices between 1983 and 1986, the study found that costs would be $7.5 billion per year. This can be compared with estimates of $3.5 billion to $6.2 billion, and of $3.4 billion to $4.3 billion by the Office of Technology Assessment (TBS 1986; all estimates in 1995$). Cost-sharing through an industry fund remained a prominent provision of this proposal.

10. For legislative histories, see Kete (1992) and Hausker (1992).

11. Electric utilities in 1985 accounted for about 70% of SO_2 emissions from point sources in the United States.

12. These estimates describe long-run costs that were expected to be obtained when the allowance bank, which was expected to build up to about 11 million tons by the end of Phase I (in 2000), would be drawn down and net contributions to the bank would be zero.

13. U.S. NAPAP (1991) (1989$). This estimate omits indirect use and nonuse benefits.

14. On March 10, 1997, EPA Administrator Carol Browner argued, "During the 1990 debate on the acid rain program, industry initially projected the cost of an emission allowance to be $1,500 per ton of sulfur dioxide.... Today those allowances are selling for less than $100" (New Initiatives in Environmental Protection, *The Commonwealth,* March 31, 1997). In testimony before the House Commerce Subcommittee on Energy and Power on the economics of the Kyoto Protocol (March 1998), Janet Yellen, chair of the Council of Economic Advisers, noted that "emission permit prices, currently at approximately $100 per ton of SO_2, are well below earlier estimates.... Trading programs may not always bring cost savings as large as those achieved by the SO_2 program."

15. See "Economists' Cold Forecast; Assumptions: Expect Their Dire Predictions about the Impact of the Global Warming Treaty on the United States. Ignore All of Them," by Elaine Karmarck, *Baltimore Sun,* December 28, 1997.

16. The cost function they estimate treats fuel type (high-sulfur and low-sulfur coal), labor, and generating capital as fully variable inputs. The econometric model consists of the cost function plus two share equations that specify the share of total costs attributed to capital and labor, and an equation for the firm's mean annual emissions rate. The study uses a translog form for the cost function, adding dummy variables for each plant in the database to measure fixed effects that vary among the plants. Costs for generating units with scrubbers are taken directly from reported data. The model does not investigate whether early commitments to build scrubbers were economical, but several studies have suggested that several of these investments were not.

17. Ellerman (2003) provides another review of the SO_2 program and an explicit comparison of Ellerman et al. (2000) and Carlson et al. (2000).

18. In their command-and-control scenario, Carlson et al. (2000) apply a uniform emissions rate standard to all facilities. GAO (1994) and Van Horn Consulting et al. (1993) allow intrautility trading, but no trading between utilities. GAO also models a scenario that requires each facility to achieve its SO_2 allowance allocation without trading, and it finds that cost savings more than double when internal trading is allowed in the command-and-control baseline.

19. The 1978 Clean Air Act regulation of sulfur emissions from newly constructed fossil fuel-fired electricity-generating facilities imposes a rate-based standard that requires a 90% reduction in a smokestack's SO_2 emissions, or a 70% reduction if the facility uses low-sulfur coal. Although nominally a performance standard, it effectively dictates technological choices and precludes compliance through the use of process changes or demand reduction. The only available technology to achieve such reductions is scrubbing, and the use of low-sulfur coal is not a permissible way to avoid the threshold for the strict standard.

20. Note that the different estimates of average control cost reflect different estimates of baseline emissions in the absence of the SO_2 emissions cap.

21. The retirement rates for coal-burning facilities and replacement with facilities using scrubbed coal technology are taken from projections by the Energy Information Administration (U.S. EIA 1996). The estimates assume that no additional retrofit scrubbers are constructed after Phase I. In fact, eight scrubbers have been installed since the beginning of Phase II; however, the cost of these scrubbers is below that of the most efficient scrubbers installed in Phase I. Carlson et al. (2000) state that if scrubber prices fall below their estimate, then theirs should be interpreted as a high estimate of the costs of the program.

22. Technological change here captures both exogenous efficiency improvements at the power plant and improvements induced by the program, but it does not capture improvements in scrubber technology and performance.

23. A CEM system continuously measures the concentration of pollutants emitted from a generating-unit. Under Title IV, the owner or operator of the unit is required to install CEM systems for all units enrolled in the SO_2 trading program.

24. In 1995, the continuous emissions monitors estimated 7% higher emissions than did the historical approach on average, although estimates vary considerably among facilities.

25. Winebrake et al. 1995; Fullerton et al. 1997.

26. This analysis excludes generating units that installed scrubbers for compliance.

27. In contrast, Bailey (1996) finds that PUC decisions *did not* impede use of the allowance market. She investigates the effect of PUC rulings clarifying the regulatory stance on allowance trading. Using state-level data on market participation, she estimates a probit model to see whether PUC guidelines and rulings affected utility participation in the allowance market within a state. The study finds that PUC rulings were positively correlated with trading

activities, but it also finds that the causal relationship may have been the reverse—that is, that desire to participate in the market often prompted utilities to request regulatory rulings to reduce uncertainty.

28. A complementary issue is the effect on the measure of benefits. Williams (2002) demonstrates that the improvement in labor productivity from reducing pollution can have sizable positive effects when measured in an general equilibrium framework.

29. The number 1.78 (178%) is the product of 1.29 times 1.35 times 1.02. The number 1.29 is the ratio of general equilibrium to partial equilibrium cost from Goulder et al. (1997) for a revenue-raising policy, such as an emissions tax. The number 1.35 is the ratio of command-and-control to efficient least-cost from Carlson et al. (2000). The number 1.02 is the ratio of general equilibrium costs for a performance standard relative to an emissions tax identified in Goulder et al. (1997).

30. In contrast, Taylor (2001) finds that capital costs have declined significantly over time.

31. Continuous emissions monitoring may not be necessary for a trading program to work. Protocols could be developed to deal with concerns about measuring emissions or emissions reductions at sources that are difficult to monitor. Such protocols have been proposed for managing trades between point and nonpoint sources of water pollution. Ellerman et al. (2000) estimate the cost of a CEM system to be about $125,000 per unit, or 7% of the total direct compliance cost.

32. In addition to surrendering allowances for a subsequent year, facilities with insufficient allowances in their accounts must pay an automatic penalty of $2,000 per ton (1990$), adjusted for inflation.

33. A related concern is that tradable permits may instill a property right that would be difficult to change. This was forestalled in the design of Title IV by explicitly stating that allowances did not constitute a property right.

34. The Jeffords (I-VT) bill (S. 366) caps annual allocations of SO_2 emissions allowances at 2.25 million tons (25% of the 8.9 million tons allocated annually under Title IV). The Bush administration's proposal (S. 1844), sponsored by Sen. Inhofe (R-OK), caps annual emissions of SO_2 at 4.5 million tons in 2010 and 3.0 million tons in 2018. The Carper (D-DE) bill (S. 843) caps annual emissions of SO_2 at 4.5 million tons in 2009, phasing down to 2.25 million tons in 2016.

35. For a comparison of the bills, see http://www.rff.org/multipollutants.

36. The Tracking and Analysis Framework model was developed to support NAPAP. The entire model and documentation (Bloyd et al. 1996) is available at http://www.lumina.com/taflist.

References

Arimura, T. 2002. An Empirical Study of the SO_2 Allowance Market: Effects of PUC Regulations. *Journal of Environmental Economics and Management* 44(2)(September): 271–89.

Bailey, E.M. 1996. Allowance Trading Activity and State Regulatory Rulings: Evidence from the U.S. Acid Rain Program. MIT-CEEPR 96-002WP (March). Mimeo. Cambridge, MA: Massachusetts Institute of Technology.

Banzhaf, S., D. Burtraw, and K. Palmer. 2002. Efficient Emission Fees in the U.S. Electricity Sector. RFF Discussion Paper 02-45 (October). Washington, DC: Resources for the Future

Bloyd, C., et al. 1996. *Tracking and Analysis Framework (TAF) Model Documentation and User's Guide*, ANL/DIS/TM-36, Argonne National Laboratory, December.

Bohi, D. 1994. Utilities and State Regulators Are Failing to Take Advantage of Emission Allowance Trading. *The Electricity Journal* 7(2): 20–27.

Bohi, D.R., and D. Burtraw. 1997. SO_2 Allowance Trading: How Do Expectations and Experience Measure Up? *The Electricity Journal* 10(7)(August/September): 67–75.

Burtraw, D. 1996. The SO_2 Emissions Trading Program: Cost Savings without Allowance Trades. *Contemporary Economic Policy* 14(April): 79–94.

———. 2000. Innovation under the Tradable Sulfur Dioxide Emission Permits Programme in the U.S. Electricity Sector. *Innovation and the Environment,* Proceedings from OECD Workshop, June 19.

Burtraw, D., and E. Mansur. 1999. The Environmental Effects of SO_2 Trading and Banking. *Environmental Science and Technology* 33(20)(October 15): 3489–94.

Burtraw, D., A.J. Krupnick, E. Mansur, D. Austin, and D. Farrell. 1998. The Costs and Benefits of Reducing Air Pollutants Related to Acid Rain. *Contemporary Economic Policy* 16(October): 379–400.

Carlson, C., D. Burtraw, M. Cropper, and K. Palmer. 2000. SO_2 Control by Electric Utilities: What Are the Gains from Trade? *Journal of Political Economy* 108(6): 1292–326.

Cronshaw, M.B., and J.B. Kruse. 1996. Regulated Firms in Pollution Permit Markets with Banking. *Journal of Regulatory Economics* 9: 179–89.

Edison Electric Institute. 1983. Evaluation of H.R. 3400—The Sikorski/Waxman Bill for Acid Rain Abatement. Washington, DC.

Ellerman, A.D. 1998. Note on the Seemingly Indefinite Extension of Power Plant Lives, A Panel Contribution. *Energy Journal* 19(2): 129–32.

———. 2000. From Autarkic to Market-Based Compliance: Learning from Our Mistakes. In *Emissions Trading,* edited by R.F. Kosobud. New York: John Wiley & Sons.

———. 2003. Ex Post Evaluation of Tradable Permits: The U.S. SO_2 Cap-and-Trade Program. MIT/CEEPR Working Paper 03-003 (February). Also forthcoming in OECD, *Ex-Post Evaluation of Tradable Permits: Case Study Reports.*

Ellerman, A., and J.-P. Montero. 1998. The Declining Trend in Sulfur Dioxide Emissions: Implications for Allowance Prices. *Journal of Environmental Economics and Management* 36(1) (July): 26–45.

———. 2002. The Temporal Efficiency of SO_2 Emissions Trading. MIT-CEEPR Working Paper. Cambridge, MA: Massachusetts Institute of Technology.

Ellerman, A.D., P.L. Joskow, R. Schmalensee, J.-P. Montero, and E. Bailey. 2000. *Markets for Clean Air: The U.S. Acid Rain Program.* Cambridge: Cambridge University Press.

Elman, B., B. Braine, and R. Stuebi. 1990. Acid Rain Emission Allowances and Future Capacity Growth in the Electric Utility Industry. *Journal of Air and Waste Management Association* 40(7): 979–86.

Fieldston Company, Inc. 1996. *Coal Supply and Transportation Markets during Phase One: Change, Risk and Opportunity.* Prepared for Electric Power Research Institute, TR-105916 (January).

Fullerton, D., S.P. McDermott, and J.P. Caulkins. 1997. Sulfur Dioxide Compliance of a Regulated Utility. *Journal of Environmental Economics and Management* 34(September): 32–53.

Goulder, L.H., I.W.H. Parry, and D. Burtraw. 1997. Revenue-Raising vs. Other Approaches to Environmental Protection: The Critical Significance of Pre-Existing Tax Distortions. *RAND Journal of Economics* 28(4)(Winter): 708–31.

Government Accounting Office (GAO). 1994. Air Pollution: Allowance Trading Offers an Opportunity to Reduce Emissions at Less Cost. GAO/RCED-95-30. Washington, DC.

Hart, G.R. 2000. Southern Company's BUBA Strategy in the SO_2 Allowance Market. In *Emissions Trading,* edited by R.F. Kosobud. New York: John Wiley & Sons.

Hausker, K. 1992. The Politics and Economics of Auction Design in the Market for Sulfur Dioxide Pollution. *Journal of Policy Analysis and Management* 11(4)(Fall): 553–72.

Heller, J., and S. Kaplan, 1996. "Coal Supply and Transportation Markets During Phase One: Change, Risk and Opportunity." Report prepared for the Electric Power Research Institute, January.

ICF. 1989. Economic Analysis of Title V (sic) (Acid Rain Provision) of the Administration's Proposed Clean Air Act Amendments (HR 3030/S 1490). Prepared for U.S. Environmental Protection Agency, Washington, DC (September).

———. 1990. Comparison of the Economic Impacts of the Acid Rain Provisions of the Senate Bill (S. 1630) and the House Bill (S. 1630). Prepared for U.S. Environmental Protection Agency, Washington, DC (July).

———. 1995. Economic Analysis of Title IV Requirements of the 1990 Clean Air Act Amendments. Prepared for U.S. Environmental Protection Agency, Washington, DC (September).

Kete, N. 1992. The Politics of Markets: The Acid Rain Control Policy in the 1990 Clean Air Act Amendments. Ph.D. dissertation, Johns Hopkins University.

Keohane, N.O. 2000. Environmental Policy Instruments and Technical Change: The Sulfur Dioxide. Unpublished manuscript. Department of Economics, Harvard University.

———. 2002. Environmental Policy and the Choice of Abatement Technique: Evidence from Coal Fired Power Plants. Paper presented at the Second World Congress of Environmental and Resource Economics, Monterrey, CA, June. Available at http://weber.ucsd.edu/~carsonvs/papers/867.pdf (accessed November 27, 2002).

Kruger, J.A., and M. Dean. 1997. Looking Back on SO_2 Trading: What's Good for the Environment Is Good for the Market. *Public Utilities Fortnightly* 135 (August).

Lile, R., and D. Burtraw. 1998. State-Level Policies and Regulatory Guidance for Compliance in the Early Years of the SO_2 Emission Allowance Trading Program. Discussion Paper 98–35 (May). Washington, DC: Resources for the Future.

McLean, B.J. 1997. Evolution of Marketable Permits: The U.S. Experience with Sulfur Dioxide Allowance Trading. *International Journal of Environment and Pollution* 8(1/2): 19–36.

Montero, J.-P. 1999. Voluntary Compliance with Market-Based Environmental Policy: Evidence from the U.S. Acid Rain Program. *Journal of Political Economy* 107(October): 998–1033.

Nelson, R.A., T. Tietenberg, and M.R. Donihue. 1993. Differential Environmental Regulation: Effects on Electric Utility Capital Turnover and Emissions. *Review of Economics and Statistics* 75(2): 368–73.

Pizer, W. 2002. Combining Price and Quantity Controls to Mitigate Global Climate Change. *Journal of Public Economics* 85(3), 409-34.

Popp, D. 2001. Pollution Control Innovations and the Clean Air Act of 1990. Working paper W8593 (November). National Bureau of Economic Research.

Portney, P.R. 1990. Economics and the Clean Air Act. *Journal of Economic Perspectives* 4(4): 173–81.

Regens, J.L., and R.W. Rycroft. 1988. *The Acid Rain Controversy.* Pittsburgh: University of Pittsburgh Press.

Resources Data International. 1995. *Phase I Databook.* Boulder, CO.

———. 1997. Implementing an Emissions Trading Program in an Economically Regulated Industry: Lessons from the SO_2 Trading Program. In *Market Based Approaches to Environmental Policy: Regulatory Innovations to the Fore,* edited by R.F. Kosobud and J.M. Zimmerman. New York: Van Nostrand Reinhold.

Rose, K., A.S. Taylor, and M. Harunuzzaman. 1993. *Regulatory Treatment of Electric Utility Compliance Strategies, Costs and Emission Allowances.* National Regulatory Research Institute, Ohio State University, Columbus (December).

Rubin, J. 1996. A Model of Intertemporal Emission Trading, Banking and Borrowing. *Journal of Environmental Economics and Management* 31(3): 269–86.

Seligman, D.A. 1994. *Air Pollution Emissions Trading: Opportunity or Scam?* Washington, DC: Sierra Club.

Smith, A., J. Platt, and A.D. Ellerman. 1998. The Cost of Reducing SO_2 (It's Higher Than You Think). *Public Utility Fortnightly* May 15: 22–29.

Solomon, B.D., and R. Lee. 2000. Emissions Trading Systems and Environmental Justice. *Environment* 42(8): 32–45.

Sotkiewicz, P.M. 2002. The Impact of State-Level PUC Regulation on Compliance Costs Associated with the Market for SO_2 Allowances. Public Utility Research Center, University of Florida (December 20).

Stavins, R.N. 1998. What Can We Learn from the Grand Policy Experiment? Lessons from SO_2 Allowance Trading. *Journal of Economic Perspectives* 12(3)(Summer): 69–88.

Swift, B. 2000. Allowance Trading and SO_2 Hot Spots—Good News from the Acid Rain Program. *Environmental Reporter* 31:19(May 12): 954–59.

———. 2001. How Environmental Laws Work: An Analysis of the Utility Sector's

Response to Regulation of Nitrogen Oxides and Sulfur Dioxide under the Clean Air Act. *Tulane Environmental Law Journal* 14(2)(Summer): 309–425.
Swinton, J.R. 2002. The Potential for Cost Savings in the Sulfur Dioxide Allowance Market: Empirical Evidence from Florida. *Land Economics* 78(3): 390–404.
Taylor, M. 2001. Legislative-Driven Innovation: The Influence of Government Action on Technological Change in Environmental Control. Ph.D. dissertation. Carnegie Mellon University, Pittsburgh.
Temple, Barker and Sloane (TBS), Inc. 1983. Evaluation of H.R. 3400: The "Sikorski/Waxman" Bill for Acid Rain Abatement. Prepared for the Edison Electric Institute (September 20).
———. 1986. Evaluation of H.R. 4567: The "Acid Deposition Control Act of 1986." Prepared for the Edison Electric Institute (April 14).
Torrens, I.M., J.E. Cichanowicz, and J.B. Platt. 1992. The 1990 Clean Air Act Amendments: Overview, Utility Industry Responses, and Strategic Implications. *Annual Review of Energy and the Environment* 17: 211–33.
U.S. Energy Information Administration (U.S. EIA). 1991. Coal Data: A Reference. DOE/EIA-0064(90). Washington, DC.
———. 1994. Electric Utility Phase I Acid Rain Compliance Strategies for the Clean Air Act Amendments of 1990. DOE/EIA-0582. Washington, DC.
———. 1995a. Coal Data: A Reference. DOE/EIA-0064(93). Washington, DC.
———. 1995b. Energy Policy Act Transportation Rate Study: Interim Report on Coal Transportation. DOE/EIA-0597 (October). Washington, DC.
———. 1996. *Annual Energy Outlook 1997*. DOE/EIA-0383(97). Washington, DC.
U.S. Environmental Protection Agency (U.S. EPA). 1995. *Acid Deposition Standard Feasibility Study: Report to Congress*. EPA 430-R-95-001a (October).
———. 2001. *EPA's Acid Rain Program: Results of Phase I, Outlook for Phase II*. EPA 430-F-01-022. Washington, DC: Clean Air Markets Division (October). Available at www.epa.gov/airmarkets (accessed November 21, 2002).
———. 2002. *EPA Acid Rain Program: 2001 Progress Report* EPA-430-R-02-009. Washington, D.C.: Clean Air Markets Division (November). Available at www.epa.gov/airmarkets/cmprpt/arp01/2001report.pdf (accessed February 3, 2003).
———. No date. GIS Analysis: Geographic Mean Centers of SO_2 Allowance Trading Activity 1995–1999: Plants Acquiring Allowance for Compliance, Beyond Each Year's Allocation; Plants Supplying These Allowances. Available at http://www.epa.gov/airmarkets/cmap/trading.html (accessed November 25, 2002).
U.S. National Acid Precipitation Assessment Program (NAPAP). 1991. 1990 Integrated Assessment Report. Washington, DC.
U.S. Office of Technology Assessment (U.S. OTA). 1983. An Analysis of the "Sikorski/Waxman" Acid Rain Control Proposal: H.R. 3400, The National Acid Deposition Control Act of 1983. Staff memorandum (revised July 12).
Van Horn Consulting, Energy Ventures Analysis, Inc., and K.D. White. 1993. Integrated Analysis of Fuel, Technology and Emission Allowance Markets. Prepared for the Electric Power Research Institute, EPRI TR-102510 (August).
White, K. 1997. SO_2 Compliance and Allowance Trading: Developments and Outlook. Prepared for the Electric Power Research Institute (EPRI), EPRI TR-107897 (April).
White, K., Energy Ventures Analysis, Inc., and Van Horn Consulting. 1995. *The Emission Allowance Market and Electric Utility SO_2 Compliance in a Competitive and Uncertain Future*. Prepared for the Electric Power Research Institute (EPRI), TR-105490, Palo Alto, CA (final report: September).
Williams, R.C. III. 2002. Environmental Tax Interactions When Pollution Affects Health or Productivity. *Journal of Environmental Economics and Management* 44: 261–70.
Winebrake, J., M.A. Bernstein, and A.E. Farrell. 1995. Estimating the Impacts of Restrictions on Utility Participation in the SO_2 Allowance Market. *The Electricity Journal* 8(4): 50–54.
Zuckerman, B., and S.L. Weiner. 1998. Environmental Policymaking: A Workshop on Scientific Credibility, Risk and Regulation. Center for International Studies, Massachusetts Institute of Technology, Cambridge, MA, September 24–25.

CHAPTER 3

Industrial Water Pollution in the United States

Direct Regulation or Market Incentive?

Winston Harrington

THE PRINCIPAL INSTRUMENT governing efforts to improve and maintain water quality in the nation's streams and lakes is the Clean Water Act (33 U.S.C. Chapter 26). Water quality became a mainly federal responsibility with the passage of the Water Pollution Control Act Amendments of 1972.

Prior to 1972, water quality was primarily a state and local concern, and the federal government's role was limited to providing grants to municipalities for wastewater treatment—the grants began in 1956—as well as information and planning assistance to the states. At the time, the states' approach to water quality was use-based; water bodies were classified according to the highest desired use, and water quality standards were set accordingly. Implicitly, disposal and transport of waste were accepted as legitimate uses of the nation's water resources. By 1970, however, strong consensus emerged that the use-based approach had not prevented the steady decline in water quality throughout the country. Several well-publicized examples of poor water quality in the late 1960s—culminating in a June 22, 1969, incident in which an oil slick on the Cuyahoga River near Cleveland caught fire—dramatized what appeared to be a growing problem. On the other hand, the first National Water Quality Inventory, conducted by the Environmental Protection Agency (EPA) in 1973, found that in general water quality had improved in the preceding decade, at least in terms of fecal bacteria and organic matter (CEQ 1976).

At the time there was discussion in Congress as to the appropriate policy instruments to be used to control industrial pollution. In November 1971 Senator Proxmire offered an effluent-charge amendment to clean water legislation then under consideration, and he and Senator Muskie debated the issue on the Senate floor. Muskie's main objection, apparently, was "We cannot give anyone the option of polluting for a fee" (Kelman 1981, *102*). As Yogi Berra might have said, if you give someone a choice, he might take it.

Aside from general skepticism about the effectiveness of effluent taxes, there was also concern that even if it might be useful in certain circumstances, it was not the right tool for the time. A command-and-control approach appeared to provide much more assurance that reductions in pollution would be achieved—and achieved quickly. Accordingly, use of economic incentives was rejected in favor of direct regulation of effluents from industrial sources, together with a mix of regulations and subsidies for publicly owned facilities used to treat domestic waste.

The new federal approach set as a national goal nothing less than the elimination of pollutant discharges into the nation's waters by 1985. This "zero-discharge" goal referred not to effluent itself but to the pollutants in effluent. But it meant that in the long run, waste disposal and assimilation would no longer be an acceptable use of water resources. Two interim goals were set: the nation's waters were to be "fishable and swimmable" by 1983, and toxic pollutants in amounts harmful to human activities or aquatic ecosystems were to be eliminated.

The Clean Water Act relied primarily on two tools to achieve these goals: First, the Construction Grants Program would provide massive federal support to publicly owned treatment works (POTWs)—wastewater treatment plants owned and operated by municipalities and local sewer districts. These grants would pay 75% of the capital costs for new wastewater treatment plants, or for expansion of existing plants.[1] The Construction Grants Program was in operation from 1973 to 1988 and, over its lifetime, paid out grants of $60 billion. It was replaced by a revolving loan fund.

The second tool was a system of technology-based regulations governing the discharge of water pollution from point sources. These point sources included both POTWs and two classes of industrial facilities: *direct dischargers*, which discharge effluent directly into receiving waters; and *indirect dischargers*, which discharge effluent into a sewer, which carries it to a POTW. The industrial standards are the focus of this investigation, and we describe them in more detail in the next section.

The Clean Water Act was in the vanguard of a major change in the federal government's regulation of economic activity. Until the late 1960s, federal regulation tended to be economic, concerned with such matters as regulating the prices of goods or services produced by industries thought to be natural monopolies and whose activities crossed state lines. These included railroads, airlines, and transmission of natural gas and electricity. Federal regulation also restricted activities of banks and sought to prevent excessive concentrations of market power. The 1970s began a period of "social regulation," concerned with workplace safety and health, environmental quality, exposure to hazardous chemicals, unsafe consumer products, and like concerns. Ironically, as social regulation waxed, economic regulation waned, with deregulation of airlines, trucking, railroads, banking, and now electricity.

Regulations for Point Sources

For point sources, the backbone of the regulation is the National Pollutant Discharge Elimination System (NPDES), which requires permits of all significant dischargers of wastewater into surface waters. These permits state the effluent discharge limits the source must meet, expressed in allowable mass loadings, pollutant concentrations, or other norms. The dischargers affected include both industrial plants and POTWs, which are mostly owned by municipalities or special sanitary districts and are designed to treat domestic waste.

The specific requirements in the permits are determined by a complex system of regulation that begins with federally established effluent guidelines.[2] The guidelines establish a set of technology-based performance standards that all point sources must meet, except where water quality considerations demand even more stringent standards.[3] To allow for the vast heterogeneity in American industry, the guidelines are very detailed, breaking industrial plants into a large number of categories and subcategories, each with its own set of pollutant-specific regulations. Among other things, the emissions limitations differ on the bases of product, industrial process, age of equipment, geographic region and size.

For water bodies where application of the technology-based effluent guidelines would not be sufficient to achieve the appropriate water quality objective, permit writers were required to set even more stringent "water-quality-limited" standards.[4] Setting these standards was a matter of allocating to individual plants the total waste load that the water body could handle. This maximum load necessarily depended on the current conditions of the receiving water body and its capacity to absorb waste.

The front-line administration of this program—that is, the writing of the NPDES permits and the routine monitoring and enforcement of permit requirements—could be delegated to appropriate state agencies that demonstrated sufficient legal and institutional capacity for the job. EPA regional offices would administer the program in other states and provide oversight to the delegated programs. At present, nearly all the states have delegated programs. The state departments of environmental quality (DEQs)[5] are supervised by the 10 EPA regional offices.

POTWs

Regulations for POTWs. Before turning to industrial point sources, we first describe the effluent discharge policy for POTWs, which also had to obtain NPDES permits. POTWs have an important influence on the industrial point source program. Because many industrial plants—the indirect dischargers—discharge wastewater into municipal sewers, POTWs have the dual role of regulator and service provider for indirect discharging plants.

Separate sets of guidelines for POTWs expressly address the treatment of household waste, which consists of about 100 gallons per person per day of organic waste rich in fecal bacteria and containing about 300 milligrams per liter (mg/l) each of biochemical oxygen demand (BOD) and total suspended solids (TSS), plus varying amounts of organic phosphorus and nitrogen. In 1968, 60% of households served by POTWs had "secondary treatment"—use of physical processes (e.g., skimming, screening, settling) followed by biological processes that together were capable of removing about 80–90% of BOD and TSS, leaving waste concentrations of about 15–30 mg/l. (U.S. EPA 2000). One of the goals and eventual achievements of the Clean Water Act was to implement secondary treatment throughout the United States and "tertiary," or advanced, waste treatment processes where needed to meet water quality standards.

The typical restrictions contained in an NPDES permit for a POTW are as follows:

- specific limitations on both conventional and nonconventional pollutants in both wastewater and sludge;[6]
- limitations designed to respect statutory limits on concentrations of numerous toxic "priority pollutants" specifically mentioned in the Clean Water Act;
- criteria on acceptable uses for sludge;
- removal efficiency requirements (e.g., 85% removal of BOD); and
- other operating requirements to ensure effective operation and maintenance.

Regulations for Indirect Discharges. Household wastes show little variation from one day to the next, at least compared industrial wastewater, and POTW designs take advantage of this characteristic. Much industrial waste is similar in important ways to domestic waste; for example, the food and paper industries have waste streams that are primarily organic. Thus many industrial wastes can potentially be treated in POTWs. However, industrial wastewater can also cause serious problems for POTWs. Toxic material or highly acidic or alkaline material can disrupt the microbial ecology of the waste treatment plant, reducing its efficiency. Other wastes, toxic or otherwise, can pass through the plant unaffected and pose a direct public health risk or a threat to aquatic ecosystems. Still other wastes, such as alcohol, might be treatable but pose a threat of fire or explosion within the sewer itself. Finally, some industrial wastes could be too clean. Cooling water, for example, would simply add to the flow of the plant, diluting the waste stream and making pollutant abatement more difficult and costly.

The pretreatment guidelines (40 C.F.R. Part 403) were designed to assist POTWs in dealing with the above problems. They contain instructions for setting up a pretreatment program, plus specific prohibitions against industrial discharge of wastes that would harm the POTW or pass through it without adequate treatment. In addition, the guidelines established technology-based pretreatment standards for the quality of wastewater sent to a POTW from certain industrial categories. For other industries, the standard for pretreatment are be set at the local level. In states where permit responsibility has been delegated, the state DEQs can further delegate responsibility for writing and enforcing permits to the local POTW. Nearly all states have done so.

In addition, the effluent guidelines for each industry contain pretreatment standards for new and existing plants discharging into sewers. These standards were designed to prevent industrial discharges from interfering with plant operations and to limit pass-through of untreated pollutants to what a direct discharging plant would be allowed.

To set the local limits for pollutant discharge by industries, the POTW conducts a "headworks analysis," or a pollutant by pollutant estimate of the total waste loading that the plant can safely accept from nonhousehold sectors. An

EPA guidance document (U.S. EPA 2001) provides detailed instructions on preparing the headworks analysis and recommends that it be revisited every year. The headworks analysis begins with an estimate of allowable waste discharge into the environment, in either the plant effluent or the sludges. The allowable effluent discharge is generally taken from the NPDES permit, but the POTW may have more discretion on sludge composition. If the plant wishes to produce sludge that is salable, for example, the permissible loadings of toxic materials is much lower than it otherwise might be. Given the permissible discharge of each pollutant, getting the permissible influent at the headworks requires knowledge of the total removal efficiency of each pollutant by the plant. Calculation of these removal efficiencies is thus central to the headworks analysis.

Next, the POTW estimates waste discharges from waste haulers, such as septic tank cleaners, and other wastes that may be delivered by truck, and from "uncontrolled sources," perhaps small commercial operations that are difficult to control. These wastes are subtracted, pollutant by pollutant, in their entirety from the plant's "maximum allowable headworks loading." From the remainder, the POTW further subtracts a safety margin to allow for contingencies and a growth allowance. What is left is the waste load that can be allocated to nonhousehold users, or in EPA terminology, the maximum allowable industrial loading (MAIL).

Now the POTW must allocate the allowable discharges for each pollutant to the industrial users. Evidently, the most common allocation method is to set "uniform-concentration local discharge limitations," which "have become synonymous in the Pretreatment Program with the term local limits," according to EPA (2001, *6–3*). In this method, the allowable discharges of each pollutant are allocated to users so that the limits, expressed in terms of pollutant concentration, are the same for each pollutant.

Although this is apparently the most common method chosen, it is not required. EPA guidance and regulations do not, for example, rule out the use of marketable permits to allocate the MAIL, much as emissions offsets and ultimately cap-and-trade programs grew out of aggregate emissions limits in nonattainment areas under the Clean Air Act. So far, however, there has been little use of tradable permits in this context anywhere in the country. Apparently the only tradable permit program currently operating for the allocation of local limits in POTWs is in Passaic, New Jersey.[7]

Regulations for Direct Industrial Discharges. In the 1972 act, Congress directed EPA to prepare guidelines for 30 designated industries,[8] including such major sectors as pulp and paper, organic chemicals, seafood, and fruit and vegetable processing. As noted above, the standards were supposed to be technology based. Congress in fact specified several kinds of standards: First was best practicable technology (BPT), which all plants in affected industries were to adopt by mid-1977. Congress did not define practicable, but EPA appeared to rely on two rules of thumb: Where applicable, BPT meant secondary or biological treatment, and otherwise it would represent the best standard of treatment currently found in the industry. For other categories, guidelines were often based on the "average of the best" plants identified in the industry survey. The "best" plants were those in the surveys that were identified as having "suitable and well-operated treatment in place." (Caulkins and Sessions 1997, *98*). More stringent standards were the best available technology (BAT) that was economically achievable, which were to be installed by mid-1983. The informal rule for BAT standards was the "best of the best" surveyed. Still more stringent were the new source performance standards (NSPS), which were to be applied to new plants seeking permits after the standards were promulgated.

Rulemaking

In this section we consider two classes of outcomes of the effluent guidelines process. After describing how the effluent guidelines were written, we discuss what may be called the rulemaking outcomes, or the administrative outputs of the process. Then we turn to the "real" outcomes—the effectiveness of the effluent limitations on the ground and their actual cost.

Process Inputs

One of the charges raised against the use of command-and-control (CAC) policy instruments is the supposed administrative effort required to implement the standards. This is a rather sweeping complaint, considering the great variety of CAC policies. In the case of the effluent guidelines, however, it seems to be a valid concern. The "phase I" guidelines, for example, required thousands of pollutant discharge regulations in hundreds of industrial subcategories. The burden fell not only on the agency but also on other interested parties, such as affected firms, trade associations, and environmental groups, in making extensive comments on proposed regulations and, frequently, challenging final regulations in court.

Writing the effluent guidelines was an enormous task for the young agency. Not only did it have to give operational meaning to words like *best, practicable,* and *economically achievable,* it also had to collect a vast amount of information—about production techniques, location, waste products, and waste treatment technology—for each industry to be regulated. The agency started with a very narrow information base on these matters. Whereas the Federal Power Commission or the Interstate Commerce Commission regulators dealt with a single industry, EPA had to address the full range of manufacturing in the American economy.

The information requirements were exacerbated by another factor. Very early in the standard-setting process, EPA became aware that the great heterogeneity in the products and processes of each of the 30 industries would preclude use of the same standards for all plants in that industry. Each industry, therefore, had to be subcategorized into generally homogeneous subsets for purposes of water pollution regulation. The number of subcategories was large—20 in dairies, for example, and 105 in fruits and vegetables. In all, EPA created over 360 industrial subcategories for the first 30 industries requiring effluent guidelines. Each required separate BPT, BAT, NSPS, and pretreatment regulations.

To collect and organize the mass of information required to set these standards, EPA hired engineering consulting firms. Naturally, the firms with the necessary expertise in the appropriate industrial and wastewater treatment technology had ties to the industries being regulated. These ties were an asset in winning the cooperation of the firms to be regulated, but they also raised concerns in some quarters about conflicts of interest.

The contractors had the task of preparing "development documents" containing information on the structure of the industry, its production and waste treatment technologies, the estimated cost of pollution abatement, and suggested effluent standards. The suggested standards were studied by EPA and often circulated among industry sources for comment. The agency then issued the proposed standards. Interested parties had the opportunity to submit additional information and make comments on the regulation during a comment period lasting several months. EPA took these comments and, after some further time had elapsed, issued final standards.

In promulgating the effluent guidelines, the agency had to follow the procedures of informal or "notice-and-comment" rulemaking as specified in the Administrative Procedures Act (5 U.S.C. 553). This act requires the agency to give notice of proposed rulemaking, propose regulations, and allow a 60- to 90-day period for interested parties to make comment before promulgating final regulations. EPA received many comments on the proposed effluent guidelines and was required to answer each comment in the Federal Register.

The regulations were often challenged in court. As of 1976, the National Commission on Water Quality (1976) reported that more than 250 lawsuits had challenged specific guidelines. Every set of guidelines faced litigation.

Outcomes: Productivity

Over time, the guidelines were part of a policy that imposed substantial costs on and achieved substantial pollution reductions from industrial sources. EPA estimates that today the guidelines reduce water pollutant discharges by 690 billion pounds per year (U.S. EPA 2004). However, the agency was unable to meet the statutory deadlines in P.L. 92-500.

Initial BPT Standards. Given the difficulties of promulgating the standards, it is not surprising that the promulgation of the effluent guidelines fell behind the schedule set by Congress. Indeed, not one of the original 30 sets of guidelines was issued within the statutory deadline of one year—probably more because of the unrealistic deadline set by Congress than because of any regulatory dawdling by EPA. It should be clear from the procedural and analytical requirements, not to mention the necessity of collecting and responding to public comments, that EPA's task was impossible within the statutory limitation and the resources available to it.

By any other standard, the effluent guidelines process was remarkably productive in terms of regulations written. Compared with other new rulemaking processes of approximately the same time, including health and safety regulation at the Occupational Safety and Health Administration, hazardous substance regulation under the Toxic Substances Control Act, and regulation of consumer product safety at the Consumer Product Safety Commission, many more regulations were written for the effluent guidelines program than for any other program (Magat et al. 1986).

BAT Standards. Amendments to the Clean Water Act in 1977 made a mid-course correction to the effluent guidelines program. In particular, the BAT standards were modified to focus on toxic discharges, and a new set of standards, called best conventional technology (BCT), designated more advanced abatement of conventional pollutants than that provided by the BPT standards. The shift toward toxics was driven by the report of the National Commission on Water Quality, together with the consent decree settling the lawsuit *Natural Resources Defense Council v. Train* 8 ERC 2120 (D.D.C. 1976). The 1977 amendments established a list of toxic "priority pollutants" and required EPA to develop BAT standards for toxic pollutants in 23 industries.

According to Adler et al. (1993), EPA had difficulties writing new BAT toxics regulations or revising existing BAT regulations in additional industries to reflect the new orientation on toxics. In 1987, further amendments to the Clean Water Act directed EPA to prepare a plan indicating how it would comply with the BAT requirements. When EPA's plan was unveiled in early 1988, the Natural Resources Defense Council (NRDC) again sued, leading to a consent decree in which EPA agreed to prepare more than 20 new BAT regulations in the next 10 years. As of 1992, NRDC estimated that only one-third of all direct dischargers had permits based on the BAT standards. The other two-thirds were in categories for which no standards were written and had permits based on the best professional judgment (BPJ) of the permit writer (Adler et al. 1993).

Zero Discharge. As a step toward meeting the zero-discharge goal, the BAT standards were to "result in further progress toward the national goal of eliminating the discharge of all pollutants ... [and] require the elimination of all discharges if ... such elimination is technically and economically feasible" (Sec. 301(b)(2)(A)). Sec. 304(b)(3) requires EPA to "identify control measures and practices available to eliminate the discharges of pollutants from classes and categories of point sources."

Inasmuch as complete elimination of all discharges would seem to be impossible, since 1972 there has been confusion and dispute over the meaning of the zero-discharge goal and how to make it operational. Adler et al. (1993) argue that the point of zero discharge was what we now call pollution prevention: the reduction, elimination, or capture of pollutants before they enter the wastewater stream. They strongly criticize EPA for failing to implement the portions of the Clean Water Act relating to zero discharge. BAT retains to this day an end-of-pipe focus, and EPA has issued zero-discharge regulations in very few industrial categories.

Outcomes: Regulatory Stringency

The original premise of the BPT, BAT, and NSPS standards was to limit the influence of politics on the process, since the standards would be based on the technological possibilities as determined by disinterested experts at EPA and elsewhere (Freeman 1978). Seemingly, this language did not permit much bargaining between the agency and the regulated sources of pollution. However, the statute allowed considerations of cost and technological feasibility in the setting of standards, opening the door for bargaining between EPA and the regulated industry.

For this reason the promulgated effluent guidelines in most subcategories were substantially and systematically different from the proposed guidelines. For the first 30 industries, the BPT standards for BOD and TSS increased on average 44% and 92%, respectively. (Of course, if a standard initially called for 90% reduction of a pollutant, a 92% increase in the standard would still result in an 81% reduction.) Only for a small number of subcategories were the final standards tighter than the contractors' suggested standards or the agency's proposed standards.

Those results might seem to suggest that the affected industries intervened to influence the content of the regulations in a way favorable to their interests. However, a statistical analysis of the BPT rulemaking experience by Magat et al. (1986) found otherwise. Magat et al. examined the effect on the BPT regulations of "external signals," including the number of comments received from industrial sources on the regulation, the political and economic power of the industry affected, and warnings or projections of unemployment or plant closures, and found none that were statistically significant.

Instead, they found that several *internal* variables, including document quality and staff turnover, were much more important in explaining outcomes. The researchers made their own *ex ante* assessments of the quality of the development documents and economic analyses, based primarily on whether information presented was internally consistent and whether calculations could be replicated. They found that when the development document failed their quality test, the resulting BPT standards were 33% weaker for BOD and 44% weaker for TSS than they would have been had the document been stronger. Turnover during the process was associated with a weakening of the standards of 33% (BOD) and 68% (TSS).

Costs

One of the most important barriers to analysis of regulatory instruments is the great difficulty of getting good estimates of the cost of compliance, especially *ex post*. *Ex ante* estimates are a little easier to come by because, since the 1970s, federal rulemakers have been required either by agency policy or by presidential executive order to prepare estimates of the cost of compliance. From the very first rules promulgated in the effluent guidelines program, the agency estimated costs of compliance. The development documents prepared to support the individual regulations contained fairly detailed data on abatement costs of "model plants" of various output capacities and levels of abatement.

Two very different cost criteria have at various times played visible roles in effluent guidelines rulemaking. One is *affordability*, a criterion that focuses on the overall impact of the regulation on the financial health of the firms and workers in the industry. To measure affordability, the regulators and their contractors

Table 3-1. Ex ante *Incremental Costs of BOD Removal ($/kg)*

Chicken, large plants	0.10
Chicken, medium plants	0.16
Chicken, small plants	0.25
Duck, Large plants	1.04
Duck, Small plants	3.15
Fowl, large plants	0.10
Fowl, small plants	0.20
Further processing	0.35
Turkey	0.60

Source: Magat et al. (1986).

combined abatement cost estimates with output demand elasticities taken from the economics literature to get the effect of the regulation on firm and industry profit. Employment and plant-closure effects were in turn based on the estimated loss of profits, using a simple model of industry entry and exit. More aggregate indicators were also calculated, such as the effect on inflation and overall unemployment.

The other important cost criterion is *cost-effectiveness.* For the effluent guidelines, whose objective is the reduction of pollutant discharges from industrial plants, cost-effectiveness refers to the incremental reduction in pollutant discharges per dollar of abatement cost. For a given resource outlay, a cost-effective policy is one that produces more environmental benefit than any other with the same cost, or alternatively, one that achieves the environmental objective at the least cost. A necessary condition for cost-effectiveness is that the incremental costs of pollutant reduction should be the same for all firms in all industries.

Incremental Costs and Economic Impact. In the 1970s, when the BPT and BAT standards were prepared for the first group of industries, rulemakers focused on affordability. From these development documents, researchers at Resources for the Future were able to construct rudimentary incremental abatement cost functions that would yield estimates of the cost of compliance with the regulations. These cost functions show that against a goal of minimizing total discharges of BOD and TSS, the effluent guidelines were far from cost-effective.

For example, among various subcategories of the poultry industry, *ex ante* incremental costs of BOD removal varied between $0.10 and $3.15 per pound. As shown in Table 3-1, the costs varied by plant size and type of bird processed, with chicken and fowl processing plants having particularly low marginal abatement costs. These regulations were remanded and later withdrawn, but nevertheless they indicate the lack of interest in cost-effectiveness criteria at this time.

Not surprisingly, given such results, examination of economic analyses conducted at the time to support the effluent guidelines give no sign that cost-effectiveness calculations were ever made. Instead, the regulators were visibly concerned about affordability. Affordability calculations did not affect the changes in the rules during the rulemaking process, but given their prominent place in the calculations, they did appear to affect the stringency of the regulation in the contractors' draft development documents or in the proposed regulations.[9]

For the BAT rules promulgated during the 1980s, however, considerations of cost-effectiveness (defined on the basis of pollutant toxicity rather than mass) apparently did play a role. Fraas and Munley (1989) examined the technologies discussed in the development documents supporting BAT rules promulgated between 1981 and 1987, using a probit model in an attempt to uncover the decision rule that would determine which technologies were included and excluded. They divided the time interval into three periods, corresponding to the regimes of Anne Gorsuch, William Ruckelshaus, and Lee Thomas as EPA administrator. During the initial period (Gorsuch), they found, a cost-effectiveness rule (i.e., constant incremental costs) did quite well in explaining inclusion or exclusion of BAT technologies. Under Ruckelshaus and Thomas, cost-effectiveness still played a role, but it was diminished relative to considerations of affordability and regulatory stringency. The threshold for inclusion increased, indicating rules of greater stringency, and there was a larger variance in the errors associated with cost-effectiveness, indicating that other considerations entered besides cost-effectiveness.

Comparison of *ex ante* and *ex post* Cost Estimates. In 1979, in response to a directive from Congress, EPA published *The Cost of Clean Air and Water*, which estimated the expected cost to governments at all levels and to all industries of complying with the Clean Air and Clean Water acts. According to this report, in the decade between 1977 and 1986, the nation's manufacturers would spend \$3.8 billion per year in capital costs and \$15.9 billion per year in operating costs (2002\$).

Table 3-2 compares this estimate with the actual costs for the years 1982 through 1994, as reported in the annual Pollution Abatement Capital Expenditure (PACE) survey. As noted in *The Cost of Clean*, delays in the issuance of rules and still greater delays caused by legal challenges to them threw the timetable off by several years. Because these delays were much more of an issue in some categories than in others, there is no exact comparison as to time period, and looking at the period 1982–1994 is probably no worse than looking at 1977–1986. Also, the PACE survey did not begin until 1982. As shown in Table 3-2, the capital costs rose sharply beginning in 1988, doubled by 1991, and then receded slightly. Examination of more-detailed data shows that most of the increase is in a few industrial categories, notably organic chemicals and petroleum and coal products. Apparently this reflects the promulgation of the costly and important BAT rules in these industries.

In the aggregate, reported abatement costs in every year from 1982 to 1994 fell far short of the abatement costs estimated *ex ante* in 1979 (Table 3-2). On average, capital costs were overestimated by 72% and operating and maintenance costs by 117%. The pattern of overestimation holds in all industrial categories; as shown in Table 3-3, in only one category (stone, clay, and glass) were costs underestimated. Table 3-3 also illustrates how the costs of abatement are concentrated in a few industries. Chemicals alone accounts for 31% of all abatement investment and 27% of operating costs.

This comparison is consistent with earlier findings by Harrington et al. (2000), which compared *ex post* and *ex ante* cost estimates for about 25 environmental and occupational health programs. Abatement costs tend to be overestimated for

Table 3-2. *Estimated* ex post *Abatement Costs, All Manufacturing (Million 2002$)*

Year	Capital	Operation and maintenance
1982	$1,629	$5,817
1983	1,314	6,326
1984	1,373	6,644
1985	1,526	6,911
1986	1,524	7,070
1987	NA	NA
1988	1,775	7,267
1989	2,420	7,765
1990	3,385	8,192
1991	3,469	7,818
1992	3,019	7,911
1993	2,696	7,790
1994	2,795	8,092
Mean, 1982–1994	2,244	7,300
Ex ante	3,870	15,893

NA = not available

Table 3-3. *Water Pollution Abatement Expenditures: Comparison of* ex ante *and* ex post *Estimates (Million 2002$)*

	1979 estimate[a]		1986 PACE[b]	
	Capital costs	Operation and maintenance	Capital costs	Operation and maintenance
All industries	3,870.6	15,893.2	1528.4	7,083.7
Food and kindred products	175.0	2,349.9	159.0	823.0
Textiles	346.9	1,107.3	15.4	139.5
Lumber	24.4	135.1	16.0	62.6
Furniture	1.2	3.1	1.0	26.9
Pulp and paper	371.6	1,594.5	142.4	831.3
Printing and publishing	NA	NA	6.3	37.9
Chemicals	1,284.4	2,999.5	478.3	1,913.7
Petroleum and coal products	361.7	2,738.7	178.6	849.4
Rubber and miscellaneous plastics	101.9	265.3	14.1	76.4
Leather	33.2	72.6	1.6	24.8
Stone, clay, and glass	13.0	101.6	20.1	100.4
Primary metal	490.6	2,541.8	109.8	748.6

[a]U.S. EPA (1979).
[b]U.S. Bureau of the Census (1989).
NA = not available

several reasons. Three are probably of special importance for effluent guidelines. First, the cost estimates do not—cannot—take account of technological innovation. Regulators were supposed to find an off-the-shelf technology that would achieve the specified abatement, and the cost estimate referred to this technology. Second, the baseline level of industrial pollution control was probably better than EPA expected when the cost estimates for the first group of industries were prepared (1973–78). As a result, the typical industrial plant found it easier to comply than EPA expected. Finally, as noted above, the regulations were relaxed considerably between the contractors' reports and the final promulgated standards. Most likely, the cost estimates reflect the agency's proposed standard rather than the more relaxed standards that were promulgated.

Patterns of Abatement Investment. One stated purpose of the zero-discharge requirement of the Clean Water Act was to encourage alternatives to end-of-pipe treatment. Although regulations were never promulgated to promote source reduction, the existing BAT regulations as well as the pretreatment requirements for indirect sources provide some incentive for source reduction. Even though technology-based standards for pollution abatement typically designate a particular end-of-pipe technology, permitees are generally allowed to use whatever technology is capable of meeting the requirements.

With this in mind, let us consider what the PACE data reveal about the choice of abatement technique. At least between 1986 and 1994, the PACE questionnaire allowed respondents to disaggregate expenditures into "end of pipe" and "process change" categories. In Figure 3-1, we show the share of abatement investment designated as process change, by sector, for the years 1986 and 1994. Process change for all industry increased from 18% to 30% of all investment in this interval. Only two industries showed declines: furniture and nonelectrical machinery. Several categories, including printing and publishing, fabricated metal products, and chemicals, could show very large gains in the importance of process-change investment. Over this period, total investment in abatement nearly doubled, so the absolute increase in process change investment was greater than these figures suggest.

Real Outcomes

We consider two outcomes of the industrial water pollution permitting process: first, the extent to which the new permits resulted in actual reductions in the amounts of pollutants entering water bodies, and second, the evidence related to changes in water quality.

Have Pollutant Discharges from Point Sources Been Reduced? According to an EPA estimate made in the early 1990s, full compliance with BAT-based permits and secondary treatment for POTWs would lead to a 97% reduction in the direct discharge of priority pollutants from POTWs and industrial point sources into the nation's waters (Adler et al. 1993, *139*, citing internal EPA documents). That assumes not only that discharge permits reflect BAT, but also that the sources comply with them.

By the mid-1980s direct dischargers had apparently achieved and thereafter sustained a high rate of compliance with permit conditions. EPA estimated in 1984 that only 6% of major direct dischargers were in "significant noncompliance" with their permit requirements. Over the next decade, the rate remained at this level, except for a brief excursion to about 14% in 1990 (U.S. EPA 1987, 1995). Thus, by the mid-1980s, direct dischargers had achieved a high rate of compliance with their NPDES permits. As noted above, however, most direct dischargers had permits that were based not on BAT but on the less stringent BPJ standards. It appears, therefore, that direct dischargers may have achieved sig-

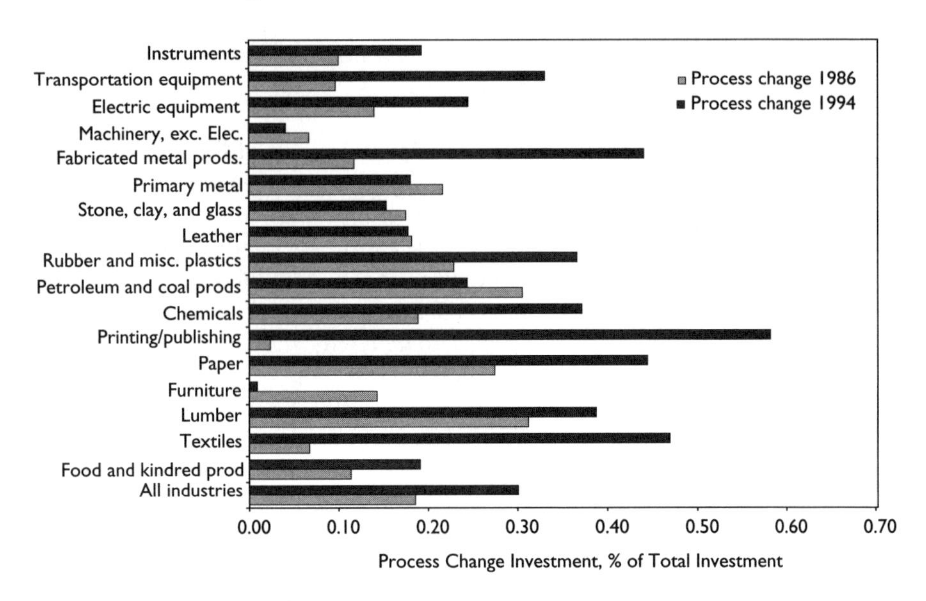

Figure 3-1. *Share of Abatement Investment Designated as Process Change, 1986 and 1994*

nificant effluent reductions by 1994, but they had not achieved the reductions anticipated in the BAT standards.

Among POTWs, the large-scale investment in wastewater treatment increased the fraction of the United States served by wastewater treatment plants from 42% in 1970 to 74% by 1985. Among all POTWs, the rate of significant non-compliance with permits was 11% in 1984 and had declined to about 9% in 1994. However, an EPA-sponsored audit of POTW pretreatment programs in 1992 found that 54% of significant industrial users were in "significant noncompliance," which means they failed to observe at least one component of the POTW pretreatment program. Of these, 35% were in violation of discharge standards and 36% were in violation of self-monitoring or reporting requirements (Adler et al. 1993).

At least until the mid-1990s, in other words, the pollutant reductions achieved by the effluent guidelines program were significantly less than what had been achieved by full implementation. This failure probably accounts for at least part of the gap between *ex ante* and *ex post* estimates of compliance costs.

Has Water Quality Improved? It is generally agreed that the Clean Water Act has had some important local successes in improving water quality. In rivers near major cities, contact recreation is now encouraged in areas where it had been forbidden in the 1970s, an outcome probably attributable to the regulation of pollutants from point sources. Knopman and Smith (1993) cite improvements in the Potomac near Washington, the Delaware near Philadelphia, and the Flint River in Georgia. State water quality control administrators' assessments of their own states' progress during the first decade of the Clean Water Act indicated that most believed there had been significant improvement (ASIWPCA 1984). A report by the National Research Council (1993) acknowledged these improvements but warned that there were still many problems.

Even today, a systematic assessment of the effects of the Clean Water Act on water quality is very difficult and uncertain because of a lack of suitable data and no consensus on the most meaningful ways to aggregate trend data across water bodies and pollutants. The best evidence, from the 424 stations in the National Stream Quality Assessment Network (NASQAN),[10] is mixed. For example, consider dissolved oxygen (DO), the water quality criterion that might be expected to show the most improvement as a result of pollution abatement from point sources. During the 1980s there was no significant change in average DO concentrations across all monitoring sites. Taken individually, 38 stations showed increasing DO, 26 showed decreasing DO, and for the rest there was no significant change. Among the 26 stations near urban areas, however, there was apparent improvement, as the percentage reporting large DO deficits (average concentrations less than 6.5 mg/l) decreased from 40% of all stations to 20%. Note the small sample size, however (Smith et al. 1991).

Smith et al. (1991) also examined loadings of several pollutants delivered to streams. For nitrate, the data show no change in loads except in two of the nation's 14 water-resource regions, where concentrations declined by 0.4% per year (lower Mississippi) and 1.6% per year (upper Mississippi). Phosphorus concentrations declined in all but one region, and in three midwestern regions they fell by more than 3% per year. The authors attributed the improvement to reductions in point-source discharges and more widespread use of low-phosphorus detergents. Finally, suspended solids declined in 11 regions (by at most 1.3% per year) and increased modestly in three regions.

Those results suggest that the Clean Water Act, and in particular its point source programs, has made at least modest improvements in water quality, especially considering the increase in economic activity during the 1980s. Incomplete implementation of the Clean Water Act during the 1980s was one reason improvements were not larger, but probably a far more important factor was the failure of the act to do much about nonpoint sources.

Long-Run Responses of Point Sources to the Clean Water Act

In the long run, it is expected that economic incentive (EI) and command-and-control (CAC) approaches to pollution abatement will lead to different results on the ground. Two hypotheses in particular address these differences. First, we hypothesize that EI approaches will exhibit greater dynamic efficiency—that is, they will encourage more innovation in waste treatment than CAC instruments, since they exert more pressure on firms to reduce pollutant discharges.

In addition, we expect EI approaches to be more adaptable to changes in conditions. Suppose, for example, economic growth in a region means that existing pollution-reduction policies cannot achieve acceptable air or water quality. If emissions permits are marketable and fixed in supply, the reallocation of waste discharges is taken care of automatically, without intervention by the authorities.[11] To accommodate growth in a CAC system, however, the entire regulatory process needs to be reopened, with new regulations established for at least some sources. In fact, one of the ways that CAC systems have accommodated change in the past has been to convert, at least in part, to economic incentive systems. Indeed, this is how tradable permits were first introduced into stationary source air pollution policy in the late 1970s (Krupnick and Harrington 1981). Below we will discuss evidence for and against the proposition that economic incentives are likewise being gradually introduced into the point source water pollution program.

Thirty years (1972–2002) is certainly enough time to observe the effects of the Clean Water Act on patterns of point source water pollution, the overall structure of water pollution abatement, and how these changes in turn have affected the nature of regulation. And as we describe below, some interesting and important developments make the policy landscape much different from what it was 30 years ago. Unfortunately, these changes are very difficult to document systematically because the relevant data, when collected at all, are scattered in EPA regional offices, state DEQs, and POTWs. What follows is mostly a journalistic account, based on conversations the author had with officials in EPA's Office of Water, regulatory authorities in the water program offices of state DEQs, pretreatment coordinators at POTWs, and engineers at regulated firms and their consultants.

The Legacy of the POTW Construction Grants Program

The Construction Grants Program appears to have had a powerful effect on industrial wastewater discharges. These federal construction grants provided 75% of the capital cost of abatement technology, more if the state chipped in. At the time, economists and others (e.g., Freeman 1978) pointed out that these subsidies created perverse incentives for POTWs. In the first place, a capital-only subsidy distorts the mix of capital and other inputs and encourages the installation of waste treatment that is excessively capital intensive. Freeman also argued that the subsidy would encourage the building of excess capacity. Perhaps there would be no such incentive if POTWs could be certain that the subsidy would always be available, but such a guarantee would be impossible to give, and in any case, it was known that only a fixed amount of money was appropriated for the purpose. In addition, communities could have viewed the excess sewer capacity as a way of attracting industry.

A study by the Congressional Budget Office (CBO 1985) found that extensive excess capacity was in fact installed, and specific examples emerged in conversations with POTW staff and others. For example, the Washington Suburban Sanitary Commission (which serves Montgomery County and Prince George's County, outside Washington, D.C.) was able to obtain construction grants for plants that would double its capacity.[12] Twenty-five years later, the commission still has so much excess capacity that it does not even monitor for conventional pollutants because it can treat anything its industrial sources send. (It does monitor for pH and toxics, however, since those pollutants can interfere with plant operations.)

The excess capacity in POTWs became an attractive alternative to pollution abatement for many direct dischargers facing major expenses in complying with permit requirements based on the new effluent guidelines, particularly in plants processing organic wastes that could easily be handled by POTWs. A case involving Smithfield Foods provides an example.[13] In the late 1980s, Virginia promulgated new effluent regulations limiting phosphorus discharges from point sources into the Chesapeake Bay watershed to 2.0 mg/l. Two plants operated by Smithfield Foods in Isle of Wight County found it extremely difficult to meet this requirement and challenged the regulation. As the suit went through the courts, the plants continued to discharge wastewater with excessive phosphorus concentrations, and the state brought an enforcement action. The case was eventually resolved in part by an agreement by the firm to pipe the wastewater from these two plants 20 miles to an underutilized POTW.

According to EPA regulations (40 C.F.R. Sec. 35.2125), construction grant monies could not be used to treat industrial wastes. But once the plant was built, there was nothing to prevent the plant from accepting waste from industrial sources, especially if the industrial user paid a share of the costs. In fact, given the excess POTW capacity exists, using it to treat industrial waste is efficient as long as there are no pass-through or interference issues, because otherwise the POTW would be underutilized. Of course, that is not to say that it is efficient *ex ante* to overbuild POTWs with the expectation of using the excess capacity to treat industrial waste.

Trends in Direct and Indirect Discharge

Many industrial dischargers, though obviously not all, have a choice of whether to seek an NPDES permit and discharge directly to receiving waters or to send wastewater to a POTW. Evidence suggests that over the past 25 years, there has been a gradual shift away from direct discharge by major industrial dischargers and toward indirect discharge. Certainly in industries treating primarily organic waste, this trend is unmistakable, but it may extend to other industries as well. In Maryland, for example, the number of major direct dischargers has slowly declined from about 50 to 35 in the past two decades.[14] According to Maryland officials, the decline is not the result of plants' switching over to indirect discharge. Rather, it is due to the gradual retirement of the existing direct discharging plants, together with the tendency of new plants to opt for indirect discharge whenever possible. Elsewhere, however, industrial plants facing difficulties meeting direct discharge standards have made the switch, the Smithfield plants discussed above being a prominent example.

At least in terms of numbers of plants, indirect dischargers now greatly outnumber direct dischargers. In 1995, EPA reported that there were about 7,000 major dischargers holding NPDES permits and discharging directly into receiving waters. Of these, more than half were POTWs. At the same time, there were some 35,000 significant industrial users discharging into POTWs (U.S. EPA 1995).

In interviews, industry representatives, consultants, state officials, and pretreatment coordinators all agreed that most plants would vastly prefer to send wastes to a POTW. The reasons are not hard to find. One plant owned by a major manufacturer of dairy products and without access to a POTW incurred incremental costs of $3.16 per pound of BOD or TSS to meet BAT requirements—an order of magnitude greater than the unit fees imposed by POTWs. Moreover, the excess capacity induced by the Construction Grants Program probably has made POTW operators eager to take industrial waste since they can charge very low fees and still more than cover short-run marginal costs.

Industrial dischargers have other, less tangible reasons for avoiding direct discharge. Because the discharge goes directly into the receiving waters, permit requirements for direct dischargers are much more stringent and more closely scrutinized than those for indirect dischargers. They are also more visible, because environmental watchdog groups pay much closer attention when there is no POTW to mediate the discharge. Direct dischargers are also more concerned about liability, although recent federal regulations making violations of indirect discharge permits a violation of federal law may prompt more concern about indirect discharges.

The apparent shift from direct to indirect discharge carries with it several interesting ramifications for point source water pollution policy. First, it moves pollutants into regimes in which effluent discharge regulations were less vigorously enforced, at least until the mid-1990s. As noted above, from 1984 to 1994 the rate of noncompliance with permit requirements was 50% greater for POTWs than for industrial direct dischargers, and the noncompliance rate with pretreatment permit requirements exceeded 50%. Second, it moves wastes out of closed systems, where all wastes that are generated are treated, and into combined sanitary and storm sewer systems. Some cities, particularly in the Northeast and Midwest, still have combined systems, and their untreated wastes can bypass POTWs and flow directly into water bodies. Third, it raises the possibility that more wastes will move into a regime in which a system of effluent charges may be emerging. This third possibility is discussed further below.

The Spread of Waste-Based Sewer Surcharges

Fees charged by POTWs for accepting industrial wastes can be fixed, based on the flow rate, based on the waste load, or some combination of the above. Some POTWs, though not many, have based fees on waste load since the 1960s—that is, before significant federal involvement in water quality. Over time, more and more POTWs have resorted to the use of waste-based fees, and they have gradually increased the range of pollutants on which a fee is collected. Today, a majority of POTWs charge for oxygen-demanding materials (usually BOD, by pound or kg of oxygen demanded) and TSS, and increasingly they charge for a variety of other pollutants, including nitrogen, phosphorus, and fats, oils, and grease.

These waste surcharges are said to have increased rapidly during the 1970s, but since the early 1980s, they have stabilized or even declined somewhat. Their nominal basis is, or at least was, the average cost of abatement at the POTW. An EPA estimate prepared in 1986 of the cost of waste treatment at POTWs was 18 cents per pound of BOD at a 25 million-gallon-per-day (MGD) plant and 35 cents per pound at 0.38 MGD and 3.3 MGD plants.[15]

A sample of fees charged industrial users for BOD and TSS discharges is shown in Table 3-4. For space reasons we have left out fees on other pollutants, such as the fee on Kjeldahl nitrogen charged by some POTWs. The third column

Table 3-4. *Waste Surcharges for Selected POTWs*

POTW	*BOD[a] ($/lb)*	*TSS[a] ($/lb)*	*Effective BOD fee for dairy plants[b] ($/lb)*
Providence, RI	0.600	0.600	0.75
Loudon, TN	0.460	0.460	0.575
Delta, CO	0.340	0.340	0.425
Salem, OR	0.301	0.222	0.357
Nashville, TN	0.293	0.140	0.328
Longmont, CO	0.270	0.230	0.325
Boulder, CO	0.260	0.285	0.33
Henderson, NV	0.230	0.220	0.285
Lakeland, FL	0.186	0.116	0.215
Mitchell, SD	0.180	0.160	0.22
Savannah, GA	0.160	0.160	0.20
Springfield, MO	0.151	0.086	0.172
Aberdeen, SD	0.150	0.012	0.18
Terre Haute, IN	0.150	0.150	0.187
Wilmington, NC	0.140	0.100	0.165
Gwinnett County Public Utilities, GA	0.120	0.120	0.15
Wilson, NC	0.109	0.152	0.147
Norman, OK	0.083	0.111	0.110
Princeton, NJ	0.060	0.076	0.079
North Davies Sewer District, UT	0.056	0.043	0.067

[a]These charges apply only to wastes at concentrations in excess of a locally determined threshold, usually 200–300 mg/l. For example, if waste strength is 1,000 mg/l and the threshold is 300 mg/l, the fee is charged for only 7/10 of the waste load.
[b]To calculate "effective BOD" for dairy waste, add 25% of the fee for TSS to the BOD fee. A rule of thumb for the concentration of TSS in dairy waste (except for cheese plants) is 25% of the BOD concentration (Roy Carawan, personal communication, North Carolina State University, Nov. 4, 2002).
Sources: Pretreatment coordinators listserv at yahoo.com, Mannapperuma et al. (1993).

in Table 3-4 is for "effective BOD" applicable to dairy waste, based on a rule of thumb given to the author by Roy Carawan of North Carolina State University, which is that TSS concentration in untreated dairy plant effluents, not including cheese plants, is usually the greater of 500–600 mg/l or one-quarter of the BOD concentration. This effective fee is a crude representation of the waste surcharge that would be facing dairy plants expressed in terms of BOD concentration.

Are Waste-Based Surcharges Effluent Fees?

In form, at least, waste-based surcharges look very much like effluent fees. Whether they perform like effluent fees depends on three questions. Are they large enough to have incentive effects? If so, do plant managers respond to these incentives by installing pretreatment or using some method of source reduction of waste loads? Finally, do regulators at the POTWs, their state overseers, and EPA use the fees to limit waste loads? This last characteristic is assumed by the "textbook" discussions of economic instruments and their use. However, casual observation of the cases cited as examples of emission fees suggests that few fee systems have this property. Rates are set not with some target aggregate emis-

Table 3-5. *Cost of Waste Disposal at Five Dairy Plants*

		Onsite wastewater treatment				
Plant (n)	Flow (million gallons/day)	BOD in (lbs./day)	BOD out (lbs./day)	Annual abatement cost ($)	Cost/lb. removed ($/lb)	Waste strength delivered to POTW (mg/l)
Plants with pretreatment						
2	157.4	973,000	176,000	193,000	0.242	649
3	116.3	2,514,000	76,000	526,000	0.216	377
181	74.0	1,285,000	193,000	134,000	0.123	1516
Plants without pretreatment						
1	153.8	1,023,999	—	—	—	3868
5	74.3	649,000	—	—	—	5072

Source: Bough et al. (1988) plus author's calculations.

sions rate in mind, but to meet other objectives, such as raising revenue. Of course, the fact that these instruments are not "pure" does not mean they are not interesting or useful.

The slender evidence available to the author on the incentive properties of the waste-based surcharges comes from scattered sources. First, a paper presented at a conference on food processing wastes about 15 years ago suggests that at least at some plants, the cost of pretreatment is less than the waste-based fees charged by many POTWs. Table 3-5 shows data on five dairy processing plants from Missouri, including the incremental costs of waste treatment, expressed in terms of BOD removal. As shown, the incremental costs at the three plants with pretreatment vary from 12 to 24 cents per pound of BOD. Comparing these results with the surcharges shown in Table 3-4, we see that the BOD waste surcharges exceed the highest incremental pretreatment costs at seven plants, and the effective BOD for dairy wastes exceeds the highest incremental costs at 21 plants. It should be noted, however, that these results may not be typical. These three plants were in a uniquely favorable position regarding the disposal of sludges. Discussions with other dairy plant operators suggest that pretreatment of dairy wastes is rarely cost-effective compared with sending the untreated waste to the POTW and paying the fees.

What *is* cost-effective, however, is source reduction. Dairy operators report that the presence of waste surcharges has caused them to take a closer look at plant operations, and as a result, they have found numerous opportunities for reducing waste loads by making changes to product specification, production process, product sequencing in a multiproduct operation, and production run lengths. For example, longer production runs mean less cleaning of equipment, and therefore less wastewater.

Nonetheless, there is some reason to question whether plant managers are fully alive to the possibilities of reducing waste surcharges by examining waste loads in the plant. One pretreatment coordinator described a dairy that was paying millions of dollars per year in waste surcharges, apparently unknowingly: the plant was a unit of a large multistate, multiplant operation, and the bills were being paid by the headquarters office, which had little idea what kinds of plant-level waste reduction economies were available. When this pretreatment coordinator brought the expense to the attention of his technical contact at the dairy, that individual contacted the home office and implemented a study of local waste-load reduction possibilities.[16]

The anecdotal evidence for the incentive effect of waste surcharges is therefore mixed. In principle, it would be easy enough to test the proposition statistically, but assembling an appropriate dataset is not easy.

Almost certainly regulators do not treat the surcharges as effluent fees, and it is not obvious how they could, given the current U.S. regulatory approach. Consider how a POTW operator might react to the increase in waste load brought on by local economic and population growth that threatens increase waste discharges—a situation analogous to the challenge that led to the introduction of pollution offsets in air quality management. The POTW has a permit that specifies a limit on its waste discharge. Some portion of the available waste load has been allocated to industrial dischargers by a headworks analysis, as discussed above. The increase in the waste load could imply a reduction in the waste load available to the industrial dischargers.

In an effluent fee regime, the way to respond to such increased waste loads is to increase the fees, which will presumably induce waste reduction among industrial dischargers. However, the POTW cannot be certain that the waste reduction will be large enough or occur quickly enough to prevent the overall POTW discharge limits from being exceeded. Without this certainty, it is unlikely that the POTW will take the approach of simply raising the fee, nor is it likely that the regulatory authorities at the state DEQ or at EPA will allow the use of fees only to control effluent, without setting new quantity limits based on a new headworks analysis. In other words, there appears to be an incompatibility between the use of fees upstream from the POTW and a nontransferable quantity instrument downstream.

Conclusion

Let us close with a discussion of the experience with the effluent guidelines in the context of hypotheses on the performance and characteristics of economic incentive instruments and command-and-control instruments.

When Congress was deliberating the shape of federal water pollution legislation, around 1970, it briefly appeared as though the use of economic incentives—namely effluent fees—to control industrial and domestic discharges might get a hearing. Ultimately, however, effluent fees were opposed by pollutant dischargers on the grounds that they were taxes and by environmental advocates who had no faith in the incentive properties of economic instruments. They were a "license to pollute": polluters would simply pay the fees and continue to pollute (Kelman 1982). The use of the police power was thought to be the only way of getting reliable results in a timely fashion.

By these lights, the performance of the effluent guidelines was surely disappointing. The NRDC's performance audit of the Clean Water Act after 20 years (Adler et al. 1993) found serious delays both in promulgating the required BAT standards and in issuing permits based on those standards, all of which required further legislation in 1987 to establish a new timetable for preparation of BAT standards. To put this in perspective, however, we should note that the Clean Water Act was among the first environmental statutes to impose significant and costly responsibilities on U.S. industry. Sources of pollution had to be convinced that environmental quality was a problem in the United States, that they were part of the problem, and that the federal government had both the power to change their behavior and the willingness to use it. To an extent probably unappreciated in 1972, the rulemaking procedures and the courts provide many opportunities for regulated parties to contest the regulations, resulting in delay and weaker standards. Still, the technology-based regulatory approach, adopted primarily because of its presumed effectiveness, was probably not as effective as its proponents had expected *ex ante*.

Today the effluent guidelines approach appears to be changing in ways that no one anticipated in 1972. At that time, the focus of the program was on the BPT, BAT, and NSPS standards for direct dischargers. Over time, direct dischargers, though still important in some industries, have gradually become fewer in number and less important in environmental terms. Among indirect dischargers, furthermore, it is likely that waste surcharges increasingly are having incentive effects as they are raised by local POTWs for revenue purposes. That is, this quintessential regulatory program is gradually evolving into an economic incentive program.

Waste-based surcharges are in fact only one of at least three ways of introducing economic incentives into industrial water pollution control. The other two, which are beyond the scope of this paper, are the use of marketable permits to allocate the maximum allowable industrial load at POTWs, a program now in use at only one POTW, and the potential use of marketable permits to allocate total maximum daily loads for direct dischargers into water-quality-limited water bodies.

As has been frequently observed, many, if not most, economic incentive programs in use around the world have evolved out of purely regulatory instruments. Perhaps it is time to look at this common observation from the other end. As time passes and new challenges emerge, CAC programs appear to have a strong tendency to adapt by incorporating economic incentives into their structure.

Acknowledgment

The author would like to thank without implication Dan Balzer, David Lankton and Kit Rogers for research assistance; Roy Carawan, Ed Stone and Gary Kelman for helpful discussions; and Paul Portney, William Anderson, Robert Bastian, and participants in three seminars at the U.S. Environmental Protection Agency for excellent comments and corrections on an earlier version of this chapter.

Notes

1. In addition to these federal funds, several states contributed matching funds to the capital costs of municipal wastewater treatment plants. Maryland, for example, contributed an additional 5%.

2. 42 C.F.R. 403. Statutory authority for the effluent guidelines is found in the Clean Water Act (33 U.S.C. Chapter 26).

3. Technology-based standards are effluent limits for dischargers that are based on the performance of a designated abatement technology, without consideration of the environmental or social problem caused by the discharges.

4. Water bodies were classified by the states according to desired use, and water quality standards were set accordingly.

5. This is a convenient generic term. Actual state names for the agencies responsible for environmental quality vary.

6. Conventional pollutants include biochemical oxygen demand (BOD) and total suspended solids (TSS). BOD is the amount of dissolved oxygen that will be consumed in the water by the pollutant. EPA has identified 15 "pollutants of concern" that are often found in sludge and wastewater from POTWs. These include the "conventional pollutants" BOD and TSS, plus 12 metals, such as arsenic and mercury. The POTW is also required to watch for other pollutants that may be local problems.

7. For a description of this program, see Industrial Economics (1998).

8. In addition, EPA concluded in 1974 that 18 more industries required effluent guidelines. The total number of industrial categories with guidelines today is about 55.

9. For effluent guidelines issued during the 1970s, EPA circulated the contractors' reports in the contractors' report covers, presumably after EPA review. Later this practice was ended and EPA issued supporting documents under its own covers and tied to the proposed regulation.

10. Although there are thousands of water quality monitoring sites in rivers, streams, and estuaries across the country, the data are not suitable for discerning long-term trends in water quality; there are relatively few stations where data have been collected over a long period by the same methods and for the same pollutants. As of 1993, the only two exceptions are two networks operated by the U.S. Geological Survey: NASQAN, 420 stations located on major rivers; and the Hydrologic Basin Network, 55 stations located on headwaters in pristine areas, designed to provide water quality baselines. Two long-term environmental monitoring projects were initiated in the early 1990s. In 1991, the Geological Survey began the National Water Quality Assessment, a long-term effort at water quality assessment that is just now beginning to yield results. EPA launched the Environmental Monitoring and Assessment Program, designed to monitor trends in ecological resources generally.

11. It should be pointed out that an effluent fee system is not quite so automatic, since it requires the authorities to raise the tax rates.

12. Personal communication, Steve Laszlo, Washington Suburban Sanitary Commission, September 10, 2002.

13. The facts of the case are summarized in *United States v. Smithfield Foods, Inc.,* 965 F. Supp. 769, 772–781 (E.D.Va. 1997).

14. Personal communication, Ed Stone, Maryland Department of Environment, September 30, 2002.

15. 51 Fed. Reg. 24986 (1986).

16. Personal communication, Lyle Milby, pretreatment coordinator, Norman, Oklahoma, September 24, 2002.

References

Adler, R.W., J.C. Landman, and D.M. Cameron. 1993. *The Clean Water Act Twenty Years Later.* Washington, DC: Island Press for the National Resources Defense Council.

Association of State and Interstate Water Pollution Control Administrators (ASIWPCA). 1984. *America's Clean Water: The States' Evaluation of Progress: 1972–1982.* Washington, DC.

Bough, W.A., E. McJimsey, and D. Clark-Thomas. 1988. Operating Costs of Dairy Pretreatment vs. POTW Facilities and the Establishment of a Waste Minimization Program. Presented at the Food Processing Waste Conference, October 31–November 2, 1988, Atlanta.

Caulkins, P., and S. Sessions. 1997. Water Pollution and the Organic Chemicals Industry. In *Economic Analyses at EPA: Assessing Regulatory Impact*, edited by R. D. Morgenstern. Washington, DC: Resources for the Future.

Congressional Budget Office (CBO). 1985. *Efficient Investment in Wastewater Treatment Plants.* Washington, DC.

Council on Environmental Quality (CEQ). 1976. *Environment Quality—1976.* Washington, DC: Government Printing Office.

Fraas, A.G., and V.G. Munley. 1989. Economic Objectives within a Bureaucratic Decision process: Setting Pollution Control Requirements under the Clean Water Act. *Journal of Environmental Economics and Management* 17(1): 35–53.

Freeman, A.M. 1978. Air and Water Pollution Policy. In *Current Issues in U.S. Environmental Policy,* edited by P. Portney. Baltimore: Johns Hopkins University Press for Resources for the Future.

Harrington, W., R.D. Morgenstern, and P. Nelson. 2000. On the Accuracy of Regulatory Cost Estimates. *Journal of Policy Analysis and Management* 19(2).

Industrial Economics, Inc. 1998. Sharing the Load: Effluent Trading for Indirect Dischargers. EPA-231-R-98-003. Washington, DC: U.S. Environmental Protection Agency.

Kelman, S. 1981. *What Price Incentives? Economists and the Environment.* Westport, CT: Auburn House.

Knopman, D.S., and R.A. Smith. 1993. Twenty Years of the Clean Water Act: Has U.S. Water Quality Improved? *Environment* 35(1): 16.

Krupnick, A.J., and W. Harrington. 1981. Stationary Source Pollution Policy and Prospects for Reform. In *Environmental Regulation and the U.S. Economy,* edited by H. Peskin, P.R. Portney, and A.V. Kneese. Baltimore: Johns Hopkins University Press for Resources for the Future.

Magat, W.A., A. Krupnick, and W. Harrington. 1986. *Rules in the Making: A Statistical Analysis of Regulatory Behavior.* Washington, DC: Resources for the Future.

Mannapperuma, J., E.D. Yates, and P. Singh. 1993. Survey of Water Use in the California Food Processing Industry. Proceedings of the 1993 Food Industry Environmental Conference.

National Commission on Water Quality. 1976. Report to Congress.

National Research Council. 1993. *Managing Wastewater in Coastal Urban Areas.* Washington, DC: National Academy Press.

Smith, R.A., R.B. Alexander, and K.J. Lanfear. 1991. Stream Water Quality in the Conterminous United States—Status and Trends of Selected Indicators during the 1980s. *National Water Summary 1990–91—Stream Water Quality: Hydrology.* Water-Supply Paper 2400. U.S. Geological Survey.

U.S. Bureau of the Census. 1989. *Pollution Abatement Costs and Expenditures, 1986.* Washington, DC: Government Printing Office.

U.S. Environmental Protection Agency (EPA). 1979. *The Cost of Clean Air and Water.* Report to Congress.

———. 1987. *National Water Quality Inventory:* 1986 Report to Congress.

———. 1995. *National Water Quality Inventory:* 1994 Report to Congress.

______. 2000. Progress in Water Quality: An Evaluation of the National Investment in Municipal Wastewater Treatment. Washington, D.C. Office of Water. EPA-832-R-00-008 (June).

———. 2001. Local Limits Development Guidance. Washington, DC: Office of Wastewater Management, U.S. EPA.

———. 2004. Fact Sheet: Extension of Comment Period for the Preliminary Effluent Guidelines Plan for 2004/2005. http://epa.gov/guide/fs2004-2005plan-ext.htm (accessed July 14, 2004).

CHAPTER 4

Industrial Water Pollution in the Netherlands

A Fee-based Approach

Hans Th. A. Bressers and Kris R.D. Lulofs

THE DEBATE OVER replacing command-and-control (CAC) pollution abatement policies with economic incentive (EI) strategies has always been rather heated, though carried on more in theoretical terms than based on empirical evidence (see Mitnick 1980). The Netherlands offers an example of an EI strategy in the form of effluent fees.

The Dutch fees on water pollution are a side effect of the Surface Water Pollution Act of 1970, which was intended to improve the greatly deteriorated quality of surface water. The need to clean up household and industrial effluent called for a large investment in sewage water treatment, and the law permitted responsible authorities to introduce fees to cover their annual costs. The fees were a mechanism to raise resources to finance sewage water treatment, not so much an effort to influence households and businesses directly.

The law delegated the task of water quality management to the provincial authorities and often, through these authorities, to established district water boards. Within a few years almost every water board had raised its rates to a level that businesses emitting wastewater into the sewage system found it cheaper to reduce pollution. Degradable organic pollution, for instance, was taxed more than twice as much per unit of pollution as in the comparable German program (Brown and Johnson 1983). Meanwhile, almost all district water boards also established fees on heavy metals in wastewater. Because the fees were earmarked, with levies related to the actual purification costs, the fees on heavy metals were relatively low because heavy-metal abatement costs are relatively low.

In this chapter, we present evidence on the effectiveness of Dutch effluent fees in the period 1986–1995 and compare it with earlier findings from the period 1975–1980. We discuss industrial wastewater and reduction of the waste load, not the improvement of the surface water quality, albeit this improvement has been impressive. The analysis focuses on the effectiveness, static and dynamic efficiency of fees.

Policy outcomes are analyzed, although we also look at implementation problems—one of the largest sources of uncertainty threatening the effectiveness and efficiency of political measures. Furthermore, by taking into account the power of the instruments to affect the decisionmaking agenda of the business manager, we anticipate that instruments may be effective by alternative mechanisms normally not taken into account in analyses of CAC and EI policies.

We have organized this chapter as follows. First we present background on the situation in the Netherlands, including the history of water management issues. Then the Surface Water Pollution Act is described. This act provides water authorities with regulatory instruments to set general rules and grant permits, to develop and set water quality plans and water quality standards, to collect levies, and to monitor and enforce implementation. The policy toward industrial wastewater involves a mix of instruments that are part of the broader context of Dutch water quality policy. Hence the effluent fees in this case interact with other instruments and aspects of the context in which they are used, and we analyze these other factors in this study.

A subsequent section reports the empirical analyses on the effects of fees. Each water board has applied different rates according to its costs and the units of pollution to be covered in its district. This opens up possibilities to assess the effectiveness of fees. The analysis covers the periods 1975–1980 and 1986–1995. Finally, we report our conclusions.[1]

Background

The Netherlands is a small country, approximately the size of the U.S. states of Maryland and Delaware combined. The population density is one of the highest in the world, however, and the country has a large industrial component and thousands of miles of waterways of all sizes. The struggle against water is a theme throughout Dutch history.[2] The Netherlands is situated in a low-lying delta. Because about 60% of the country is below sea level, drainage is necessary. Floods—in particular, the catastrophic 1953 flood that killed 1,800 people—prompted efforts to seal up sea inlets and reinforce the dikes. Now a new threat has emerged: climate change might cause the sea level to rise. In addition, the capacity to drain river water from the continent might be exceeded and require new investments. Besides such safety-related big issues, small issues dominate quantitative aspects of water management, such as the ongoing battle between cattle farmers, who want a high groundwater level, and farmers, who demand a low groundwater level. Authorities cannot drain the land in a manner that satisfies both at the same time.

Some industry sectors unintentionally played leading roles in setting the agenda for a cleanup operation because their prominent and visible pollution caught public attention in the 1960s and 1970s. For instance, the paper and cardboard industry emitted untreated wastewater into surface water. The pollution caused bad odors, the natural environment was visually and ecologically damaged, and fish died. The media publicized the problems, which were especially bad in the province of Groningen, where the sugar industry and the potato industry contributed additional pollution to the same surface water systems. Initially, pipes were constructed that transported wastewater from Groningen into the Waddenzee, an extremely vulnerable ecosystem connected by open water to the North Sea. Public conscience soon led to the invention of the concept of "dirt-pipes." Later, the companies were required to purify their wastewater before emitting it into the pipes, and their water treatment equipment was optimized and stabilized. The possibility arose of emitting the now-purified wastewater directly into the surface water. In other provinces, emitting wastewater into the sea was not an option, so it was emitted into the sewage system or directly into surface water.

Situations such as the hotspot in Groningen were important drivers behind the Surface Water Act Pollution Act of 1970. The act was first proposed in 1964 but took years to become law because many government entities—central government agencies, provinces, water boards, and municipalities—lobbied to become the principal actor. In the final version, the law distinguishes between national surface waters that are managed by the national government, and other waters, which are the responsibility of subnational governments. The national waters include the main rivers, the Waddenzee, and the North Sea. The water quality tasks of the national government consist of giving management directives, planning water management, granting permits to discharge into state waters (including the discharge of effluent from treatment plants of adjacent water boards), collecting fees for the same category of discharges, measuring surface water quality, and subsidizing abatement measures (typically the construction of treatment plants by water boards).

The authorities responsible for regional waters have the same function, with the exception of providing subsidies, but with the important addition of building and operating water treatment plants. The law assigns responsibility for regional waters to the provinces. However, the provinces are authorized to delegate these tasks to district water boards or municipalities. These provinces and water boards manage nearly all treatment plants. Municipal treatment plants hardly exist and are regarded as an anachronism. The prime responsibility of municipalities in water quality management is to manage sewage systems and, to some extent, control pollution of discharges into their sewage systems.

By 1970, the year the act took effect, industry and private households were producing roughly 45 millions of "population equivalents"[3] of oxygen-consuming organic pollution, which along with heavy metals is a principal subject of our analysis. As a result, overall water quality in the country had become rotten—both figuratively and literally. Since 1970, the organic pollution problem has decreased, thanks to the creation of a treatment infrastructure and reduced pollution. Meanwhile, however, emissions of toxic substances have become a more urgent issue because of new information on the magnitude and seriousness of this pollution. Given the hydrological characteristics of the country, surface water and groundwater interact and affect drinking water quality. Although the initial concern was organic water pollution, toxic substances such as heavy metals were also addressed by the Surface Water Pollution

Act of 1970 and related fees. The act addressed both aspects of surface water management.

Who Is Responsible for Industrial Wastewater?

The Surface Water Pollution Act applies to all surface waters in the Netherlands. Responsible water authorities are supplied with regulatory instruments to set general rules and grant permits, to develop and set water quality plans and water quality standards, to collect levies, to monitor and enforce compliance, and to compensate for efforts beyond reasonable impacts. Within the national government, the responsibility for industrial wastewater is mandated to the Public Works Agency, which consists of a central directorate, a large research institution, service departments, and regional directorates. When granting and enforcing permits, the Public Works Agency and its subagencies deal with the industries that discharge into national waters. The subagencies are not responsible for indirect discharges (industrial waste in sewage systems that subsequently discharge into national waters) once a public wastewater treatment plant has been built. That is because only the district water boards, established by the provinces, are entitled to build sewage water treatment plants. The district water boards also handle subnational waters. In the Netherlands, 30 district water boards operate about 450 wastewater treatment installations. Consequently, these district water boards themselves become dischargers into national waters with the effluent of their treatment plants and are subject to permit granting from the Public Works Agency. This system introduces constraints to pollution that sometimes are similar to a bubble concept, and for some categories of indirect discharge, the district water boards are responsible instead of the Public Works Agency.

Actors such as the Ministry of Traffic and Public Works, Agency of Public Works, and district water boards have teams of supervisors and control officials. These teams control the compliance to the rules, especially in case of surface-water pollution.

Given the long history of water issues in the Netherlands, the surface water policy system is well organized and can take early and significant action. The central and regional actors cooperate in a multiscale, multiactor system of governance in which both private and public parties play a role. The emphasis is on consent and "voluntary," often negotiated, cooperation. An example is the Commission on Integral Water Management, formerly known as the Commission on the Implementation on the Pollution of Surface Water Act.[4] This consultation organ of all parties involved in water quality management currently has three functions: coordinating, guiding research, and giving advice on subjects such as administrative and juridical matters, implementation aspects of collecting effluent fees, measurement of water quality, and emissions standards. It no longer confines itself to water quality, which is regulated by the Surface Water Pollution Act.

Water Quality Policy: Quality Standards for Surface Water

Water quality standards have been set for several hazardous substances. In most cases, these standards express allowed concentrations of harmful substances in surface water. The process of writing such quality standards is as much as possible based on scientific information.

Based on risk limits, "environmental quality standards for surface water" are issued by government. The status of these standards differs greatly. Those regulated by law are obligatory, "hard" standards; the "soft" standards" are often guidelines based on the stricter negligible risk limits that imply an effort toward attainment. Standards for metals take into account natural background levels and differ by region. For nutrients and other natural parameters, area-specific standards are in use because of naturally wide regional variations and the large number of water types.

Emissions Policy

The relationship between *water quality policy* and *emissions policy* is as follows: The quality standards, both hard and soft, are considered "immission" standards. The water authorities, as the coordinating actors, take into account the characteristics of the water system and calculate the total amount of pollution that can be tolerated by all emitting sources—the immission. This determines the size of the pollution "bubble." It is the duty of the water authorities to use regulation, incentives and other instruments to ensure that the sources in a region do not exceed this maximum. The bubble principle means that water quality cannot be allowed to deteriorate, and that the functioning of the water boards' treatment plants may not be disturbed by irregular discharge, heavily diluted wastewater, or obstructing substances.

The bubble principle is not valid, however, for especially harmful substances, such as the substances on the "black list," "gray list," and "priority list." For these substances, a "standstill principle" is in effect: the discharge may in no case increase.

Regulation for Individual Companies

In operational terms, industrial emissions can take two paths: directly into the surface water, or through the sewage system into the public wastewater treatment plants and thence into the surface water. Any direct emission into the surface water, even after pretreatment, whether by companies or by wastewater treatment plants, must be licensed. Licensing is a responsibility of the district water boards or the Public Works Agency.

Emissions into the sewage system have to be licensed on the basis of the Surface Water Pollution Act under the following conditions:

1. If the emissions of a company exceed 500 m^3 and/or the pollution of oxygen-demanding substances exceeds 5,000 population-equivalents.
2. If listed toxic substances or categories of firms are involved. A 1981 amendment to the Surface Water Pollution Act gave the district water boards and the Public Works Agency direct responsibility for these emissions. This was a victory for the water boards, which had sought to assume that responsibility

from the municipalities. International agreements and treaties on gray-list and black-list substances made such a change inevitable. For practical reasons, it was decided to list not substances but categories of firms, as the new responsibility of the regional water boards.

For all other emissions into the sewage system, a license to connect to the sewage system, which municipalities usually own, is necessary and sufficient. The municipality itself also needs a license to connect its sewage system to the public wastewater treatment plant. Requirements can also be related to the efficient functioning of the public wastewater treatment plant.

The target for reducing pollution in wastewater was raised considerably in the period 1985–2002. However, the goal was limited to hazardous substances like heavy metals, chemicals, and fertilizers. Practitioners explained to us that research obligations are usually imposed on companies to limit or reduce the discharge of certain substances. The company has to prove that a discharge has no negative influence; previously, the permitting organization had to prove that a substance was harmful. However, the impact of more ambitious licenses differs by industry sector. Some sectors are way ahead of the licensers; other are more passive, in which case regulation might have substantial impacts. In the past, these companies almost always paid attention only to substances that determined the fees they had to pay.

Charges on Companies' Wastewater

The district water boards collect nearly 100% of the fees. About 80% of the amount is used for purification management and 20% for remaining tasks related to water quality and quantity. The levy returns are more than 99% based on the discharge of oxygen-demanding substances. Heavy metals contribute only a tiny proportion (CIW 1999, *28*).

Each firm pays a fee roughly based on the amount of pollution it produces. Pollution is calculated in a unit called a population-equivalent (PE), which is equivalent to the amount of organic pollution in wastewater normally produced by one person. There are roughly two ways to determine the number of PEs: a model approach for small businesses, and measurement for large companies. The charge per PE has risen sharply during the period the fee system has been in effect and varies regionally depending on the district water board. These regional differences are not based on different environmental conditions or quality objectives, but rather reflect regional costs of building and operating treatment plants. This might, of course, also reflect economies of scale. Because of expanding treatment capacity, the fees involve large amounts of money.

By 1995, the average fee was approximately €36 per PE, with a household paying more than €100 per year. For companies, the sum could easily reach €1 million. The average annual quantity of unpurified water that companies discharge to public water treatment plants is 10,000 to 15,000 PEs. Thus, the average company pays a water board €500,000 yearly. It can be cheaper for companies to treat the wastewater themselves. Inflation in the period 1980-1999 was 25%, but the average district water board fee level increased from €16 to €40 per

PE, and the annual costs for public wastewater treatment increased from €368 million to €682 million (CBS 1999).

The total revenue increased from €155 million in 1975, to €355 million in 1980, €490 million in 1985, €608 million in 1990, €842 million in 1995, and almost €1 billion in 2000 (CBS 1998).

How is the levy calculated for individual companies? A distinction is made between households and businesses, and the latter involves a further distinction:

- Buildings with a discharge smaller than 5 PEs pay a levy fixed at 3 PE or 1 PE.
- Companies with discharge smaller than 1,000 PEs determine their emissions by using a chart for wastewater coefficients; no measurements or calculations are necessary. If such a "chart company" believes it is being overcharged by the assigned coefficient, it can ask for temporary measurement and have the coefficient changed accordingly.
- Companies with discharge greater than 1,000 PEs are obliged to determine the discharge by measurement and taking samples; they are known as measurement companies.

Normally, the burden of measurement is imposed on those companies by means of their license, but if actual measurements are necessary or requested by the company, the water quality manager (in a water board or the Public Works Agency) takes an independent set of measurements. The calculation of the number of PEs is done the same way. The tariff per PE, however, differs significantly between district water boards.[5]

Performance of Policy Instruments

As in the United States, the quality of industrial water discharges is regulated by a licensing regime. The implementation of the licensing system under the Surface Water Pollution Act of 1970 proved far from perfect in the 1970s and 1980s. As in other environmental sectors, substantial backlogs of applications for licenses and inadequate enforcement were familiar phenomena (Audit Office 1987)—even for heavy-metal pollution, one of the district water boards' top priorities. Since the early 1980s, regional water quality managers have been empowered to issue licenses directly so that they can impose restrictions on firms that discharge gray-list and black-list substances—in particular heavy metals—into the municipal sewage systems. A survey conducted by DHV among water quality managers from 1984 to 1987, for example, showed that just over half of these firms had been issued licenses. Moreover, the requirements contained in these licenses were more moderate than initially envisaged. The company's economic position was often the overriding concern, even to the extent that the "best existing techniques" criterion was often interpreted as "the best practicable or affordable techniques," a formulation that the central government had expressly avoided in relation to these substances. Nevertheless, Table 4-1 indicates that industry did reduce pollution in the period 1975–1995.[6] Before 1975 a substantial decrease had already taken place. An *ex ante* estimate for 1969, before the law was enacted,

Table 4-1. *Gross Discharge of Oxygen-Demanding Substances (Million PEs), 1975–1998*

	1975	*1985*	*1995*
Consumers	13.7	14.5	15.5
Industry	15.3	5.9	3.3
Miscellaneous[a]	4.2	3.9	4.1
Total gross discharge	33.2	24.3	22.9

[a]Includes agriculture, energy services, construction, and waste disposal.
Source: www.rivm.nl.

was that industry was responsible for 29 million PEs, compared with 15.3 in 1975, the initial time period of Table 4-1.

Table 4-1 illustrates what has become common knowledge: Thanks to the almost total reduction of gross discharge realized by industry, households now are responsible for the bulk of discharge of organic pollution.[7] By 1980, organic pollution caused by industrial production had declined by two-thirds since 1969. Almost half the remaining organic pollution, from both industry and households, was removed in sewage treatment plants, many of which had been newly built.

Table 4-2 gives information on some benchmarks. Expenditures are nominal; the table contains a price index to indicate the level of inflation. The selected sectors of industry are the companies covered by Dutch sector codes 10–41. We can observe that in 1986 the cost for private wastewater treatment exceeded the fee level, which indicates overinvestment in private wastewater plants in the previous years; by 1997 this situation had changed dramatically.

Impact of Fees, 1975–1980

Why did Dutch industry reduce its water pollution? The dramatic decrease in pollution occurred between 1969 and 1980, when the economy was generally booming, and thus pollution per product unit decreased even more sharply than total pollution. The effectiveness of water quality policy has been underestimated.

Our data set includes 17 district water boards that set their own emissions charges. Although we know the fees of the districts and the emissions of organics and toxics for the years 1975, 1980, 1986, and 1995, a straightforward comparison is nevertheless problematic. Between 1980 and 1986, water district boards merged, the definition of a population equivalent (PE) changed, and sectors of industry were redefined. Thus we must analyze two periods, 1975–1980 and 1986–1995.

The 17 large district water boards that were part of the 1975–1980 analysis cover 97% of the total volume of organic pollution; the 23 smaller areas cover only 3%.

Theoretically, changes in pollution per product unit might be accounted for by other factors than policy outputs. Alternative explanations could include (1) independent technical developments, (2) an increase in the value of waste matter as raw material, (3) an increase in environmental awareness within the companies or the population as a whole, and (4) information by nongovernmental institutions. Applying a "modus operandi" method (Scriven 1976) to the examination of these different factors suggested that the observed pollution reduction cannot be attributed to these other factors,[8] and logically speaking, policy measures must account for it. As the next arguments will demonstrate, this has indeed been the case in the Netherlands: the substantial reduction in industrial wastewater pollution has been the result of Dutch water quality policy. Since that policy contains a mix of measures, the question remains, which of the various instruments applied has contributed most to the policy's apparent effectiveness?

First, we offer a few words on the technique.[9] Differences in goal attainment among the 17 districts can be accounted for by two sets of factors. One is the differences in the economic structure of the water districts and the relative weights of particular industries; the other is the degree to which different policy instruments are used.

Regional Economic Structure and Differences in Pollution Reduction. Using data on the contribution of industrial sectors to regional pollution in 1975, as well as data on the average decrease in pollution per industry in the Netherlands between 1975 and 1980, we calculated the amount of pollution decrease that could be expected on the basis of the regional economic structure for each water district: how would pollution in each region have changed if the industries in that region had reduced their discharges at the same rate as they did nationally? These expected values were closely correlated to actual decrease of pollution. In the case of organic pollution, the Pearson's r was +.79 (r^2 = .62) and in the case of heavy metals pollution, it was +.74 (r^2 = .55).

At first sight, it would seem that only the unexplained share of regional differences in pollution decrease is a relevant dependent variable when we attempt, statistically, to assess the impact of various policy instruments on the level of goal achievement attained. This is, however, not entirely the case. Variations in organic pollution between industries can also be examined and, as it turns out, provide important information on the effects of effluent fees in particular.

The 14 industries that account for 90% of the pollution form the research units in the analysis described. Since they account for such a large part of the pollution, this analysis is not so much a sample study as a semipopulation study. Nevertheless, in view of the small number of units involved, some caution is in order for analyzing and drawing conclusions based on this sample of industries.[10]

Most obviously, decreases or increases in the level of production of the different industries could affect the volume of pollution produced in the period under investigation. A second cause of differences in abatement rates can be found in the fact that in some industries, it is much more difficult, and thus more expensive, to produce "cleanly" than in others. Therefore, relative abatement costs can also play an important role in determining the level of pollution reduction achieved.

Finally and (in connection with the focus of this study) most importantly, it is also conceivable that a policy output, namely effluent fees, could lead to differences in abatement between industries. The more pollution units per unit of production that an industry generates, the greater the impact that the fees (which are related to the number of pollution units produced) will have on production costs. This can mean that some industries will bear significantly higher charges. Efforts to reduce the number of pollution units will be greater to the extent that a given industry is liable for higher fees compared with its production value. In the discussion that follows, this factor will be referred to, for convenience, as the "fee factor."

The extent to which organic pollution of industrial wastewater decreased between 1969 and 1980 appears to be related to the three above-mentioned factors as follows: production increase, $r = -.21$; abatement costs, $r = .50$; and fee factor, $r = .73$. The signs of these relations are all in the anticipated direction. The strong correlation with the fee factor is very striking. When the three factors are related simultaneously to pollution decrease, they account statistically for 63% of the decrease.[11]

Policy Instruments and Regional Differences. Having concluded that an important part of the regional variation in pollution abatement cannot be explained by differences in the industrial structure of the regions, we can explain the remaining variation by the degree to which different policy instruments are used. The degree to which these different instruments are actually used varies greatly from region to region.

The dependent variable for this part of the analysis is "relative success of abatement." This variable has been calculated as the difference between the actual percentage of abatement and the percentage of abatement expected in view of the industrial structure of the region. Regional variation in the fee level cannot, however, straightforwardly be related to this variable. The reason for this is as follows.

The relative success of abatement has been calculated for the period 1975–1980. Before this time, however, pollution of industrial wastewater had already declined substantially. The above analysis of interindustry variations makes it plausible to assume that abatement in this period was also largely the result of fees. The 1975 amount of pollution was already more or less in equilibrium with the fee level of that time. For this reason, the fee factor influencing the abatement of pollution of industrial wastewater is indicated better by the difference between the rate before 1975 and the rate after 1980 than by the level of the fee in any one of the preceding years.[12]

The increase in the rate charged in 1974–1980 has a very strong positive correlation with the relative abatement success for the years between 1975 and 1980 ($r = +.86, n = 15, p = .000$ for organic).[13] This relation is not the result of one or two extreme values for the units analyzed, although the two water quality districts with respectively the most and the least relative success weaken the strength of the relationship somewhat. With these two observations omitted, the correlation is even stronger: $r = +.92$ ($n = 13, p = .000$).[14] Such high correlation between the fee factor and pollution decrease makes it natural that the relationships with the indicators of the various other policy instruments used are absent or weak. The influence of many other policy instruments examined, with 30 indicators, was assessed. Among these indicators, only water board permit-granting to municipalities and abatement schemes drawn up with companies show relatively substantial (but still modest) positive correlations with the relative abatement success.

The relative abatement success—the dependent variable in all these analyses—has been defined as the percentage decrease of oxygen-consuming pollution of industrial wastewater, adjusted to take the regional industrial structure into account. It should therefore be possible to explain the variation in the percentage decrease of organic pollution by:

- the decrease expected as a result of differences in the regional structure of industry;
- the increase in the rate charged;
- the number of municipalities with a permit issued by water boards (thus giving these authorities indirect influence on sewage permits issued to individual plants); or
- the weighted number of abatement plans.

As it turns out, 96% of the percentage decrease of organic water pollution can be explained, statistically, by the first two factors (r^2 after correction is .95).[15]

Explaining Effectiveness. Our statistical analyses were accompanied by assessments of insiders,[16] drawn from different sets of data: the opinions of negotiators from the water boards and company representatives who were responsible for water quality. In questionnaires and interviews, fees emerged as the most influential policy instrument for dealing with organic pollution, but for heavy metals, respondents attributed equal amounts of influence to a broader range of instruments. There are some small but interesting differences between the results of the questionnaire and those of the statistical analyses.

According to the statistical analyses, fees were most important in decreasing oxygen-consuming pollution, followed closely by heavy-metal pollution (Bressers 1983a,b). Why did the regional water quality administrators recognize the importance of fees for the abatement of organic pollution but not for decreasing pollution by heavy metals? The explanation seems to be as follows. These officials can clearly see that fees will prompt most companies to reduce their organic pollution. There are, however, few companies that are going to abate their heavy metal pollution solely because of effluent fees: discharging heavy metal is simply too cheap. Abatement plans and inspections are almost always necessary to persuade heavy-metal polluters to take appropriate action. What these water quality administrators do not see (indeed, what they individually cannot see) is that the success of these abatement plans is much greater in districts with higher fees. In both cases companies have to be persuaded by the water authorities to get started; in some cases, however, officials must continue to exert pressure to ensure that the agreed-upon measures are in fact carried out.

Even when they cannot motivate companies to take abatement action, fees apparently play an important role in facilitating the task of the water quality administrators in getting abatement measures implemented. This conclusion opens interesting perspectives on the possibility of applying fees not only as an alternative for, but also as a complement to, direct regulation.

Here, we limit ourselves to some implications of the fee system for interorganizational relationships. A very important aspect of the system is that it makes the regulator-regulated interaction a constructive form of consultation because it lets pollution prevention pay. Furthermore, it gives the water boards an entrée to the firms; permit-granting responsibilities did not provide that because so many firms discharged their wastes indirectly, into a municipal sewage system. On that basis, water board officials also often tried to discuss pollution from heavy metals and other pollutants. Indeed, they often claimed that certain discharges would not be allowed in the near future. Legally, they had no basis for such statements.

Another point meriting attention is the role played by informal negotiations. It is clear that water board officials attached greater importance to informal contacts than we would have anticipated on the basis of the statistical analyses. Negotiations do not constitute a separate policy instrument, but rather represent a manner in which the total policy mix is applied. Informal negotiations serve as a sort of lubricant for the machinery of the other policy instruments discussed above; compare the case in Germany (Hucke 1978) and for Great Britain (Vogel 1983). The following statement by a water board official illustrates this function: "When I'm going to have a talk with a company about the abatement of their discharges, I always take my pocket calculator along. I calculate their potential savings on fees and invariably get an interesting conversation started."

This primarily informative function of negotiations is especially important when the policy instruments make abatement cost-effective. Regulators can not only provide information about the exact consequences of the charges, but also indicate directions to improve performance of the company. This situation tends to occur more often in the case of organic pollution than with heavy metals because the effluent fees make reductions in discharges economically attractive.

Impact of Fees, 1986–1995

In 1986 the concept of "population equivalent" was redefined.[17] The analysis for this period is limited to organic pollution and involves 17 of the 18 large district water boards, covering about 90% of the organic pollution.[18] We also confined ourselves to the major waste-producing industrial sectors: nine subsectors of the food industry (processing of meat, fish, fruits and vegetables, food oils, dairy, potato starch, animal destruction, beverages, and others), and the textile, paper, printing, chemistry, basic metal, and galvanic sectors. Industrial sectors that have much smaller impacts on organic water pollution are thus not considered. Together, the selected sectors in our sample produced some 70% to 75% of all organic pollution produced by industry. The remaining one-quarter consists to some extent of still-remaining discharges in national waters (for example, from the Rhine, though untreated discharges are phased out) and to some extent are not related to industrial production, but to facilities for the personnel. So the analysis below deals with by far the greatest share of organic waste water pollution by industrial processes.

The ecological relevance of industrial organic water pollution changed dramatically during the 1970s and early 1980s. In 1969, the 14 most-important sectors produced some 29 million pollution units (of approximately 180g of oxygen demand). By 1975, this number had already dropped to some 15 million—more or less equaling the contribution of households—and in 1980, to 9 million. By the time our analysis in this section begins, the 15 sectors in our analysis produced less than 4.4 million PEs (of 136g oxygen demand). In the same period, the collective treatment capacity of the district water boards increased to some 25 million PEs, sufficient in most situations to treat all discharges of organic pollution. Given that situation, the further decrease of industrial organic waste load was increasingly seen by the water boards as not a benefit to the environment but as a weakening of their tax basis, without any environmental benefit, leaving the other parties (such as households) with higher fees.

Whereas in 1975–1980, water board officials actively warned industry to reduce pollution as much as possible,[19] and pollution became expensive, 1986–1995 presents a different administrative context. Only when complete renovations required a new capacity assessment, or when growing pollution was caused by population growth was the decrease in industrial organic pollution good news. In many other cases, ecological arguments for pollution reductions were neutral—and economic arguments were negative for the district water boards.

One might expect that after the initial investment period in the 1970s, major fees increases would end. But this proves not to be true. Between 1980 and 1995, the average fee of the major water boards increased from €15.80 to €36.33 (then levied in Dutch florins), a 130% increase. In our research period of 1986–1995, the increase was €12.91, or 55%.

In the first part of the research period, 1986–1990, the average fee level rose much less steeply than before or after, from €23.45 to €27.24. This €3.79 (16%) increase compares with an €9.10 (33%) increase in the period 1990–1995—almost double the rate. These differences cannot be explained by inflation rates. Generally, slowdowns in the increase in fees are observed from 1983 to 1985 and from 1987 to 1989. The explanation is that after the first large investment period, when the treatment plants were built, a second period of huge investments began in 1990, when the installations were equipped with devices to remove phosphor (see Table 4-2).

As for 1975–1980, calculations were first made for each district water board and for each sector to estimate the pollution in 1990 if all sectors in the region performed according to the average of their sector. By adding all these sector estimates for each district water board, we calculated what the 1990 pollution in each district water board would have been if all sectors in a region behaved according to the national average.

These estimates can be seen as a way to eliminate all national developments from the variance, leaving only the variance that should be accounted for by regional differences. It also eliminates the consequences of the different sector

Table 4-2. *Expenditures and Performance of Public and Private Wastewater Treatment*

	1985	*1990*	*1995*	*1996*	*1997*
Public wastewater treatment					
Investment in treatment facilities (million €)	221	206	331	416	177
Annual expenditures on public wastewater treatment (million €)	365	475	648	690	679
Annual capital costs (million €)	238	304	360	391	384
Annual operation and maintenance costs (million €)	127	171	288	299	295
Capacity (million PEs)	22.7	23.7	24.4	—	24.8
Average expenditure for public wastewater treatment per PE (€ based on actual performance)	20	23	34	34	32
Average expenditure maximal potential of capacity performance per PE (€)	16	20	28	—	27
Average fee of district water boards per PE (€)	23	27	37	38	40
Fees paid by selected industrial sectors (million €)	83	96	107	101	103
Private wastewater treatment					
Investment in treatment facilities (million €)	70	273	112	92	102
Annual expenditure by companies to reduce wastewater load (million €)	213	288	364	364	370
Annual capital costs (million €)	121	140	170	170	175
Annual operation and maintenance costs (million €)	92	148	194	194	95
Plants operated by companies	330	567	610	—	624
Capacity (million PEs)	8.3	14.5	15.7	—	15.9
Average expenditure maximal potential of capacity performance per PE (€)	26	19	24	—	23
Price index	100	116	120	122	125

Source: CBS (1999).

composition of regional industry. This can be important, because some sectors can decrease pollution much more easily and cheaply than others, perhaps because of certain new technologies.

Although the calculated expected pollution in 1990 is very strongly correlated with the actual emissions in 1990 ($r = .995$ $p = .000$, of course mainly a big correlation with big and small correlates with small artifact), the actual pollution decrease percentage correlates only moderately with the calculated expected pollution decrease ($r = .567$ $p = .009$). The variable "relative success" of pollution abatement is the number of percentage points that pollution has declined more in the region than one could expect on the basis of the national averages per sector and the share of the various sectors per region.[20] For instance, in Groningen the observed percentage PE decrease was 17.12, the calculated decrease was 15.78, and thus "relative success" was +1.34.

Table 4-3 gives an overview of the correlation between the rise in the fee level and the relative success of abatement. "Relative success of the region" is indeed correlated to the increase of fee levels per region. If one takes the same period of fee increase as the period of pollution decrease, then the correlation is not in all cases statistically significant with this small sample. It is, however, quite possible that there was some anticipation of fee increases or—alternatively—some delay in taking additional measures in industry. So the fee-increase period

Table 4-3. *Correlation between Relative Success in Reducing Emissions and Fee Increase (n = 17)*

Period-variable relative success in reducing emissions	*Period-variable fee increase*	*Pearson's r*	*Significance*
1986–1990	1986–1990	.226	.191
	1987–1991	.425	.044
1990–1995	1990–1995	.340	.091
	1988–1993	.529	.014
1986–1995	1986–1995	.459	.032
	1986–1994	.478	.026

Table 4-4. *Correlation between Relative Success in Reducing Emissions and Fee Increase, Controlling for Expected Decrease (n = 17)*

Period-variable relative success in reducing emissions	*Period-variable fee increase*	*Partial r*	*Significance*
1986–1990	1987–1991	.453	.039
1990–1995	1988–1993	.601	.007
1986–1995	1986–1994	.524	.019

Table 4-5. *Model Summary and Coefficients*

Period	*r*	*r^2*	*Adjusted r^2*	*F*	*Sig.*	*B*[a]	*Beta*[b]	*Sig.*
1986–1990	.679	.461	.384	5.977	.013	2.868 (S)	.555	.020
						1.502 (FI)	.374	.013
1990–1995	.799	.638	.586	12.330	.001	1.386 (S)	1.009	.000
						1.997 (FI)	.572	.014
1986–1995	.764	.584	.524	9.807	.002	1.290 (S)	.920	.001
						1.250 (FI)	.478	.037

[a]Unstandardized coefficient.
[b]Standardized coefficient.

can best be derived inductively.[21] The periods of fee increases with the strongest connection to the emissions decrease are 1987–1991, 1988–1993, and 1986–1994 (see Table 4-3). This indicates that in the first subperiod—after the steep rises in the previous period, and with some low-hanging fruit yet to pick, anticipation was somewhat more influential than delay (a one-year lead). In the later part of the full period, after relatively low fee increases and with higher abatement costs, lag was stronger than lead. In the analysis of the full period, these two influences prove to be balanced.

Instead of calculating the degree of relative success, it is also possible to compute the partial correlation between emissions decline and fee-level increase, controlling for the emissions reduction that could be expected on the basis of the shares of various sectors in the region's industries. The results are summarized in Table 4-4. Table 4-5 gives the outcomes of a linear regression model that includes both sector shares (S) and fee increases (FI) as independent variables.

The outcomes presented for 1986–1995 seem a reasonable combination of the trends observed in the two partial periods. The overall assessment of the impact of the fees in 1986–1995 can be that fees are still important in influencing the behavior of firms, but less so than during the 1970s.

In 1986–1990, fees still had a considerable influence on the behavior of firms regarding their organic pollution load, but again, less than in the 1970s, when direct regulation was absent or lax and steep fee increases created a shock in many sectors.

In the 1990–1995, the impact of the fees was somewhat stronger than in the previous five years, but still not as strong as in the 1970s. In this period, fees rose at double the pace of the previous period. Another factor is that more and more water boards began to anticipate the economic consequences if firms treated their wastewater themselves. Particularly if industry handled the easy part (the "thin water"), the public treatment system would be left with a lower tax basis, while not improving the environment, since sufficient treatment capacity already existed. This interesting phenomenon gives empirical evidence for the possibility that fees can drive industry to go beyond what is regarded efficient. But it should be emphasized that only a small share of the tax increase during 1986–1995 can be attributed to this effect. The decrease in industrial pollution was only some 10% of the total PEs taxed in 1986 because of the much larger and stable share of households, while fees rose 55%.

Fees in the 1986-1995 period were important and influenced the behavior of firms, but not as dominantly as they did during the 1970s—a conclusion one might expect. The fee instrument's secondary forms of efficacy—putting the subject on management's agenda, communicating the urgency of the pollution problem, suggesting possible abatement measures, activating industry associations and consultancies, enabling informal negotiations by waterboard officials—had all ceased. The shock effect had disappeared and the possible cost savings achieved by decreasing pollution was simply one more factor taken into account in making decisions about new products or production equipment.

In the remainder of this sector we will deal with two special topics: the influence on competitiveness and the role of fees in multi-instrumental "policy mixes."

Competition: The Dutch Paper and Cardboard Industry. A possible distortion of economic competitiveness is an often-used argument against unilateral national environmental regulations. It typically leads to acknowledging the necessity of multiscale arrangements. Sometimes this concern about competitiveness can be genuine, but it can also lead—even deliberately—to a complete halt to further incentives for more sustainable development. Therefore, it is interesting to consider an empirical study that illustrates that the many economic science theories have an uncertain empirical basis (de la Fuente 1994; see also Bressers 1995). In fact, the following example of the paper and cardboard industry shows that strong environmental policies are not necessarily detrimental to a competitive position in global markets.

What kind of test would one like to have? Assume two countries of similar development, size, and population, both with very open economies and exporting more than half of their production. Further, assume that they are adjacent, so both are in the same region and share the same international economic and governance incentives. Further, suppose that one of these countries has a very vigorous policy for a certain environmental objective, while the other completely lacks such policy, and that this situation lasts for 20 years. Additionally, the one country's strong policy includes the dreaded instrument of pollution fees, often accused of having a detrimental effect on competitiveness because of the money transfer to government. Here, then, is competition hell in one country, and paradise for another—an attractive setting for a real-life experiment.

This is precisely the situation that existed in the Netherlands and Belgium from 1970 to 1990. The Netherlands started in 1970 with a very ambitious water quality policy, while Belgium during most of this period was in the process of federalization, splitting the Flanders and Walloon communities and discussing at what level future water quality policy should be placed. To make the comparison even stronger, we can further refine "hell." What would be the most disadvantaged industrial sector in the regulated country? It would be a sector that produces bulk goods and is thus unable to defend itself by brand images (such as the case with, say, perfume); is unusually dependent on exports (some 75%); faces a lot of competition from many other countries and is therefore vulnerable; and generates relatively high water pollution per unit of production, so effluent fees and environmental measures exert maximum pressure on its profitability. Following these criteria, the paper and cardboard industry is ideal for this test.

At the beginning of the period (1970–1974) the profitability of the Belgian paper and cardboard industry tended to be somewhat higher than its Dutch counterpart. In the following period, when the first energy crisis struck the Dutch paper industry somewhat harder and Dutch effluent fees—by far the highest in the world[22]—came into full effect, this profitability gap widened. At the end of the 1970s, the Dutch sector responded with a very high investment rate (Porter 1990). After 1980, the gap in profitability closed rapidly, and from 1985 onward, profits were even somewhat higher in the Netherlands. The relative export position started as equal, but declined in both countries. Nevertheless, in Belgium the decline was greater. Although the export quota stabilized from 1980 onward in the Netherlands, it kept declining in Belgium until 1986 and remained considerably lower at the end of the period. The rate of utilized production capacity was about equal until 1976, led by Belgium from 1977 to 1979, equal again from 1980 to 1985, then led by the Dutch from 1986 onward.

Even though losing competitiveness in an ever more integrated world economy is one of the main arguments for internationalizing environmental governance the empirical basis for this is far from self-evident. Belgian and Dutch water policies illustrate that even rather extreme differences in the economic burden of environmental policy, including the use of fees, did not necessarily mean a competitive advantage for the nation with the less rigorous environmental regulations.

Policy Interactions: The Dutch Dairy Industry. This case involves two production sites of a large dairy food corporation[23] that reduced phosphates in wastewater because of the interactions between different policy instruments. Hofman (2000) did the primary analysis of this case concerning factories in Beilen and Bedum.

The increase in fee levels probably influenced the development of organic water pollution by the dairy sector (correlation for the 1986–1995 period is .63, sign.=.003, n = 18). The dairy industry entered in a environmental covenant with a combination of governments in 1994. The covenant covers water pollutants but does not affect the obligation to pay effluent fees. [24]

Pressure to reduce phosphorus emissions was driven by difficulties the Netherlands faced to comply with its international obligations. In the early 1980s, the issue became serious. This led to a 1988 covenant between the national government and water quality authorities, including the district water boards. A 75% decrease was agreed upon. The covenant aimed at adding technology to public wastewater treatment plants to abate phosphates.[25] In 1998, this led to a regulation that water treatment plants should abate 75% of phosphates.

Both dairy production sites (Beilen and Bedum) discharged a lot of phosphorus, a byproduct of whey processing. The district water board concluded that the emissions of the factory in Beilen should be reduced. Initially, the company planned to build its own treatment plant—an option that threatened the water board's public wastewater treatment plant, which would receive 80% less organic polluted wastewater and lose the corresponding tax revenues. One of the district water board officials had previously worked for a company that developed "green" technology. This official also preferred consultation and cooperation to a situation of conflict. He brought together the authorities, the company, and a technology engineering company to devise a solution. In a 1989 agreement, the company promised not to built its own treatment plant, the district water board promised not to obstruct business, and both agreed to seek (and eventually found) a solution.

Meanwhile, the plant in Bedum in the province of Groningen got into legal procedures about other issues, including comparable emissions of phosphates into the Waddenzee. Solutions to reduce phosphorus emissions were found in a jointly financed research project. The project received a subsidy from NOVEM, an implementing agency of the Ministry for Environmental Protection, largely because the technology would be integrated into the production process rather than have the end-of-pipe character of other potential solutions. The technology enabled a marketable by-product, which was implemented in Bedum in 1997 and in Beilen in 1998.

The case proves that a careful mix of instruments can prevent unproductive decisions. Had the company built its own wastewater treatment plant, the financial basis for the public plant would have disappeared. But the company could not tolerate regulatory uncertainty or be forced to invest in obsolete end-of-pipe technology. Command-and-control policies often do not provide the time necessary for the trial and error that innovation needs, and they often impose technology that is available now but suboptimal in the long run.

Conclusions about Effluent Fees

A comparison of Dutch and U.S. water quality policy shows some contrasts and some similarities. For the Dutch case we can draw some conclusions and lessons.

General Assessment

Generally speaking, the effluent fees have given good results compared with other instruments, both in 1975–1980 and in 1986–1995 (Bressers 1983a,b; Schuurman 1988; Bressers and Lulofs 2002). Industry reduced its emissions of oxygen-demanding substances by roughly 78%, prompted by charges that were intended to cover public wastewater treatment but acted in practice like regulatory fees. The fees impose no additional costs on industry (unless one would ask households to pay for the cleaning up industrial wastewater): the costs of treating wastewater would have to be paid one way or another. The revenue from the fees was used to build collective treatment plants, and in that sense, it is an example of a retribution. Because of the increased capacity of public wastewater treatment plants and the increased number of companies emitting their waste to the sewage system, direct emissions to surface water decreased sharply. These results were surprising because the official policy was presented as a CAC instrument, with emphasis on the permit system. In the implementation phase the EI elements were successful; the CAC elements failed.

Fairness

The effluent fee distributes the costs of water quality fairly on the basis of the polluter-pays principle. In fact, effluent fees encourage more cleanup in cases where more production value can be realized per pollution unit than in another sector or plant. This can be seen as particularly fair: the dirtiest have to clean up the most. Also, there will be more cleanup where the costs to do so are lower, which also can be considered fair. It is not easy to avoid the fee, and monitoring costs are not exceptional.

When CAC policies were in effect, an implementation gap led to a situation in which the regulatory pressure toward industry stayed low for a long time, especially for organic pollution. Even in 1970–1995, implementation of other environmental policies was characterized by obsolete permits and total regulatory failure, leading to inequalities that could be regarded as a serious disturbance of competitive equity. In water quality policy, the permits were of little importance; most companies achieved more abatement efforts under the fee system. This is a striking result, considering that it was the permits, not the fees, that were officially designed to manipulate the environmental behavior of businesses. With one exception—a hotspot situation—in the northern part of the country, the effluent fee system created no serious implementation deficit.

Effectiveness

The effectiveness of the fees was slightly lower in 1986–1995 (Bressers and Lulofs 2002) than in 1975–1980 (Bressers 1983a, b). In the earlier period, water board officials tried to assess the necessary capacity for wastewater treatment and warned industry to decrease pollution as much as possible. Industrial pollution decreased, and the urgency of the issue of organic pollution in surface water vanished. In fact, in the later period, the technical and economic functioning of wastewater treatment plants of water district boards was threatened by a further decrease, which might cause overcapacity in public wastewater treatment and erode the tax base of the district water boards, in turn leading to a steep increase in fees. Meanwhile, other polluting substances became more important. This simply implied that the effluent charges were in the period 1986-1995 to a lesser extent accompanied by campaigning water board officials aiming for emissions reduction.

Static Efficiency

The short-run pollution elasticity of companies confronted with effluent fees was not small, as economic theory might suggest. In the Dutch case, the influence of effluent fees on organic waste load reductions was prompt and extremely large. Whereas CAC policies often lead to lengthy negotiations with violators (if they are even caught), EI policies, if well communicated, can present much more inevitability to company decisionmakers. Bressers (1983a) attributed part of the fees' success in the 1970s to the drastic change they brought about in the relationship between water regulators and industry. Because industry can achieve significant savings by reducing pollution, the keynote is collaboration rather than conflict. In practice, therefore, environmental fees do not function as a purely economic mechanism.[26] They do not replace consultation between authorities and industry but actually increase its beneficial effects on environmental conservation.

Given the effluent fees' large effects and the fairness of the polluter-pays principle, we offer a positive conclusion about their static efficiency, but we found some sources of inefficiency. One might theoretically assume that companies abate pollution to the extent that the marginal costs for abatement equal the fee. The fee, however, is determined by the actual costs of a district water board and differs by region. Some allocative inefficiency might then occur: the fees do not necessarily reflect the cost of ensuring water quality but rather the actual costs of district water boards. These boards might overcomply with reasonable water quality standards or otherwise operate inefficiently. We also observe that companies usually do not know how much of their expenditure is environment-driven. However, when a firm's expenditures are triggered by fees, as part of calculations in collaboration with water authorities, the company's understanding of its own efficiencies is greater.

Nevertheless, we think that the effect of fees is not solely the result of a continuous balancing of costs and benefits; it also depends on the communicative power of the fees to place the issue on the decisionmaking agenda of business managers. So it might well be that the reaction of business managers does not equal the marginal costs but overshoots it in one period (when, for instance, media attention occurs) and lags behind in others. Our analysis of the 1986–1995 period in the Netherlands shows some indications of this. We base this both on the analysis of the statistics presented in Table 4-2 and on our analysis of the performance of the fees.

In building the public water treatment plants, the boards do not necessarily take into account the optimal scale (although the scale of such plants is more likely to be efficient than when separate companies treat the wastewater for themselves only). In some cases there is overcapacity. Nevertheless, the industrial sectors' varying degrees of pollution reduction corresponded in the 1970s to the estimated average pollution prevention costs (and to the pressure of fees on production costs). Also, recent qualitative inference indicates that the fees are now considered a cost factor to be minimized, leading to cost-effective solutions. Compared with an equal reduction target, this approach has been no doubt more efficient. Still, our analysis concerns reduction of the waste load, not the quality of surface water. Finally, one might argue that the public wastewater treatment plants might be left with the worst cases—the emissions that can be treated only at extreme cost. Practitioners often say that companies operate low-cost primitive wastewater treatment equipment and leave the high-cost, difficult-to-remove pollution to the public treatment plants. Yet even this cost disadvantage might be counterbalanced by economies of scale, which favor public plants.

Dynamic Efficiency

Theoretically, the more flexible time horizon of EI policies might stimulate innovative solutions, while the CAC time limits might in some cases lead to timely but obsolete solutions. CAC conceivably discourages innovation by pollutant dischargers by means of the "regulatory ratchet," since discovering ways to reduce emissions can become the basis of ever more stringent standards.

Although much of the Dutch industry's first measures were just good housekeeping, process-integrated measures were also taken, even before the term came into use. This can be regarded as innovative. The dairy industry example supports this hypothesis, although we have to stress that a more subtle policy mix was at work in this situation.

Furthermore, the Netherlands developed new purification technology, including nitrate bacteria and membranes. Many engineering consultancies entered the market, and industry associations were active in accelerating diffusion of process-integrated measures and purification techniques. This resulted in clear progress in terms of dynamic efficiency. Although the field is large and exceptions exist, there are studies that report significant progress in some sectors. One example indicates that, roughly, 1996 costs for 90% abatement of oxygen-demanding substances equals 1985 costs at the 45% abatement level: thus average abatement costs were halved (Krozer 2002).

Potentially the full advantages of economic incentives are realized only over time, but it is also clear that EI policies have their optimal effect only if communicated strongly and accurately. A policy mix that includes EI is sometimes necessary to achieve the full advantages of dynamic efficiency, not because the economic mechanism depends on that, but because an additional impact comes from constructive communication with administrators, from making the issue important to business managers, and from activating other parties, such consultancy firms and industry associations. A comparison of the 1975–1980 and 1986–1995 periods makes it clear that economic incentives interact with situational aspects. A clear-cut comparison with CAC policies is not possible in this case, however, because of the large implementation deficits of the latter.

Lessons from the Netherlands

The Dutch case, involving the highest effluent fees in the world, shows the enormous potential of this policy instrument. We found no evidence that this result can be explained by specific national characteristics of the Netherlands. Rather, success is strongly determined by the fit between the exact characteristics of the system of EI and (1) the setting it was used in, which included other policy instruments, and (2) the changing characteristics of the setting, which included changing preferences and interests of the involved water authorities.

To judge the effectiveness and efficiency of fees, one cannot limit the analysis to just the economic mechanism. The EI might not be exclusively or even principally effective because of the economic mechanism itself, but instead because it is helped by other mechanisms. The exact shape and intensity of the instrument and its context might change during the policy's lifespan, and thus no absolute statements can be made about the feasibility and effectiveness of policy instruments. The effectiveness of an instrument or mix depends very much on the circumstances, and no instrument will be effective in all.[27] In this case two other mechanisms were found influential: the changing administrative context and the exact manner in which the EI policies were fine-tuned and communicated to industry.

In the real world, "satisfying" is often far more realistic than "optimizing." This might be true for all performance indicators, whether they deal with the effectiveness, static efficiency, or dynamic efficiency of a program. For instance the fees led companies to decrease organic pollution loads to an extent in which the marginal abatement costs exceeded the fee level substantially. But the inefficiency of this overcompliance might be well worth it, given the larger inefficiencies or even ineffectiveness of alternative devices.

The atmosphere in which interactions between authorities and companies take place under EI policies is different, and more positive compared with CAC. Firms that are regulated, monitored, and perhaps forced to comply acquire a negative image, and monitoring is more difficult and less trustworthy under CAC policies. A company that emits an extra 10 units knows that the consequences are much more dire under a CAC policy than with fees.

Implementation of effluent fees requires an administrative entity dedicated to actually collecting the fees. Central government policies are often frustrated by lack of cooperation by local authorities appointed to implement them (Bressers and Honigh 1986). This problem did not arise in the Netherlands because the effluent fees function as a revenue-raising device. Through the fee system, the water authorities collect the money they need to build and operate treatment plants, and it was a task they were eager to undertake.

Another major problem with effluent fees is the massive amount of information that is required to assess the fee each company has to pay. Some authors consider this reason to reject fee instruments altogether. In the Netherlands this problem was reduced by charging the millions of households and small industrial polluters flat fees. Because they have relatively few opportunities to limit pollution, these categories of polluters are of minor importance to the instrument's regulating power. The amount of information required was also substantially reduced by basing the assessment of medium-sized polluters not on samples of their effluent but on a coefficient table. On the basis of easily obtainable data, such as the amount of water used by the firm or the amount of certain raw materials it processes, the probable amount of pollution can be established accurately. Yet the incentive to reduce pollution remains intact: companies that believe they are cleaner than the coefficient table indicates can have their effluent sampled and be charged on the basis of the results.

Finally, the political feasibility of financial instruments is always a difficult point. The charges on water pollution were acceptable because they were introduced as charges for sewage water purification—that is, as payments for services rendered. Industry associations in the Netherlands still maintain that other forms of fees are unacceptable to them, precisely because of the effectiveness of this instrument. In the words of Senator Pete Domenici (R-NM): "Fees have only one problem: they lack political support" (Domenici 1982). Perhaps the real question is, Who's afraid of effective government?

Notes

1. This chapter is partly based on Bressers and Lulofs (2002), *Charges and Other Policy Strategies in Dutch Water Quality Management* (Enschede) and J.Th.A. Bressers, D. Huitema, and S.M.M. Kuks (1994), Policy Networks in Dutch Water Policy, *Environmental Politics* 3(4).

2. This concern with water is illustrated by the fact that the Dutch crown prince, Willem Alexander, practices water management.

3. A population equivalent is the amount of organic pollution equivalent to the average organic pollution caused by one person in a normal household; the number of PEs is thus equal to the population. One PE equals an amount of oxygen-demanding substance that uses 136 grams of oxygen. For industrial pollution, the number of PEs is calculated predominantly on the basis of chemical and biochemical oxygen demand.

4. The conclusion of Bressers et al. (1994) is that the institutional and administrative structure of surface water management in the Netherlands is very complex, but the agencies that are directly involved share a common belief system and evidence a substantial degree of commitment and interrelatedness. This concerns only the inner core of the administrative actors, however. To characterize the broader network, it should be seen in action.

5. The charge is determined by the so-called public formula:

VE = Q (CZV + 4.67 Kj-N) / 136
VE = Discharge (expresses the amount of pollution in PE)
Q = Amount of discharged water in 24 hours
CZV= Chemical oxygen demand
Kj-N = nitrogen proportion, determined by the Kjedahl-method
136 = average disposal produced by one person in one day
Formulas for black-list substances and remaining metals vary (CIW 1999, *260*) but resemble:

VE black-list substances: = (load in kg/year – 0.0006 × number of discharged PE) × 10
VE remaining metals = (load in kg/year – 0.04 × number of discharged PE) × 1

6. As an aside, we have to explain that the basis for the levy was revised in 1986; the calculation was based on 136g instead of 180g of oxygen use.

7. For individual households there is no incentive to reduce water pollution. The levy is differentiated by only two categories, on the basis of size of the household. If one is willing to accept that pollution by households correlates with the amount of drinking water used, there is some incentive in the price of drinking water. However, Dutch drinking water is, within the European context, relatively cheap.

8. This method is also referred to as the detective paradigm. Scriven proposes (1) to make an inventory of all possible suspects that might explain the phenomena; (2) to reconstruct the characteristic pattern of event for every suspect; and (3) to empirically assess which of the characteristic patterns or causal chains occurred. This should lead to the analytical confession of the perpetrator.

9. The analysis of the period 1986–1995 is done in a similar manner.

10. A more extended version of this analysis can be found in Bressers (1983b). Methodological problems, especially those in connection with the operationalization of the relevant theoretical notions to a model of analysis, are dealt with there in greater detail. Here we have limited ourselves to a survey of the findings and a reflection on the conclusions drawn from the analysis.

11. In the case of 2 of the 14 industries (potato starch and livestock), fees had less influence than expected on the basis of their fee factor. In anticipation of ultimately being hooked up to a *smeerpijp* (literally, a dirt pipe), the potato starch industry had been allowed to continue emitting untreated discharges; during this time, it was paying much lower rates than were other industries. The livestock industry comprises thousands of relatively small farms. Here the fees are seldom calculated according to the amount of pollution actually produced. Both phenomena were dealt with and can be isolated in the statistical analysis. When these two industries are left out of the analysis, the correlation of the fee factor with the decrease in pollution is even stronger: $r = .84$. The correlation with the other two factors remains about the same, with production increase showing $r = .24$, and abatement costs, $r = .48$. These three factors, taken together, account for 76% of the differences in the decrease of organic pollution in industrial wastewater among these 12 industries.

12. However, it cannot be determined on theoretical grounds for which period the fee increase should be used as an independent variable. On the one hand, there is a time lag between the stimulus to abatement efforts and the installation and use of the relevant control measures. On the other hand, it is not implausible to expect the companies to anticipate charges and rate increases (Bressers 1983b). It is better, therefore, to decide on empirical grounds which period should be considered in determining the rate increase. This period starts in the year in which the rate of the charge has the highest negative correlation with the relative success of abatement between 1975 and 1980. The rate in that year can be taken as the best indicator for the extent to which abatement efforts before 1975 negatively affected abatement activities in the subsequent period by the resulting increase of marginal costs of further abatement. The rate charged in that year should then be subtracted from the charge at the end of the period with the highest positive correlation in order to get the best possible indicator of the fee factor that stimulated abatement of pollution in the 1975–1980 period.

13. And $r = .65$, $n = 13$, $p = 0.008$ for heavy metals, because two district water boards did not charge heavy metals.

14. Further analysis has shown that this correlation cannot be attributed to water authorities' raising rates to maintain revenues despite industry's abatement efforts.

15. The combination of policy outputs that accounts for the greatest part of the variation in relative success in abating pollution by heavy metals includes the increase in the charge rate 1975–1980, the proportion of heavy metal dischargers with an abatement plan, and the weighted number of official reports on noncompliance. Together, these policy outputs explain statistically 82% of the difference in relative abatement success. These three policy outputs together with the abatement that could have been expected in view of the regional structure

of industry, and the shares of the various kinds of metal statistically account for 91% of the variation in decrease in heavy metal pollution (R2 after correction is .87).

16. Both approaches have their strong and weak points (Reichardt and Cook 1979). However, if both methods led to the same conclusion, our confidence in it would be increased.

17. Data of earlier years can in principle be transformed into the new values, but only approximately. Furthermore, data collection was taken over from the Ministry of Public Works by the Dutch Central Statistical Office in the early 1980s. Consequently, the quality of data in these years is rather doubtful.

18. For details, see the full report (Bressers and Lulofs 2002).

19. The district water boards themselves needed to be able to assess the necessary dimensions of the treatment works they were planning to build.

20. By subtracting the calculated percentage pollution decrease (amount of pollution in 1986 minus the calculated amount of pollution in 1990 divided by the actual pollution in 1986, multiplied by 100) from the actual pollution decrease percentage, one gets the relative success of pollution decrease in that area.

21. In the first instance, one would be inclined to take the same period for the measurement of the independent variable (the fee increase) as for the dependent variable (the pollution decline). But there is no certainty at all that the period of fee increase should be measured in the same period as the dependent variable. Instead, many causes and forms of delayed response are probable; awareness of the possible future fee increases in one's region will be high, however, considering past experience from the 1970s onward. The combination of anticipation and delay makes it impossible to determine the period of measurement of the independent variable deductively. Even per period of analysis (and per region and per sector), it can be expected that the purest measurement of the independent variable will vary. Possibilities to lower emissions have different time trajectories and consequently lead to different delayed responses. Similarly the degree of anticipation can be expected to vary with, for instance, the experience with past fee increases. All in all, then, the period of measurement of the fee increase as an independent variable can be best assessed inductively, of course within reasonable limits.

22. By 1995, the fee was some 36.33 per inhabitant equivalent (136g biochemical oxygen demand). During the 1970s and 1980s, the exchange rate was, however, very different, and the Dutch currency (45% of the value of the €) rated much higher against the dollar.

23. Friesland Coberco Foods Holding employs about 11,843 people.

24. For detailed information, see the full report (Bressers and Lulofs 2002).

25. This led to an increase in fees.

26. The argument derives from the case of the reduction of heavy metals in the period 1975–1980. In view of the damage that heavy metal pollution causes to the sewage water treatment process and to the quality of the resulting purification sludge, most water boards also introduced a charge on heavy metals in effluents. But because the fee was relatively low, the water boards felt they had little to do with the 50% reduction of heavy metals in industrial effluents achieved between 1975 and 1980. Without negotiations and licensing regulations, they thought, industry would be unwilling to budge. Statistical analysis (Bressers 1983a,b) showed, however, that negotiations in districts that had substantially raised the fees were much more successful than those in other districts. So the regression analysis revealed that the fees, far from being insignificant, were in fact the most powerful policy instrument. Nevertheless the fee on heavy metals is relatively low because the law requires that fees be related to the costs of cleaning up pollution. The additional (sludge) treatment costs because of the heavy metals are relatively low compared with the costs of abating organic pollution.

27. It is therefore essential that theories on the feasibility and effectiveness of instruments take these circumstances into account. Building on our experience with research into the effectiveness of large parts of Dutch environmental policy, a contingency theory has been

developed at the University of Twente. This contingency theory assesses the feasibility and implementability of policy instruments, taking into account the circumstances in which the instruments are applied. It therefore enables us to make global predictions and statements about the feasibility and implementation of the various policy instruments, singly and combined, in different situations.

References

Audit Office. 1987. (Algemene rekenkamer), *Milieubeleid oppervlaktewateren*, vergaderjaar 1986 19–87, 20020, nrs 1-2.

Bressers, J.Th.A. 1983a. *Beleidseffectiviteit en waterkwaliteitsbeleid* (Policy Effectiveness and Water Quality Policy). Dissertation. Enschede: University of Twente.

———. 1983b. The Role of Effluent Charges in Dutch Water Quality Policy. In *International Comparisons in Implementing Pollution Laws,* edited by Downing and Hanf. Boston: Kluwer-Nijhoff, 143–68.

———. 1995. Water Management and Global Environmental Change Policies. In *Global Environmental Change and Sustainable Development in Europe*, edited by J. Jäger, A. Liberatore, and K. Grindlach. Brussels: Luxembourg: European Commission, 77–82.

Bressers, J.Th.A., and M. Honigh. 1986. A Comparative Approach to the Explanation of Policy Effects. *International Social Science Journal* 38(2): 267–88.

Bressers, J.Th.A., and K. Lulofs. 2002. *Charges and Other Policy Strategies in Dutch Water Quality Management.* Enschede.

Bressers, J.Th.A., D. Huitema, and S.M.M. Kuks. 1994. Policy Networks in Dutch Water Policy. *Environmental Politics* 3(4).

Brown, G.M. Jr., and R.W. Johnson. 1983. *The Effluent Charge System in the Federal Republic of Germany and Its Potential Application in the US.* Seattle: University of Washington.

CBS. 1998. *Kosten en financiering van het milieubeheer,* 1995–1996, Heerlen/Voorburg: CBS.

———. 1999. *Kwartaalbericht Milieustatistieken 1999–4,* Jaargang 16 no. 4 Heerlen/Voorburg: CBS.

CIW. 1999. *Financiering Zuiveringsbeheer* (Financing Pollution Treatment and Management). Den Haag.

Domenici, P.V. 1982. Emissions Trading: The Subtle Heresy. *The Environmental Forum.* December.

Fuente, M. de la. 1994. *De relatie tussen waterkosten en concurrentiepositie.* Master's thesis. University of Twente.

Hofman, P.S. 2000. *Innovation by Negotiation. A Case Study of Innovation at the Dairy Company Borculo Domo Ingredients.* Enschede. CSTM research series no. 145.

Hucke, J. 1978. Bargaining in Regulative Policy Implementation: The Case of Air and Water Pollution Control. *Environmental Policy and Law* 8(4): 109–15.

Krozer, J. 2002. *Milieu en innovatie* (Environment and Innovation). Dissertation. University of Groningen.

Mitnick, B.M. 1980. *The Political Economy of Regulation: Creating, Designing and Removing Regulatory Forms.* New York: Columbia University Press.

Porter, M.E. 1990. *The Competitive Advantage of Nations.* New York: Free Press.

Reichardt, C.S., and T.D. Cook. 1979. Beyond Qualitative versus Quantitative Methods. In *Qualitative and Quantitative Methods in Evaluation Research*, edited by T.D. Cook and C.S. Reichardt. Beverly Hills: Sage, 7–32.

Schuurman, J. 1988. *De prijs van water.* Dissertation, Arnhem.

Scriven, M. 1976. Maximizing the Power of Causal Investigations: The *modus operandi* Method. In *Evaluation Studies Review Annual.* Beverly Hills: Sage, 101–19.

Vogel, D. 1983. Comparing Policy Styles: Environmental Protection in the United States and Great Britain. *Public Administration Bulletin* August: 65–78.

CHAPTER 5

NO_x Emissions in France and Sweden

Advanced Fee Schemes versus Regulation

Katrin Millock and Thomas Sterner[1]

BOTH SWEDEN AND FRANCE impose special charges as supplementary instruments to command-and-control (CAC) regulation of emissions of nitrogen oxides (NO_x) from energy sector and industrial boilers. The revenues from the Swedish charge (which dates to 1992) are automatically recycled through payments to industry based on the energy produced. The French tax, which in some respects is similar, applies to four categories of air pollution but at a much lower level, and the refunding of tax revenues is more targeted. The origin of the French tax system goes back to the mid-1980s and the debate on acid rain. The Swedish tax was also prompted primarily by concern over acid rain and introduced as a specific policy tool to accelerate the reduction of NO_x emissions from industry. Both systems build on existing CAC legislation, and at least in the French case the tax complements rather than substitutes for this legislation.

In this chapter, we use the term *charge* or *refunded emissions payment* (REP) for the Swedish program, since the money collected does not go to the treasury but is returned virtually in its entirety to the firms by the Swedish Environmental Protection Agency (SEPA); 0.5% of the revenues generated by the fees is used to cover the agency's administrative costs. We will use the term *tax* for the French program, since it is legally referred to as a *taxe parafiscale*, a special kind of tax. The difference is in practice fairly small, since the majority of the funds collected by the French tax are used to subsidize abatement investments among the firms that emit NO_x. There is no automatic formula for the refunding as in Sweden; rather, firms apply for subsidies to finance abatement investments.

Background

Although NO_x emissions are a serious environmental issue and many countries have official goals for emissions reductions, in relatively few places economic policy instruments are used. There are fledgling tax systems in Italy, some districts of Spain, and a few Eastern European countries, and the Netherlands is setting up an emissions trading scheme. However, it is fair to say that of the NO_x control instruments that have been in place long enough to permit analysis, the most important are found in France, Sweden, and the United States.

The modern acid rain debate started largely in Scandinavia, where acidification of lakes and fish kills were noticed in the 1960s. Sweden, Norway, and other countries had to convince the main polluters, such as Germany, Poland, and the United Kingdom, that this serious problem was due to the latter countries' emissions. Gradually the evidence was accepted, and acid rain then became a major policy issue in Europe in the 1980s, following evidence linking forest death to emissions of sulfur and nitrogen oxides. In addition, deposition of nitrogen contributes to eutrophication of watersheds. Germany in particular demanded action on a concerted European Union level. In France, sulfur emissions per unit of energy use were already among the lowest in Europe because of its investment in nuclear power. Existing regulation of pollution from industrial production units is based on regionally enforced production permits, issued by the *préfet,* the governor of a *département,* an administrative level of government below the regions. Since the office of the *préfet* also involves economic regional development, enforcement could be compromised by a company's contribution to local employment, and it would have been politically difficult to strengthen the emissions limits in the *arrêtés préfectoraux*. Hence, administrators at the French Agency for Air Quality argued for a small tax to be imposed on industrial plants as a complement to the traditional command-and-control regulation. The objective was to provide incentives for industrial plants to invest in further abatement of sulfur dioxide (SO_2) emissions.

In France, the introduction of the tax was closely linked to European Community policy on acid rain, and it initially covered only sulfur emissions. Sweden, on the other hand, introduced its tax in the 1990s specifically to accelerate reductions of NO_x emissions. A major driver of current policy in both countries is the United Nations Economic Commission for Europe Convention on Long-Range Transboundary Air Pollution and its 1999 Gothenburg Protocol. The Gothenburg Protocol sets specific limits for emissions of SO_2, NO_x, volatile organic compounds, and ammonia (NH_3) to be attained by the parties by 2010. Following the ratification of the convention, the European Union developed a directive that sets national emissions ceilings, agreed upon by member states on June 22, 2000. For France the national emissions ceiling on NO_x implies a 54% reduction from 1990 emissions levels; for Sweden it implies a 56% reduction.

Sweden has long had a very aggressive policy on the precursors to acid rain. The Scandinavian region has very old geologic structures whose bedrock is deficient in calcium and thus has little buffering capacity. Sweden is among the countries that have been most affected by acid rain and its effects on lake and forest ecosystems, and thus this nation has a very determined policy on the precursors to acid rain, notably SO_2 and NO_x emissions. Nitrogen oxides cause eutrophication of lakes, rivers, and even marine coastal areas and are precursors to tropospheric ozone, which contributes to smog and diminishes air quality in cities. In Sweden eutrophication and acidification are the major problems; tropospheric ozone is less of a concern (partly because of the limited sunshine).

For SO_2 Sweden imposes a tax of $3,000 per tonne (1,000 kilograms, or a metric ton), which has led to dramatic reductions and very low emissions levels by international standards. For NO_x the ambitions were equally high but emissions reductions were much more difficult. The Swedish charge was implemented to fulfil Sweden's commitment to reduce NO_x levels by 30% by 1995 compared with 1980 levels. Although mobile sources are an important contributor to NO_x emissions, additional reductions from the energy production and industry sectors were necessary to meet the objective.

To understand the issues related to NO_x, some technical and chemical explanation is necessary. SO_2 emissions can easily be calculated based on the sulfur content of the fuel, and thus a sulfur tax based on fuel content (with refunds for abatement) is an appropriate instrument (Yohe and MacAvoy 1987). NO_x emissions, however, are only partly due to the nitrogen content of the fuel. They are produced largely from an unintended chemical reaction between nitrogen and oxygen in the combustion chamber. The extent and speed of this reaction is highly nonlinear in temperature and other parameters of the combustion. This has two implications.

First, there is a large scope for NO_x reduction through various technical measures,[2] which we can call abatement (although some of them may also have other effects). They include changing the shape, temperature, or oxygen and moisture content of the combustion chamber. Fuel switching will automatically have a large effect on NO_x, as will exhaust gas recirculation. Other abatement strategies include adding reduction agents (ammonia) or passing exhausts through catalytic converters.

Second, there is a very strong argument for physical measurements and detailed monitoring. Engineering studies carried out in connection with the Swedish program have shown that it is not just the installation of certain pieces of equipment but, to a large extent, the fine-tuning of their operation that leads to NO_x reduction. Because of the complex, nonlinear nature of the chemical processes involved, both outside inspectors and plant managers must monitor the emissions so that they know which measures are successful. Basing fee payments on actual physical measures (typically continuous or hourly) ensures that the payments are fair and create a real incentive for abatement efforts.

Basing payments on rules of thumb or standardized emissions factors for different technologies introduces a number of risks. Such measures are bound to be inaccurate when it comes to NO_x emissions, and it is a reasonable hypothesis that such practices will reduce incentives for fine-tuning and thus lose one of the most important mechanisms for abatement.

The Policy Response

Sweden: The Refunded NO_x Charge

The Swedish charge covers NO_x emissions from all industrial boilers, stationary combustion engines, and gas turbines with a useful energy production of at least

25 gigawatt hours (GWh) per year. Initially the charge was levied on only 120 combustion plants (182 boilers) producing at least 50 GWh of useful energy per boiler. Because of its success in reducing emissions and a decline in monitoring costs, the system was later extended to include all boilers producing at least 40 GWh, and in 1997 the limit was lowered to 25 GWh. Today the approximately 250 plants (365 boilers) subject to the law emit about 14,000 tonnes of NO_x a year, which represents approximately 5% of the total NO_x emissions in Sweden.

SEPA manages the scheme at a small administrative cost amounting to 0.5% of revenues. The remaining revenue—more than SEK 500 million ($50 million) per year—is refunded to the sources that paid the charge but in proportion to output of useful energy. A refunded emissions payment scheme needs some relevant and neutral yardstick on the basis of which to refund revenues. Gross revenue or value added could be used, but this might create perverse effects, such as the inclusion of financial operations, consulting, or other nonenergy-related output that would be irrelevant for a program that targets technical abatement measures for combustion (see Sterner and Höglund 2000). "Useful energy produced" has been accepted as the yardstick for measuring output from paper mills, power plants, and other heterogeneous activities that use large boilers, engines, or gas turbines. For power plants and district heating plants, it is equal to the energy sold. For other industries, the energy is defined as steam, hot water, or electricity produced in the boiler and used in production processes or heating of factory buildings.[3]

The tax rate is SEK 40 per kilogram ($4,000 per tonne NO_x).[4] The charge is based either on actually monitored emissions, or on calculated emissions according to fuel data, fuel consumption, and oxygen concentration in the stack. All units self-report the measures in a yearly NO_x declaration submitted by January 25 of the year following the accounting period. The fee is levied on the sum of nitrogen oxide (NO) and nitrogen dioxide (NO_2) counted as NO_2. One of the reasons for this is that much of the NO can be converted to NO_2 (in the furnace or stack, or later in the atmosphere).[5] Nitrous oxide (N_2O) is normally a very small component and is not monitored.

The refund varies from year to year, but in recent years it has been just under $1 per megawatt hour (MWh) of useful energy. This implies that the average emissions factor has been 0.25 kilograms NO_x per MWh.

France: The Air Pollution Tax

The French tax on air pollution (*taxe parafiscale sur la pollution atmosphérique,* or *la TPPA*) was originally administered by the French Agency for Environment and Energy Management (ADEME).[6] It applied to all units with a power capacity of at least 20 MW or with emissions exceeding 150 tonnes per year of SO_2, NO_x, hydrochloric acid (HCl), or volatile organic compounds. Household waste incineration plants were subject to the tax if their capacity exceeded 3 tonnes per hour. The use of a different cutoff level for the French tax prevents direct comparison with the coverage of the Swedish NO_x charge. However, the French threshold for taxation is significantly higher than the Swedish one, which was lowered twice during the 1990s. In 2000, more than a third of the furnaces in the Swedish program were smaller than the French cutoff of 20 MW. The French tax covered a large number of sources (from 1,200 in 1990 to nearly 1,500 in 1999), but in proportion to the size of the French economy, this is somewhat less than the 365 boilers (in 250 plants) included in the Swedish NO_x charge. The level of the French tax, though, was only about 1% of the Swedish one. When the TPPA was expanded to encompass NO_x emissions in 1990, the level was FF 150 per tonne (about $23 per tonne NO_x). In 1995, the tax rate was increased to FF 180 per tonne, and then on January 1, 1998, to FF 250 per tonne ($38 per tonne). Plants self-reported their emissions in the preceding year by March 1 to the French regional industry board, which upon verification of the report forwarded it to ADEME by April 1. The declaration form allowed for either direct measures or (apparently more common) calculations of NO_x emissions based on fuel consumption data. If the total tax due was less than FF 1,000 for a unit, no tax was levied. Firms were not invoiced but made their payments directly to ADEME with the emissions declaration.

Whereas the Swedish charge is automatically refunded to industry according to plants' energy production, the French tax revenues were recycled through subsidies for abatement measures. In this sense, it was inspired by the water charges administered by the French regional water agencies. The water charge system has often been likened to a form of mutual insurance, under which the levies are used to finance necessary abatement operations. Another determinant for the design of the French NO_x tax was the government decree on parafiscal taxes.[7] Such taxes, which can be used to finance mutually beneficial objectives for the tax subjects, are not voted by parliament, but decided upon in government orders. As a parafiscal tax, the TPPA automatically had to redistribute its revenues. ADEME received 6% of total tax revenues to cover its administrative costs, but 75% of the tax revenues was earmarked for abatement subsidies, with the rest being allocated to the financing of air pollution surveillance systems. Any company subject to the TPPA could apply for a subsidy, which was awarded according to percentage rates of the additional fixed capital investment for emissions reductions: 15% for standard abatement technologies, 30% for innovative technologies. There was also an additional 10% subsidy to small and medium-sized companies. Individual decisions on the allocation of subsidies were taken in the directory committee of the TPPA, whose 18 members represented government, industry, and nongovernmental organizations. A study of ADEME data shows that almost all applications were funded, and in this sense, there was to a certain extent an automatic refunding, although the distributional impact depended on whether a company took the initiative to ask for an abatement subsidy. The few rejections of subsidy requests concerned plants whose abatement objectives were not considered ambitious enough in fulfilling a French decree or a European Union directive whose implementation date had already passed.

In the early years of the tax, revenues were used to cross-subsidize abatement action for other pollutants. Before NO_x emissions were covered, however, a decision limited the use of cross-subsidization by stipulating that 90% of the revenues from the tax on a certain pollutant had to be allocated toward emissions reductions of that pollutant.

Ex ante Analysis

No cost-benefit analysis was performed for the French tax on sulfur emissions when it was introduced in 1985. Although industry called for such a study, the administration seems to have been wary about the possible disparity in regional effects that such an analysis would show; it was argued that the tax was very small and a complement to existing CAC regulation in the form of operating permits. Industry raised two principal objections to the tax. First, in 1985, it contested the allocation of the tax revenues, arguing against using them to finance air pollution surveillance systems and saying that the entire amount should subsidize abatement. Since 75% of the tax revenues was allocated to abatement subsidies and tax rebates for industry financial contributions to surveillance systems were part of the design, however, resistance faded. Industry's second objection, concerning the definition of NO_x emissions, led to a legal battle that lasted through the 1990s. Major chemical producers took the stance that N_2O emissions did not qualify as air pollution under the (old) air pollution law of 1961.[8] As the court battle proceeded, companies continued to receive subsidies for abatement of all NO_x emissions, including N_2O. By 2000, the controversy was settled in the favor of the chemical companies, and ADEME refunded tax payments it had received based on N_2O emissions.

There was no formal *ex ante* analysis of the NO_x scheme in Sweden, either. The Swedish charge is based on a public inquiry concerning environmental taxes (SOU 1989). This inquiry considered a large number of possible environmental charges and taxes and suggested a package that was part of a "green tax reform" in 1991 when income and other taxes were lowered and restructured, creating an opportunity to introduce some new environmental taxes. The Swedish parliament in 1988 had decided on a 30% reduction in the aggregate level of the country's total NO_x emissions from 1980 to 1995. The environmental tax commission did not base the charge level on estimates of damage but instead took the reduction goals as given and based its proposals on estimates of the costs of abatement; those estimated costs ranged from SEK 3 to 84 per kilogram and were used to motivate the charge level of SEK 40 per kilogram. The investigation explored the general socioeconomic, distributional, and environmental consequences of all its proposals but did not conduct a detailed cost-benefit analysis, particularly not of individual components of the proposal, such as the NO_x charge. It did, however, consider the risk of increases in unregulated emissions (such as NH_3 and N_2O) as well as the risks of distorted competition between large and small energy producers. This was, in fact, the main practical reason for refunding the charges.

Ex post Analysis

France: The Air Pollution Tax

We record the development of taxable NO_x emissions in Table 5-1. For 1990 and 1995, emissions were recorded only for an eight-month period and have been

Table 5-1. *France: NO_x Emissions Subject to Tax on Air Pollution*

Year	NO_x emissions (t)	Plants	Average emissions (t)
1990	339,613	1,206	282
1991	404,797	1,230	329
1992	431,898	1,288	335
1993	360,754	1,298	278
1994	364,672	1,308	279
1995	395,699	1,423	278
1996	386,245	1,456	265
1997	378,435	1,455	260
1998	394,538	1,472	268
1999	363,699	1,474	247

Note: Emissions for 1990 and 1995 were recorded for only eight months and have been rescaled on a one-year basis.

readjusted to a yearly basis. The sectors most responsible for NO_x emissions are electricity, gas, and heating (35%), chemicals (23%), and other nonmetallic products (13%) (see Table 5-2). Gross taxes were computed as the product of the unit tax times total emissions. Net taxes are taxes effectively paid by the plants. Through the system of deductions for contributions to air pollution surveillance systems, the chemical sector received a reduction in net taxes compared with gross taxes (from 14% of total taxes to 9%). Otherwise, the distribution of net taxes across sectors follows closely that of gross taxes. The regions with the largest emissions are the Upper Rhine (Haut Rhin) and Seine-Maritime; France's largest emitter of N_2O, Alsachimie, is in Haut Rhin, and Seine-Maritime is home to a major electricity plant, several oil refineries, a cement factory, and major chemical companies.

In yearly evaluations, ADEME (1996, 1997) reports that the tax has led to annual reductions of more than 27,000 tonnes of NO_x. In the published reports on the tax (to 1997, inclusive), annual spending on abatement investments is reported, and in later years also the anticipated tonnes of abatement. Thus, sums

Table 5-2. *France: Sectoral Distribution of Emissions, 1990–1999*

Sector	NO_x emissions (t)	Percentage of total
Electricity, gas and heating	1,329,296	35
Chemicals	870,345	23
Other nonmetallic production	494,812	13
Coke	310,643	8
Iron and steel	274,826	7
Waste management	184,986	5
Food and agriculture	114,950	3
Paper	92,520	2
Plastic and rubber	58,575	2
Extraction	19,695	0.5
Other industries	69,702	2
Total	3,820,350	100

of FF 18.379 million, 26.591 million, and 40.592 million were allocated for NO_x abatement investments in 1995, 1996, and 1997, respectively. For the two years where anticipated reductions exist, they are 3,758 tonnes per year in 1996 and 56,216 tonnes per year in 1997. The emissions reduction data should be taken with some caution, since they come from the statements that companies filed when requesting subsidies for an abatement actions. No clear pattern emerges for the cost per tonne of NO_x removed, which varies significantly by year. In the early 1990s, ADEME's subsidies for NO_x abatement implied an environmental benefit of FF 4,000 to 5,000 per tonne per year. Furthermore, no attempt has been made to separate the individual effect of the TPPA from other factors, such as emissions standards in French national decrees or in European directives.

Recent econometric work aims at estimating the NO_x reduction from French stationary sources subject to the TPPA. Millock and Nauges (2003) find a significant negative effect of the tax for NO_2 emissions in a simple fixed-effects model controlling for fuel input price, economic activity (proxied by regional GDP), industry sector, and location. Estimations are performed only for plants that were not exempt from the tax (those with more than FF 1,000 of gross taxes). Furthermore, N_2O emissions are excluded from the analysis, since the major chemical companies refused to pay the tax for this pollutant. The dependent variable is yearly emissions per unit of combustion capacity, in order to avoid size effects. Based on that analysis, the upper bound for the elasticity of NO_2 emissions with respect to the unit tax rate is –0.15. This estimate constitutes an upper bound on the elasticity of emissions with respect to the tax rate because the available data did not allow researchers to control for the technology standards defined in operating permits for each plant.

There was no fine for failure to pay the tax. ADEME officials say that fines were not necessary because of high compliance—around 97%. Since the regional departments for industry oversee the self-reporting by plants, and since independent verification is required for companies not using the standard ADEME formula based on fuel consumption data, it is considered very unlikely that firms would underreport their actual emissions. Fuel consumption data reports could easily be checked against bills from fuel suppliers.

The difficult part of an *ex post* evaluation is instead the effectiveness of the subsidies. The timing of the applications for subsidies suggests that companies were anticipating a European directive or a French national decree. The supplementary abatement effects from subsidies could thus be considered nonexistent. Rather, the role of the subsidies could be interpreted as a policy instrument to accelerate abatement action that had to be taken anyway.

The sectoral distribution of the subsidies for NO_x abatement is shown in Table 5-3. The heaviest polluters were also the main recipients of subsidies. The highest subsidies went to the electricity sector (42% of total subsidies) and the chemical sector (41% of total subsidies). Abatement projects in the other nonmetallic products sector received 8% of total subsidies, followed by coke (3%) and iron and steel (3%). The effectiveness of the subsidies in reducing emissions can be questioned; econometric estimations show that it varies with the sector. Whereas abatement subsidies led to a significant decrease in NO_x emissions in the iron and steel sector, the effect in the coke sector seems to have been the

Table 5-3. *France: NO_x Abatement Subsidies by Sector, 1990-1999*

Sector	*Number*	*Million 1990 FF*	*Percentage of total*
Electricity	8	82	42
Heating	7	4	2
Chemicals	21	79	41
Other non-metallic production	11	15	8
Coke	1	5	3
Iron and steel	5	5	3
Plastic and rubber	6	1	0.5
Other industries	2	2	1
Total	61	193	100

opposite. This points to an important drawback in refunding revenues through abatement subsidies: much depends on the administrative procedure by which the subsidies are allocated, and on the effectiveness of the control measures installed. The French tax refunding system creates no continuous incentive for firms to reduce emissions, given the low tax level. Using an automatic basis for refunding, such as "useful energy output" as in Sweden, gives a firm direct incentives to reduce its emissions through whatever means prove cost-effective.

Sweden: The Refunded NO_x charge

Although the Swedish parliament wanted a 30% reduction in NO_x emissions by 1995, this objective was not met until 1998. The failure to meet the target date was primarily due to NO_x emissions from mobile sources, which reduced emissions by only 13% from 1980 to 1997. Stationary sources, in aggregate, are estimated to have achieved a reduction of more than 50% during this same period, and a large part of this reduction is due to the NO_x charge (even though it did not apply to small stationary sources). Thus the instrument used has generally been considered a success.

The Swedish Environmental Protection Agency has evaluated the NO_x charge several times (SEPA 1993–2002). The charge has also been evaluated in Höglund (2000) and Sterner and Höglund (2000). Specific emissions have fallen by 40%, from about 0.40 kilograms NO_x per MWh useful energy to 0.24 kilograms NO_x per MWh (see Table 5-4). Smaller plants, which entered the system as it was expanded, have higher emissions and have not yet reduced their emissions by more than about 20%. This implies that the reduction for the boilers that entered the system in 1992 is actually somewhat greater than 40%—a substantial reduction, since NO_x abatement is technologically difficult.[9] Data from newcomers to the system in 1996 and 1997 reinforce the impression that emissions reductions are mainly due to the REP scheme, since their emissions rates on entry are similar to those of the original plants in 1992 but immediately start to fall in response to the charge. The relative success of the Swedish charge system can be further illustrated by comparing plants between countries.

The U.S. Environmental Protection Agency compared NO_x emissions from selected plants with similar technology (coal-fired boiler plants with selective catalytic reduction) in different countries, including Sweden. The Swedish plant

Table 5-4. *Sweden: Combustion Plants Subject to Charge, 1992–2001*

Year	Production units (combustion plants)	NO_x emissions (tonnes)	Produced energy (MWh)	kg NO_x per MWh produced energy	Charges levied (and refunded), (million SEK)
1992	181	15305	37465	0.41	612
1993	189	13333	41158	0.32	533
1994	202	13025	45193	0.29	521
1995	210	12517	46627	0.27	501
1996	274	16083	57150	0.28	643
1997	371	15107	54911	0.28	604
1998	374	14617	56367	0.26	585
1999	375	12827	48956	0.26	513
2000	363	12644	51073	0.25	506
2001	393	14160	58141	0.24	566

had about half the emissions rates of the plants in the other countries. The report ascribes this discrepancy largely to the REP scheme: "The Swedish retrofitted unit, in contrast, demonstrates that NO_x levels well below the Swedish standard (and also below the German or United States standards) are achievable. ... The Swedish regulatory system, incorporating an economic incentive, clearly motivates [the Swedish plant] to achieve minimal NO_x rates rather than just comply with the applicable emission standard" (U.S. EPA 1988, *37*).

A detailed empirical comparison of emissions rates is complicated by the age, size, and structure of the plants. Fuel and other factor prices are different, and the markets for outputs are quite distinct. U.S. plants are typically bigger and produce mainly or exclusively electricity. The plants studied by Burtraw and Evans (Chapter 6) all burn coal. In Sweden coal is very rarely used, and our data cover not only power plants but all kinds of industries with large furnaces, many of which use biofuel, sometimes in combination with fossil fuels. Furthermore, Swedish power plants are typically catering to a joint demand for power and heat. Our figures for specific emissions are based on the concept of useful energy output, which includes heat that is sold commercially or used *productively* within the plant. To make our comparison as clean as possible, we have selected the four production units in 2000 in Sweden that (a) were in the power sector and (b) exclusively used coal (100% in two cases and 98–99% in the other two). Average emissions for 2000 were 0.246 pounds NO_x per MWh_{th} useful energy as defined in the Swedish regulation. Their average thermal efficiency (useful energy as a percentage of input energy) was 91%. Since these plants are not optimized for maximizing electricity production, their electricity production figures (which were not available) would not be very useful for a comparison with U.S. plants that are optimized for electricity only. We may, however, assume that a plant would have an efficiency of around 40% if it were optimized to produce electricity only. Correcting our emissions figures by multiplying by 91/40 should thus give a good approximation the NO_x emission figures on a comparable basis with the U.S. figures. The result is average emissions of 0.56 pounds NO_x per MWh_e, or about 35% of the U.S. standard. Even after this correction, the emissions figures are very low by U.S. standards. According to Burtraw and Evans (Chapter 6, Figure 6-3 and Table 6-1) most plants emit about 5 pounds NO_x per

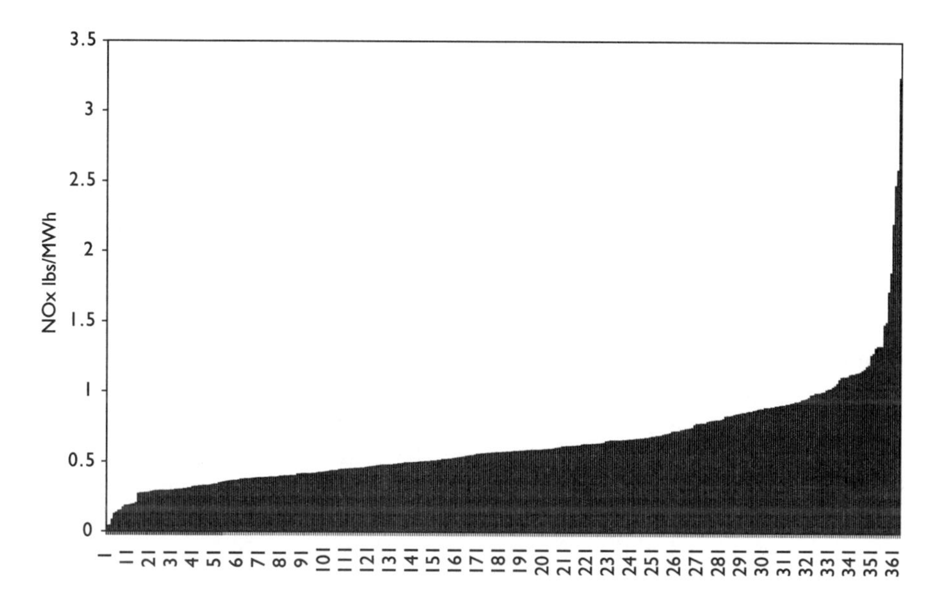

Figure 5-1. *Sweden: NO_x Emissions for Furnaces, 2000*

Note: The measure of energy used includes heat. To get approximate values corresponding to the production of electricity only, multiply by a factor of 2.5.

MWh, and the latest new source standard is 1.6, which so far has been attained by only a small number of units. Figure 5-1 shows all the emissions factors for the plants in the Swedish system in 2000 and can be compared with the figures in Chapter 6.

As an extra check we have also calculated the emissions factors for the Swedish plants in milligrams of NO_x per MJ of energy input. Converting these to pounds of NO_x per million Btus gives average emissions factors for the four plants of 0.066 pounds per Mbtu. These should be compared with the new U.S. standard, which is roughly equivalent to 0.15 pounds per Mbtu in terms of energy input. Again we see that the Swedish plants emit less than half this standard, confirming our comparison in terms of emissions in relation to energy output.

Despite the limitations of our data, we are inclined to see the Swedish charge system as effective. To judge whether or not it has been *efficient* in any overall sense is more difficult. This would mean proving not only that it was more cost-effective than other instruments but also that the level of abatement induced was appropriate, taking into account the environmental benefits compared with the allocative, administrative, and other costs of the charge. No study has undertaken to evaluate the system's efficiency fully. We do, however, have interesting partial information on the order of magnitude of some costs associated with the program.

In Sweden the main costs for the firms were for monitoring and abatement. Both of these were typically high, but no cost is due to the tax per se, since it was refunded. This applies at least at the aggregate level of all firms. For approximately half the firms there was an extra cost of net taxation (tax minus refund), while the rest realized a net gain. These intrafirm transfers will be briefly discussed below.

It would be logical to assume that the abatement costs were set equal to the charge level. (Höglund 2000) finds that average abatement costs vary between SEK 12 and 25 per kilogram—about half the level of the charge—but it appears firms had considerable difficulty in correctly calculating the opportunity cost of labor (particularly management time). They typically calculated only equipment costs, which causes an obvious downward bias. Some firms made estimates of marginal costs that were *above* the charge. These may be errors, or they might be explained as the result of proactive behavior, if the firms saw marketing or other goodwill value associated with overcompliance. Some firms with high marginal costs of abatement were municipally owned plants that provide district heating. Because they have local monopolies on this market, it is reasonable to believe they might see political considerations that have little to do with profit maximization. Nevertheless, these plants are only a fraction of the total sample.

The second largest cost was that of emissions monitoring, which was initially as high as SEK 4–6 per kilogram (Höglund 2000). Note that this corresponds to roughly $500 per tonne. A recent study by SEPA (2000) puts the average monitoring cost at about half this figure. Presumably there have been considerable technical progress and cost reduction in monitoring (partly due to the scheme itself and the experience generated), and although this cost per kilo may be particularly high, since some of the Swedish plants are fairly small, it still shows why such detailed (real-time) monitoring is not likely to occur in the French system, with fees that are of the order of $38 per tonne.

Other costs include calibration and administrative costs at the plant, estimated at about SEK 1.2 per kilogram. There may also be environmental costs through increases in unregulated emissions. Although these are uncertain, controls undertaken to reduce NO_x may also reduce temperatures or the supply of oxygen, which may lead to increased emissions of carbon monoxide, nitrous oxide, and ammonia. Economic valuation for these is very uncertain, but using a valuation of emissions that is quite high for each of these pollutants, the estimated combined total is SEK 6 to 19 per kilogram.[10] Finally, there are administrative costs for the regulator.[11] According to SEPA (2000), the costs of the agency's official administration of the system is five man-years at a cost of SEK 3.3 million in 1999. This amounts to roughly $300,000, or 0.6% of the fee collected that year. Note, however, that these costs appear low here partly because the charge level is so high.

The refundable charge system was designed to be as neutral as possible regarding aggregate competitiveness. Still, there are some possible effects on individual producers that need to be studied. The sectors affected by the charge are food and beverages, wood and wood products, paper and paper products, metal products (machinery and equipment), chemicals, energy, and waste combustion. How industries are influenced by the NO_x charge varies widely. For 1998, the paper and paper products industry faced a total net cost of approximately SEK 47 million, and the power industry had a positive net revenue of SEK 49 million (Figure 5-2).

The net revenues, both positive and negative, for the other industries are significantly smaller—less than SEK 10 million in absolute values, except for the

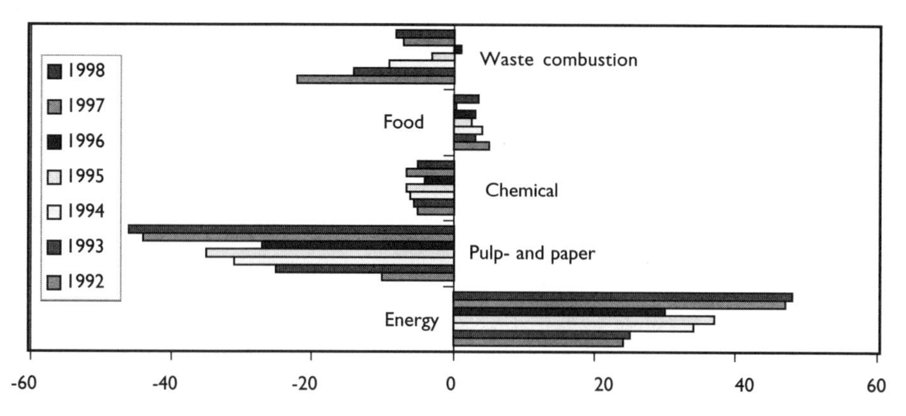

Figure 5-2. *Sweden: Net Payments by Industry Sector, 1992–1998*

Source: SEPA (2000)

waste combustion industry in 1992–1993 (SEPA 2000). In 2000, 45% of the individual firms saw net benefits (refunds were larger than the charge). Another 30% had refunds that were at least 70% of what they paid (the net fee was less than 30% of the gross fee). Naturally the resistance to this charge was quite different from what the political resistance would have been to a tax of the same magnitude.

We have no means of determining the actual efficiency of the instrument or proposing a new optimal charge level based on these data. Nevertheless, we note that the Swedish goals for NO_x reduction have still not been met, which would seem to indicate either that the goals were completely unrealistic or that the level of abatement has not been high enough—even though it certainly is high by international standards. According to SEPA (2000), the following proposals are under consideration:

- Broaden the scheme to include more sources (while also simplifying it for smaller plants).
- Raise the charge level.
- Make monitoring costs tax-deductible (to help smaller plants).

Conclusions

Both the Swedish and the French tax schemes aimed at limiting the regulatory burden on firms by imposing a threshold level for taxation. In the Swedish case, as the policy proved effective, this limit was subsequently lowered to encompass more units. The low level of the French tax meant a light regulatory burden but also less economic efficiency. Rather than driving technological development, the French tax on NO_x emissions can be regarded as a complementary policy instrument to give firms additional incentives to implement command-and-control regulation by investing in subsidized equipment.

The Swedish NO_x charge is different: it creates a strong incentive for fuel switching, modifications to combustion engineering, and the installation of specific abatement equipment, such as catalytic converters and selective noncatalytic reduction. Most important, it has also created a strong incentive to use the equipment and fine-tune combustion and other processes to minimize emissions. Swedish experience suggests a strong connection between monitoring and emissions reductions by fine-tuning, and it was the high charges that prompted firms to monitor their emissions carefully.

In both cases, the administrative costs of the tax instrument have been kept to a minimum. In France, administrative costs were fixed at 6% of total tax revenues, and in Sweden, central administrative costs are approximately 0.6% of total tax revenues. Monitoring requirements are an order of magnitude higher. In the French case, however, monitoring relied to a large extent on existing regulatory structures. Individual firms were granted a fair amount of flexibility and could choose whether to use direct measures or apply the emissions coefficients set by the regulatory agency. The French tax revenues were also used partly to finance investment in better measures of pollution.

With its levels fixed for five years by ministerial order, the adaptability of the French tax was not very quick. We believe that the environmental efficiency of the French system was probably low, but it allowed government and regulatory agencies to collect and improve information on emissions levels and abatement actions undertaken by firms in different industry sectors. In this sense, it gave government a distinct advantage compared with the existing CAC regulation.

One of the important effects of high levels of payments is that the emissions become more visible to both managers and regulators. The importance of accuracy in emissions measurement increases because significant monetary payments are based on the numbers. One of the main discoveries of the Swedish program is that emissions vary strongly with fine-tuning of plant operations. Detailed monitoring is the only way such fine-tuning can actually be undertaken, since it is the only way plant engineers themselves will realize the effects of various small changes in temperature and other combustion conditions. Accurate monitoring is thus crucial. In this respect, the French scheme has a drawback: its flexibility in the use of real monitoring versus emissions factors. The Swedish program is unique in that its very high fee level has led to the common adoption of sophisticated and detailed monitoring. As the price of monitoring equipment drops, it may be possible for other programs to get firms to adopt monitoring through other instruments, such as information disclosure, labeling, or competition mechanisms. A high fee also makes quite sophisticated pieces of abatement equipment profitable. This important effect may, however, be partly achieved through other instruments, such as subsidies for abatement, or through an emissions trading program—as long as the price of permits is sufficiently high.

The disadvantage of the refunded charge is that the output and revenue-raising effects of a tax are lost. (For the revenue-recycling effect, this is rather analogous to the loss due to grandfathering rather than auctioning off permits; see Sterner and Höglund 2000 or Goulder et al. 1999.) Although this is a loss in a simple model of a closed economy with no information asymmetries and no issues of competition or political feasibility, matters may be rather different in a real economy—particularly a small, open economy. In the Swedish economy, international competition is a major issue and politicians value an instrument that can reduce NO_x emissions without having any effect on product output. A tax as high as the Swedish REP charge would be politically impossible even in a tax-tolerant economy like Sweden's. The charge would impose a very significant burden on the energy industry and be unacceptable and impracticable where there is competition from both international firms and small domestic firms. The considerable advantage of the REP is that it makes a high charge politically feasible. Thus this instrument is promising for situations in which the technical abatement possibilities are abundant but fairly expensive and where the output and revenue-raising effects are less important.

Notes

1. We wish to thank the participants at the RFF workshop in December 2002 for useful comments and also Céline Nauges for collaboration on the French case study. We also thank Alain Milhau and Claude Siméon of ADEME for information on the French tax system. The conclusions and views expressed in this study represent those of the authors only.

2. The highest emissions coefficients are typically 20 to 30 times higher than the lowest coefficients.

3. The "productive" energy is carefully monitored, and some uses (such as cleaning stacks with excess steam) are outside the system.

4. The exchange rate with the U.S. dollar has during the writing of this chaper varied from SEK 7 to 11. We will for convenience use SEK 10 = $1 throughout the paper. This gives $4,000 per tonne.

5. The (approximate) atomic weight of nitrogen is 14 and that of oxygen is 16, so the molecular weights of NO and NO_2 are 30 and 46, respectively. One kilogram of NO will thus weigh 46/30 when further oxidized to form NO_2; this is the conversion factor used to express NO in NO_2 equivalents.

6. We analyze the tax system that was in place from 1990 through 1999. As of January 1, 2000, ADEME is no longer responsible for levying the tax, which is now under the administration of the customs authority.

7. Government decree number 59-2 of January 2, 1959.

8. The current French law on air pollution *(loi sur l'air et sur l'utilisation rationnelle de l'énergie)* dates from December 30, 1996.

9. One might ask what the reduction would have been in the absence of the charge. There are standards, and according to SEPA (1997), these standards alone would have given less than 40% of the reduction in emissions. The calculations are very uncertain, however. According to staff at SEPA (Lindgren 2003), there is now a tendency to set standards by working backward from the charge, but without doubt, the policy instrument that mattered was the charge, not the standards.

10. Values of 10, 108, and 40 SEK/kg for CO, N_2O, and NH_3, respectively.

11. Höglund (2000) also estimates welfare losses from refunding, but these are small, about SEK 1 per kilogram.

References

ADEME. 1995–97. Taxe parafiscale sur la pollution atmosphérique. Rapport d'Activité 1990/95, 1995, 1996, 1997, Agence de l'Environnement et de la Maîtrise de l'Energie. Paris.

Goulder, L.H., I.W.H. Parry, R.C. Williams III, and D. Burtraw. 1999. The Cost-Effectiveness of Alternative Instruments for Environmental Protection in a Second-Best Setting. *Journal of Public Economics* 72(3): 329–60.

Höglund, L. 2000. Essays on Environmental Regulation with Application to Sweden. Ph.D. Dissertation. University of Gothenburg, memorandum series 99. Göteborg University, Department of Economics.

Lindgren, M. 2003. SEPA. Personal communication, February 5, 2003.

Millock, K., and C. Nauges. 2003. The French Tax on Air Pollution: Some Preliminary Results on its Effectiveness. FEEM Nota di Lavoro 44.2003. Milan: Fondazione Eni Enrico Mattei.

Swedish Environmental Protection Agency (SEPA). 1993–98. Memorandum 733-3720-93 Mt, 733-4143-94 Mt, 733-6828-95 Mr, 733-4238-96 Mr, 733-4677-97 Mr, 713-3713-98 Rt. Stockholm.

———. 1997 *Svavelskatt och NO_x-avgift—utvärdering.* Report 4717. Stockholm: SEPA.

SEPA (Swedish Environmental Protection Agency). 2000. The Swedish charge on nitrogen oxides. Stockholm. Available at http://www.environ.se/.

———. 2002. Continuously updated information on NO_x. Available at http://www.environ.se/.

SOU. 1989. Ekonomiska Styrmedel i Miljöpolitiken. Delbetänkande av miljöavgiftsutredningen.

Sterner, T., and L. Höglund. 2000. Output Based Refunding of Emissions Payments: Theory, Distribution of Costs, and International Experience. Discussion paper 00-29. Washington, DC: Resources for the Future.

U.S. Environmental Protection Agency (EPA). 1998. *1997 Compliance Report—Acid Rain Program.* Washington, DC.

Yohe, G.W., and P. MacAvoy. 1987. A Tax cum Subsidy Regulatory Alternative for Controlling Pollution. *Economic Letters* 25: 177–82.

CHAPTER 6

NO_x Emissions in the United States

A Potpourri of Policies

Dallas Burtraw and David A. Evans

EMISSIONS OF NITROGEN OXIDES (NO_x) are precursors to secondary pollutants, including particulate matter and ozone, and contribute to nitrogen deposition and thus to environmental problems, such as acidification of ecosystems. Particulate matter is associated with both morbidity and mortality. Ozone has a widely recognized effect on human morbidity and potentially on mortality, although the latter effect is not firmly established. NO_x is also associated with other environmental problems such as reduced visibility.

Ozone is primarily a summertime problem because of the photochemical process that leads to its formation. Both particulate matter and ozone pollution are widespread problems in the United States, and many metropolitan areas are not in compliance with the National Ambient Air Quality Standards (NAAQS) for these pollutants.

Like ozone and particulate matter, NO_x is, because of its ubiquitous nature, also one of the six criteria air pollutants for which there are NAAQS standards. However, it was essentially unregulated through the 1970s and 1980s, especially since emissions from stationary sources were so loosely regulated until the 1990 Clean Air Act Amendments. NO_x is the only criteria pollutant that has increased nationally since 1970 (U.S. EPA 2002a).

The electricity sector is an important focus of NO_x policies for two reasons. First, it contributes about 20% of NO_x emissions in the United States. Second, these emissions are often emitted through tall stacks at high velocity, causing wide dispersion and contributing to regional pollution problems.

This chapter surveys the important NO_x programs affecting the electricity sector. These programs have followed a path beginning with traditional command-and-control (CAC) regulations leading to expanded use of flexible, market-based approaches. Today both approaches play an important role. We describe the programs and provide measures of their effectiveness, paying particular attention to the effect of these policies on coal-fired boilers subject to Title IV of the 1990 Clean Air Act Amendments.

Background

This section reviews more than three decades of policy at the federal level affecting NO_x emissions from coal-fired power plants.

Traditional CAC Regulations

Emissions of NO_x from electricity generation had little binding regulation until the 1990s. The 1970 amendments to the Clean Air Act implemented performance standards for new sources, as well as those undertaking major modification, based on emissions per unit of heat input. Collectively, these standards are known as New Source Performance Standards (NSPS). The new source standards were more specifically defined in the Clean Air Act Amendments of 1977, and the standards for NO_x were modified again in 1998. All new sources and existing sources that make major modifications located in areas that are in compliance with the ozone NAAQS must prevent significant deterioration of air quality by installing the "best available control technology," which is defined as the best technology considering energy, environmental, and economic impacts. Such sources in areas that are not in compliance with the air quality standards must install the theoretically more stringent "lowest achievable emissions reduction technology." Additionally, these sources must also offset their pollution increases through reductions at existing sources. Thus there is an overlay of a regional emissions cap of sorts coupled with a specific technology standard in nonattainment areas.

Existing sources were virtually exempt from NO_x regulations until the 1990 amendments to the Clean Air Act mandated a significant reduction in NO_x emissions at existing electricity-generating facilities. Although its contribution to particulate concentrations is usually identified as the most important impact of NO_x on the environment (Burtraw et al. 1998), the regulatory handles for NO_x emissions reductions have stemmed primarily from concerns about acid rain and nonattainment of the NAAQS for ozone.

Acid Rain Provision of 1990 Clean Air Act Amendments

The contribution of NO_x emissions to acid rain was addressed for the first time in Title IV of the 1990 Clean Air Act Amendments (CAAA). The most widely recognized focus of Title IV was emissions of sulfur dioxide (SO_2) from existing electricity-generating units, and the establishment of the emissions allowance trading program for SO_2. Yet Title IV also applied to existing sources of NO_x a traditional prescriptive approach similar to the type used for new sources. The requirements specified emissions rate limitations expected to correspond to specific abatement technologies, which typically meant combustion modifications such as low-NO_x burners.

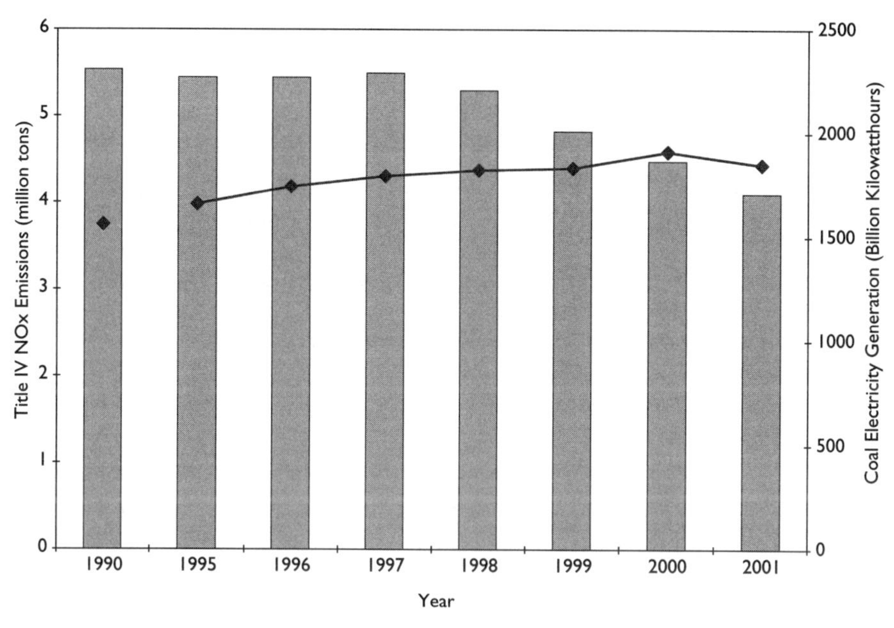

Figure 6-1. *NO_x Emissions and Electricity Generation from Coal-Fired Power Plants*

Sources: EIA (2002, Table 8.2b); EPA (2002b, Figure 11).

Figure 6-1 illustrates the emissions of NO_x from coal-fired boilers affected by Title IV, along with a measure of generation from electricity-dedicated coal-fired boilers, from 1990 to 2001. The units affected by Title IV represented about 85% of total NO_x emissions from the electricity sector. Over this period, NO_x emissions from these sources fell by 26%, while electricity generation from coal-fired plants grew by about 16%. Although this picture provides no information about the efficient level of emissions or cost-effectiveness of the programs, it does illustrate that a reduction in emissions has been achieved over a period of increased utilization of coal-fired power plants and tremendous economic growth in general.

Those reductions were achieved using a two-phased strategy with different technology-based emissions rate standards for each major coal-fired boiler configuration. Phase I of the NO_x reductions under Title IV was originally intended to begin in 1995, but litigation delayed the start until 1996. The first phase targeted 265 older coal-fired generating units and required tangentially-fired coal boilers to meet an emissions limit of 0.45 pounds of NO_x per million Btus of heat input, and dry bottom wall-fired units to meet an emissions rate of 0.50 pounds of NO_x per million Btus. Phase II, which took effect in 2000, required further rate reductions for Phase I units and required additional types of coal boilers to reduce emissions levels to rates between 0.40 and 0.86 pounds per million Btus, depending upon the boiler type.

In most cases, units affected by Phase I were retrofitted with low-NO_x burners or similar modifications that control fuel and air mixing to limit NO_x formation. However, a considerable amount of flexibility was introduced to ensure that these requirements did not impose unreasonable targets. The annual basis for the rate standards introduced the ability to average over hours in the year, providing much more flexibility in compliance than the hourly based standard that characterizes new source standards. Second, firms were given the flexibility to average emissions among commonly owned and operated facilities. Third, the law allows a unit to obtain a waiver, called an alternative emissions limit, if it could not meet the performance standard even after installing the control expected to attain that standard. Further, if units exempted from Phase I voluntarily chose to meet Phase I standards by 1997, they were exempt from Phase II rate requirements until 2008.

Ozone Policy

In addition to acid rain, the second major regulatory handle for controlling NO_x emissions has been the failure in many jurisdictions to attain the ozone NAAQS. Currently, an area meets the ozone NAAQS standard if the daily maximum one-hour average concentration measured by a continuous ambient air monitor does not exceed 0.12 parts per million more than once per year, averaged over three consecutive years. From 1998 to 2000, 30 of the 98 regions that in 1991 were determined to be in nonattainment with the ozone standard continued to fail to meet the standard while six additional areas had fallen into nonattainment since 1991. The Environmental Protection Agency (EPA) issued revised standards for ozone as well as particulate matter in 1997, and in 2001 the Supreme Court upheld the standards. The new ozone standard averages air quality measurements over eight-hour time blocks, increasing the difficulty of compliance with the ozone standard.[1,2] If the new ozone standard had been in effect during 1998–2000, 329 counties would have been found in nonattainment.

The main sources of the precursor emissions of ozone—NO_x and volatile organic compounds (VOCs)—are fossil fuel–fired electricity generating units, industrial boilers, and internal combustion engines. Before 1990, VOCs were indicted as the limiting type of emissions in the formation of ozone. This is often true in urban areas with high NO_x emissions where there are not many trees, which are a natural source of VOCs. As a result, past efforts that focused on controlling ozone in urban areas emphasized VOC control. However, in areas with high VOC concentrations, such as rural regions with high biogenic emissions of VOCs, ozone production is NO_x limited. When high NO_x concentrations from cities (or power plant plumes) disperse in suburban and rural regions, they cause major ozone episodes. In fact, it was recently determined that ozone concentrations reach their maximum levels outside cities and have a regional characteristic (Mauzerall and Wang 2001).

In 1989, a federal scientific report suggested NO_x controls as a "new direction" for the Clean Air Act and discussed the state of the science on VOC versus NO_x controls at the time, which included model results for a half-dozen cities and from EPA's Regional Oxidant Model for the Northeast (U.S. Congress 1989). The results for urban areas were mixed, but studies in the late 1980s were fairly consistent in finding that reducing NO_x would be more effective than reducing VOCs for rural and transported ozone. Chameides et al. (1988) and Trainer et al. (1987) offered convincing demonstrations of the need for NO_x controls in areas with high biogenic VOC emissions. Recently, most science and policy analysts have recommended a strategy for achieving attainment of the ozone standard that focuses on reducing NO_x emissions.

Provisions to address nonattainment of the ozone standard are found in Title I of the 1990 CAAA. Title I required emission rate limits consistent with Reasonable Available Control Technology (RACT) for large point sources of both NO_x and VOCs in nonattainment regions and the Ozone Transport Region (described below). These controls were to achieve a 15% reduction in 1990 levels of both ozone precursors by 1996. As such, this policy is referred to as the 15% rate-of-progress plan. EPA suggested RACT rates to states that were essentially the same as rates under Title IV for coal-fired boilers. Implementation of Title I came sooner and allowed firms less flexibility than Title IV.

For regions that may not achieve the ozone standard even with the 15% rate-of-progress rule, the 1990 CAAA provides a schedule for an acceptable rate of progress toward attainment. Beginning in 1997, these regions must achieve an average 3% annual reduction in NO_x or VOC emissions, or a combination thereof, over every three continuous years until the standard is achieved. This policy, known as the 3% rate-of-progress rule, applies to summertime emissions in some regions but annual emissions in others.[3]

However, atmospheric modeling demonstrates that precursor emissions from sources outside the boundary of nonattainment areas—in particular, NO_x emissions from large, rural, fossil fuel–fired power plants—have a great effect on ozone concentrations in urban areas. Indeed, many jurisdictions found that they would not be in compliance with the ozone standard even if their own emissions were reduced to zero.

Interstate Atmospheric Transport

To address interstate transport of air pollution, one major element of the 1990 CAAA was the creation of the northeastern Ozone Transport Region (OTR) and its associated commission.[4] The role of the commission is to promote cooperation among the member states in ozone abatement strategies. In 1994 the states, with the exception of Virginia, signed a memorandum of understanding to establish a coordinated three-phase effort to reduce NO_x from large stationary sources, primarily electric utility and large industrial boilers.[5] Annual control requirements mandated by the 15% rate-of-progress requirement (RACT rules) were viewed in the memorandum as Phase I of the plan.[6] Unlike most of the country, where the tightening of NO_x RACT standards was required only in regions that had not attained the ambient ozone standard, the states in the OTR were required to submit RACT plans for the entire state. These control requirements became effective in the OTR between 1993 and 1995. However, the memorandum anticipated that these requirements would not be sufficient to bring the region into attainment with the ozone standard.

Introduction of Flexibility

Phases II and III of the NO_x abatement plan in the Northeast are collectively referred to as the Ozone Transport Commission NO_x Budget Program. These phases of the program establish emissions "budgets" for each state, which collectively serve as a cap on total allowable emissions for the period when ozone is commonly a problem, May 1 through September 31.[7] A cap-and-trade approach ensures that the regional emissions target is not exceeded, while attempting to minimize the cost of achieving it. Unlike the Title IV provisions for NO_x, under which emissions are allowed to increase with economic growth, a central feature of the NO_x cap-and-trade program is that emissions are fixed and it is the emissions allowance price that fluctuates.

Sources affected by the trading program were primarily electricity generating facilities. States allocate their budgets to their affected sources and trading is allowed among sources in different states. The planned Phase II seasonal NO_x budget for the region, which applied in 1999 to 2002, was 219,000 tons (U.S. EPA 1997). In Phase III, which was scheduled to begin May 2003, the summer allocation was to be reduced further to 143,000 tons. These budgets represent a substantial reduction from the 490,000 tons of summer emissions in the region in the baseline year, 1990. As discussed below, the actual cap was somewhat different in Phase II because of changes in source participation. Phase III never began, as the OTR program was subsumed by a geographically larger one.

Unlike the SO_2 trading program, which is administered at the federal level, implementation of the OTR cap-and-trade program required a high level of coordination among state rules. To move this process along, a model rule was developed that identified the elements that should be consistent among the participating states' regulations.[8] Each state would then include these regulations in its State Implementation Plan (SIP), which is a set of regulations subject to approval by the EPA that the state enforces to meet its obligations under the Clean Air Act. Experience with the trading program is described below.

Trading in the NO_x SIP Call Region

The OTR in the Northeast is a subset of the larger eastern U.S. region that is subject to substantial transboundary drift of NO_x. Because the states in this larger region were concerned that they could not achieve the one-hour ozone NAAQS without a coordinated effort, they participated along with EPA in a structured process known as the Ozone Transport Assessment Group to study the issue. The goal of the process was to develop consensus among the states for a coordinated effort to reduce ground-level ozone in the eastern United States. The group's final report was released in June 1997 and provided few actionable recommendations. For example, although there was considerable effort to promote an emissions trading program to control NO_x emissions, no such recommendation materialized from the process (Keating and Farrell 1999). However, it did contribute to a consensus understanding of the problem at a regional level.

EPA has the authority to require states to impose restrictions on sources of NO_x emissions to help downwind states comply with the ozone standard via Section 110 of the 1990 CAAA. In late 1997 EPA published a proposed rule relying on this statute and in 1998 formalized the proposal. This rule is widely referred to as the NO_x SIP Call, because it required states to submit revisions to their SIPs outlining their strategies for complying with state-level summertime NO_x emissions budgets.[9] The revisions were to be submitted by 1999 and take effect by May 1, 2003. Although the rule allows states to develop their own strategies and rules for reducing NO_x emissions, participation in a regional cap-and-trade program similar to the OTR program was strongly encouraged by EPA.[10]

The NO_x SIP Call originally targeted 22 states and the District of Columbia, where nearly 90% of the boilers covered in the Title IV program are located. After various court proceedings, the date for submitting revised SIPs was delayed until October 2000, and the date for achieving the reductions was postponed until May 31, 2004.[11,12] The courts also truncated the SIP Call region to 19 states and the District of Columbia.[13] States that participated in the OTR program began complying with the SIP Call on May 1, 2003, following their original plan of a third phase for that program.

At the national level, the NO_x SIP Call would lead to reductions of 22% from an annual baseline level of 5.4 million tons in 2007 to a new annual level of 4.25 million tons, according to EPA estimates. Summer-season emissions in 2007 would fall by 40% from 2.4 million tons to 1.45 million tons. In the SIP Call region, the program would lead to annual reductions of 34%, from projected baseline levels of 3.51 million tons to 2.33 million tons in 2007. In the summer season, the program is expected to reduce emissions by 62%, from 1.5 million tons to 0.56 million tons.[14] Most of these reductions will come from coal-fired power plants.

Proposals for Nationwide Trading Proposals

The NO_x program in the 19 eastern states and the District of Columbia comes at a time when calls for more drastic reductions of several pollutants are taking shape as part of a possible reauthorization of the Clean Air Act. In the 107th Congress, three Senate bills proposed market-based approaches to regulating multiple pollutants. These legislative proposals seek reductions in emissions of SO_2, carbon dioxide, and mercury as well as NO_x from the electricity sector.[15] An important difference in the bills is whether carbon dioxide is included. The electricity industry is already switching from coal or oil to natural gas as the preferred fuel for new generation facilities, and the proposals have implications for the rate at which that transition will continue. Emissions of all the mentioned pollutants are much greater from coal or oil than from gas; emissions of SO_2 and mercury are virtually zero for gas. Taken in isolation or as part of a moderate multiple-pollutant proposal, the NO_x emissions reductions under the SIP Call are expected to prompt the installation of post-combustion controls at coal-fired plants and many gas-fired facilities as the primary means of compliance. They are not expected to accelerate significantly the transition to natural gas. However, the inclusion of strict mercury reduction requirements, and especially the inclusion of carbon dioxide emissions targets, will accelerate a transition from coal to natural gas.

The Performance of Prescriptive Approaches

In this section, we analyze prescriptive regulations affecting investments at new power plants and major modifications to existing plants, and subsequently we review prescriptive regulations affecting existing sources. These regulations are usually characterized as an emissions rate-based standard that might be interpreted as a performance standard. However, in practice, these regulations often take the form of technology standards because the allowable emissions rates are usually written to correlate to a specific technology choice. Firms may have the latitude to deviate from a specific technology that is approved by EPA or the state, but in doing so the firm assumes the burden of proof to demonstrate compliance, and that can be a difficult hurdle.[16]

Prescriptive Regulation for New Sources

The New Source Performance Standards (NSPS) constitute traditional source-specific prescriptive regulations for new units and any existing units that undertake a major modification. The first generation of the standards followed the 1970 amendments to the Clean Air Act and applied to generating units that were constructed or modified between August 1971 and September 1978. The standards were modified in the 1977 amendments, which were applicable from September 1978 until July 1997. The limits specified emissions per unit of heat input to the electric boiler, and they varied based on the type of fuel used and, in the case of coal, the quality of the fuel. However, these standards were not constraining for most projects. In 1998, EPA issued new standards that were retroactive to 1997. These revised standards are based not on the ratio of emissions to heat input but on emissions to electricity output.

The standards are summarized in Table 6-1. An important observation is that the current standards are neutral among fuels; in earlier versions, the standards were more lenient for coal generation than for gas and oil. The current standard for new sources is 1.6 pounds of NO_x per MWh, without regard to fuel type. Hence, it does not distinguish among fuels or the efficiency of the generation technology of different types of facilities. Along with existing combustion controls, the standard is expected to require some form of post-combustion control, typically selective catalytic reduction, at new steam boilers. The standard is estimated to result in a decrease in emissions of 26,000 tons in 2000, primarily from coal-fired electricity-generating boilers, relative to the previous standard (63 FR 49450).

These standards appear to have had an effect on the emissions rates from coal-fired power plants. Figures 6-2 and 6-3 illustrate the 1999 NO_x emissions from different vintages of coal plants. In each figure, the vertical axis represents the 1999 emission rate of the units; the reduction in emissions rates over vintage is clear. The first three groups include units brought into service before 1970 and roughly correspond to the units built before implementation of the 1971 standards. The fourth group of plants, built in the 1970s, corresponds roughly to the original NSPS that were in effect between 1971 and 1978. The group of plants

Table 6-1. *New Source Performance Standards for NO_x*

Affected boilers	*Standard (lbs. NO_x/MMBtu)*	*Date for compliance*
Units greater than 73 MW	0.8 for lignite from North Dakota, South Dakota, or Montana; 0.7 for solid fossil fuel; 0.6 for other as lignite; 0.3 for oil; 0.2 for g	August 17, 1971
Units greater than 73 MW	0.8 for lignite from North Dakota, South Dakota, or Montana; 0.7 for solid fossil fuel; 0.6 for other lignite, bituminous, and anthracite and 65% NO_x removal; 0.5 for subbituminous and 65% for NO_x removal; 0.3 for oil; 0.2 for gas	September 18, 1978
Units greater than 25 MW	0.15 lbs./MMBtu for modified sources; 1.6 lbs./MWh for new sources	July 9, 1997

Sources: EIA (1998); Krolewski and Mingst (2000).

built after 1980 corresponds roughly to those affected by the 1977 NSPS that applied from late 1978 until 1997. The bar on the far right of Figure 6-2 corresponds to the group of plants affected by the standard that took effect in 1997.

In Figure 6-2, the horizontal axis represents the share of 1999 generation by coal-fired units. The graph is a histogram, with the areas within each rectangle representing the percentage of total NO_x emissions from 1999 attributable to these cohorts. The average emissions rate of units built before 1970 was about 6 pounds of NO_x per MWh; the figure drops to 5.37 for boilers built in the 1970s, 4.09 for those from the1980s, and 3.55 for those from the 1990s. Absent other considerations, if units built after 1970, approximately when new source standards came into being, had the same emissions rate in 1999 as units built before 1970, total NO_x emissions from coal-fired boilers would have been about 14% higher.

In Figure 6-3, the horizontal axis represents the share of installed capacity. The areas in the rectangles represent relative potential emissions, but since units are utilized differently, this does not correspond directly to the emissions actually observed in Figure 6-2. A comparison of the two figures shows that older units have a somewhat larger share of capacity than generation, reflecting the fact that newer units operate for more hours because of their greater efficiency and reliability.

Nonetheless, when the 1970 amendments were written, the implicit assumption was that the power plant capital stock would turn over on a schedule that roughly approximated a typically plant's book life (30 years). In fact, many older power plants have received extensive maintenance as well as new electronics and other innovations, which have increased their lifetimes (Ellerman 1998). The NSPS have been ineffective at reaching these plants. New plants are subject to regulations for NO_x and other pollutants that put them at a competitive disadvantage, at least as far as expenditures for pollution control. Many authors have argued that this has contributed to the extension of the lifetimes of existing plants, a result that is both uneconomic and harmful to the environment (Swift 2000). Furthermore, although the assumption implicit in the NSPS was that it

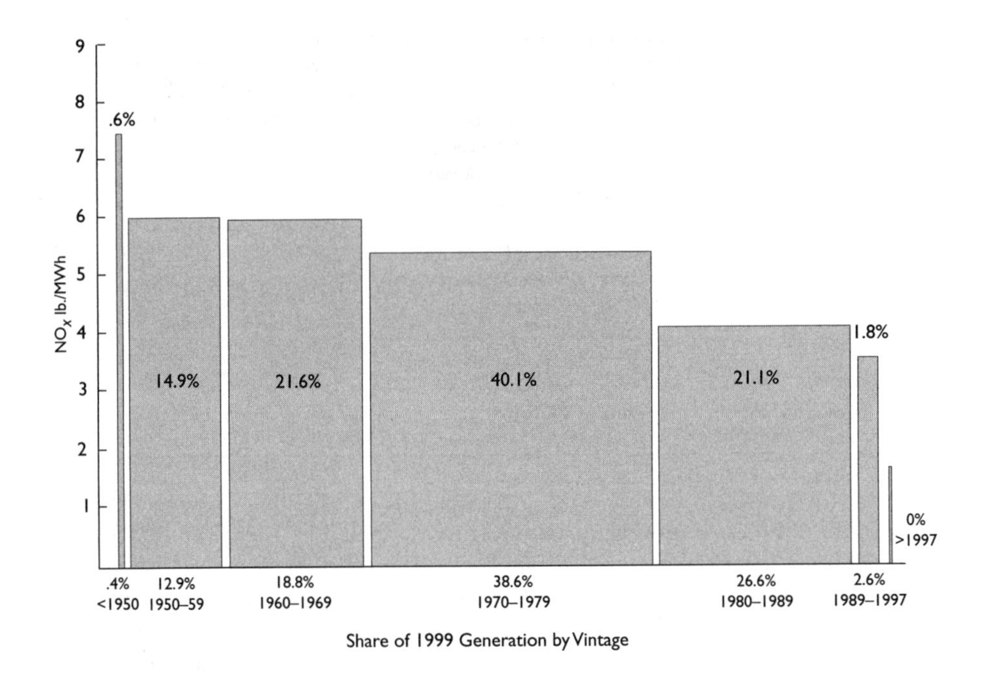

Figure 6-2. *NO_x Emissions and Share of Generation from Coal-fired Boilers*

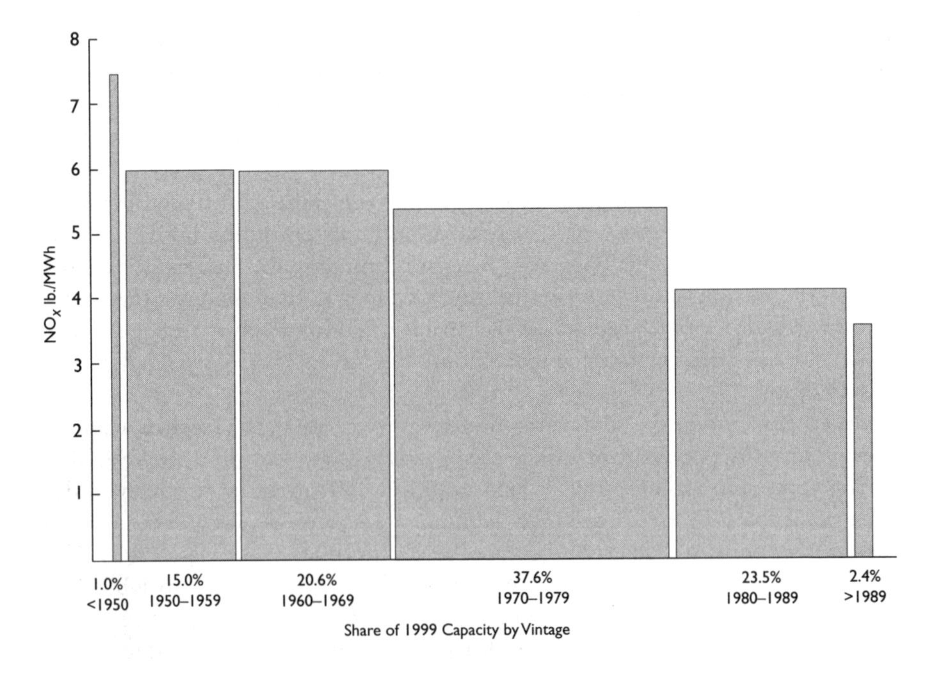

Figure 6-3. *NO_x Emissions and Share of Capacity for Coal-fired Boilers*

would be cheaper for new sources to reduce emissions, today it is often cheaper for existing sources to reduce pollution because of the strict standards applied to new sources. The cost per ton of NO_x reduction at new sources is about $565 for new coal, with the installation of selective catalytic reduction. The cost per ton at existing uncontrolled units ranges from $0 for cyclone units, to $161 for wall-fired units, to $631 for tangentially fired units (Swift 2001).

Prescriptive Regulations for Existing Units

Ex ante, EPA projected annual NO_x reductions during Phase I of the Title IV program of 0.6 million tons (ICF 1990). The anticipated direct costs of achieving these NO_x reductions are difficult to ferret out because they were combined with expected costs of SO_2 reductions, but they appear to be about $100 million. By 1995, the abatement estimate was updated to reflect the expectation that only 0.5 million tons would be reduced because of changes in the expected baseline against which emissions reductions were measured (ICF 1995). At that time, the average cost per ton of emissions reductions was expected to be $180.

In Phase II, beginning in 2000, annual emissions of NO_x were expected to fall by 2 million to 2.2 million tons compared with a baseline forecast in 1990 (ICF 1990). Whether these emissions reductions were achieved depends again on one's interpretation of the baseline against which emissions are measured.[17]

NO_x emissions in 2000 from all Title IV–affected sources totaled 5.1 million tons. This represents a reduction of 3 million tons relative to the forecast 2000 baseline of 8.1 million tons offered in EPA (2002b). However, relative to 1990 NO_x emissions, 2000 emissions from these sources are 23%, or only 1.55 million tons, lower. If we limit our comparison to only those sources affected by the NO_x provisions of Title IV (i.e., coal-fired units), emissions in 2000 were 4.48 million tons and only 19%, or 1.05 million tons, lower than in 1990. This smaller percentage reduction relative to all Title IV affected facilities is explained by the disproportionate decline in oil-fired electricity generation. The reduction from sources affected by Title IV NO_x provisions was not as great as expected because of lower than expected electricity consumption and the compliance flexibility allowed by the NO_x program. The number of units adopting each form of compliance (as described above) in Phase II is identified in Table 6-2.

We conjecture that the cost of the program was lower than expected because many units avoided installation of low-NO_x burners, the technology typically anticipated to achieve the standard, and took advantage of the flexibility afforded by the multiple compliance options. There also were more low-cost opportunities for abatement than originally anticipated. At units that complied in the standard way—by achieving emissions rate reductions at the unit—the initial response to the regulation by plant operators was to optimize the operation of boilers by adjusting air-fuel mixtures and temperatures (Swift 2001). Many units also reduced NO_x emissions through operational modifications, sometimes called trimming, that typically lower combustion temperature slightly and incur a penalty in the facility's heat rate.[18] Optimization incurs almost no cost and may result in savings, whereas operational modifications incur fuel costs. Of the 265 coal-fired units affected under Phase I, only 175 met the emissions rate limits by

Table 6-2. *2001 Compliance Strategies for Title IV*

Compliance option	*Units*
Standard emissions limitation	140
Early election	274
Emissions averaging	638
Alternative emissions limitation	27
Total[a]	1,079

[a]The total does not equal the number of affected units, 1,046, because 28 units have both early election and emissions averaging compliance plans, and 5 units have both alternative emissions limitation and emissions averaging plans.

Source: EPA (2002b, Figure 14).

installing low-NO_x burners, which involved greater capital costs as well as a slight heat rate penalty. Emissions averaging allowed 52 units to continue to operate above the average emissions limit. Overall, units lowered their average NO_x emissions rates to 0.4 pounds per million Btus, below the average rate of 0.7 pounds per million Btus in 1990 (Swift 2001).

Cost and operation data available from the Federal Energy Regulatory Commission Form 1 allow us to see the impact of both the RACT regulations required in the OTR and Phase I of Title IV of the 1990 CAAA. The plants represented in Figures 6-4 and 6-5 are those that completed the Form 1 every year in the sample period and contain boilers that must comply with Title IV. Plants with units that installed scrubbers to comply with the Title IV SO_2 provisions are removed from the sample. The plants are divided into two categories: those affected by OTR policies and those affected only by the Title IV provisions (non-OTR)[19] Recall that the RACT rules are typically very similar to the unit-specific Title IV requirements, except that they do not provide as much compliance flexibility as Title IV.

Figure 6-4 illustrates the effect on variable costs of trimming and combustion control modifications, such as the installation of low-NO_x burners, which typically lowers plant efficiency and thus raises a plant's heat rate. Although Title IV standards were established earlier, they were implemented later than similar standards in the OTR. Hence, one would expect to find an increase in the average heat rate in the region before similar increases at the national level. This is evident in Figure 6-4. Beginning in 1992 and continuing through 1995, when initial experimentation with operating controls occurred in the OTR, corresponding to the introduction of RACT regulations. Figure 6-4 indicates that a similar increase in heat rates occurred in the remainder of the country in 1996 as Phase I of Title IV came into effect. The heat rate increase outside the OTR is dampened by the inclusion of units that were not affected by the Title IV provisions until Phase II. Nonetheless, one also sees an increase in the heat rate in the region in the first year of compliance.

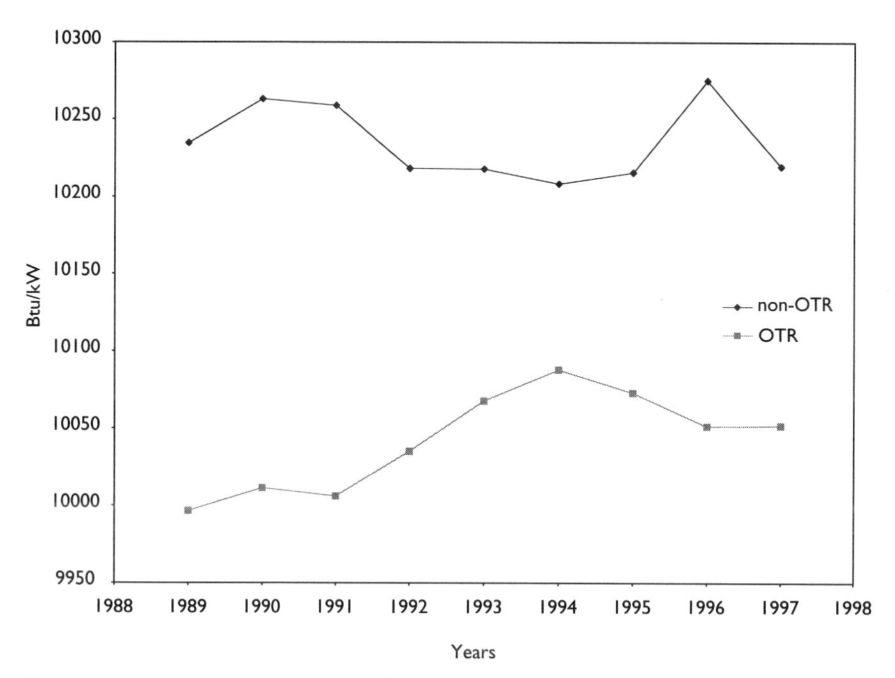

Figure 6-4. *Average Heat Rate of Title IV Coal-Fired Boilers*

Also evident in Figure 6-4 is a learning process at individual facilities as the boiler operations were modified in a period of experimentation in the search for compliance strategies. By 1996, heat rates had stabilized at levels higher than those prior to implementation of the OTR program, but well below levels during the initial years of compliance. One sees a similar decline after learning had occurred for the Title IV–affected sources in the remainder of the country. Inevitably, there was fleet-wide learning, and as companies took advantage of the opportunity to average emissions over commonly owned units under Title IV, production moved from inefficient to efficient units.

Modifications necessitated by RACT standards and the Title IV NO_x program also imposed capital costs. Figure 6-5 tracks the average change in physical plant (equipment and structures) capital costs at coal-fired boilers affected by Title IV of the 1990 CAAA. Significant increases in new capital expenditures in the OTR coincide with the introduction of RACT performance standards for NO_x. These costs reflect the installation of low-NO_x burners and some post-combustion controls.

EIA (1998) reports that the cost of retrofitting a boiler with low-NO_x burners is about \$9.30 to \$44 per kW.[20] A direct comparison between this cost and the changes in capital cost is difficult given that the signal in Figure 6-5 is confounded by the inclusion of expenditures not related to NO_x controls. Also, the flexibility in Title IV gives firms the ability to avoid capital costs. Nonetheless, the capital cost changes in Figure 6-5 are consistent with what would be expected from these policies.

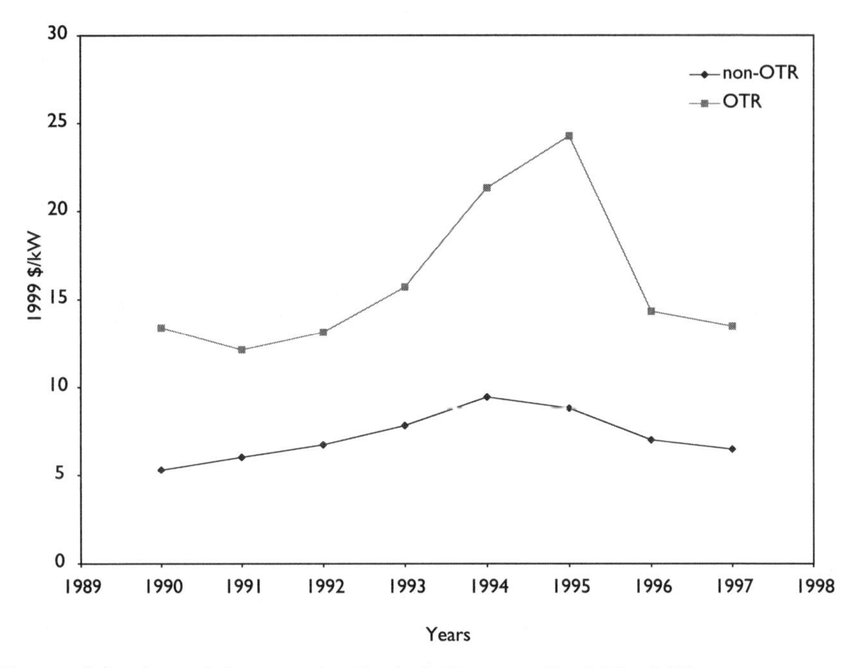

Figure 6-5. *Annual Increase in Capital Costs at Coal-Fired Plants*

Flexible Regulation: Cap-and-Trade in the OTR

Beginning in 1999, the second phase of the OTR achieved further reductions in summertime NO_x emissions through a tightened cap on aggregate emissions coupled with trading of emissions allowances. In general, the trading program can be considered a success, especially when one considers that it was developed before many of the lessons of the Title IV SO_2 program, the only trading program of similar scope, had been learned. Trading between economically unrelated sources and across state lines has been robust, suggesting that trading provided an opportunity to realize cost savings. Enforcement has been very good. Between 1999 and 2001, only eight sources were in violation of their allowance holdings. Meanwhile, emissions reductions have exceeded targets, and the program also served as the model for the NO_x SIP Call program.

Although the OTR trading program can generally be considered a success, it got off to a rocky start, as can be seen in the wild fluctuation in allowance prices in the first half of 1999. From January to April 1999, prices for 1999-vintage allowances increased from around $3,000 to about $7,000 per ton, albeit on low volumes of trade relative to the total allocation. The price of these allowances then fell considerably in May, the onset of the compliance period. By August, 1999-vintage allowances were trading under $2,000 per ton, and they fell to less than $1,000 by year's end. The price of allowances with a 2000 vintage also fluctuated over this period but not with the same magnitude. Their price spiked at just over $3,000 for a short period in May. The prices of 2001, 2000, and 1999-vintage allowances converged in the late summer of 1999.

The greatest source of uncertainty at this time can be attributed to the difficulty in achieving the coordination necessary to implement such a program (Krolewski and Mingst 2000). Participant states finalized their allocation and participation rules less than six months before the season began, and early reduction credits, which represented about 10% of the total allowances allocated in 1999, were not distributed until April and May. The rulemaking delay also raised questions about whether certain states would participate in the program at all. The most notable case in this regard is Maryland, which in fact did not participate in the 1999 trading season.[21,22]

Another source of uncertainty and market volatility resulted from the use of compliance strategies, such as load shifting and trimming, that did not rely on a significant number of retrofits. The cost and performance of retrofits could be fairly well anticipated by the market; the success of operational strategies was uncertain but ultimately exceeded expectations (Farrell 2000). Once these uncertainties were resolved, prices dropped dramatically as it became apparent that there would be sufficient allowances to cover emissions, and even a substantial number to be banked.

Although there was considerable heterogeneity in the state laws and regulations implementing the trading program (even with a generally agreed upon model rule), this variation seemed to have little effect on how it functioned—in part because the heterogeneity was in allowance allocation mechanisms and allowance set-asides, not in anything that would affect compliance or the trading process. States avoided micromanaging the allowance market through trading restrictions (Rhode Island being the notable exception). The delayed implementation of state rules, and the difference in the timing of their ratification, proved more of a problem than the actual text of the rules.

It is interesting to compare the administrative demands required to implement the RACT rules with those required of the emission trading rules. The experience of the OTR states provides an opportunity to do so. It is often argued that performance standards are more difficult to develop than a trading program because the standards must address a potentially wide variety of sources that may be affected. Despite EPA's simplifying recommendation that the RACT standards mimic the Title IV NO_x standards, their implementation was difficult. The states were required to show that RACT compliance was feasible and also to contend with sources that lobbied for rule exceptions or applied for rule waivers (because of retrofit difficulties and the like).

While this suggests an advantage of market-based approaches over prescriptive ones, analogous issues also arose for the OTR emissions trading program. Given that allocations would be made gratis, regulators had to establish and verify a baseline upon which allocations would be made. Sources recognized the value of the asset being allocated and lobbied for rules that would yield favorable allocations. Regulators also had to prove that attainment with the cap would be feasible (ESS, 1996).[23,24] Although we suspect that the creation of the RACT rules posed a greater challenge to regulators, establishing rules for the OTR trading program was not notably less burdensome.

One important regulatory concern with the OTR program was the potential for emissions to concentrate in a particular location. To address this potential problem, trading restrictions between different zones of the OTR trading region were considered. Such restrictions were not adopted because simulation modeling showed that they would be unnecessary (ESS, 1996). Swift (2004) contends that little evidence of spatial hotspot problems in the OTR from 1999 to 2002 can be found, as emissions were uniformly below allocations in all states except Maryland. An additional consideration is that, although the OTR trading program confers considerable flexibility in achieving abatement requirements, sources must still comply with RACT standards during the trading season. In effect, the concentration of pollution that can be released from any one source is limited, thus reducing the potential for emission hotspots.

A temporal concentration of NO_x emissions is of greater concern than a spatial one. Violations of ambient standards for ozone typically occur in climatic episodes when the temperatures are very high and the air is stagnant. Such climatic events correlate with high demand for electricity, when air conditioning use increases. Exploring the potential for episodic ozone events with the OTR trading program, Farrell (2003) notes that the program design is not particularly well-suited to prevent ozone episodes, because allowances can be used at any time during the ozone season. He also suggests that rate-based prescriptive policies would not necessarily perform any better. However, we should note that RACT rules do act as a backstop to total daily emissions from affected sources because they allow an averaging period for compliance—typically between 15 minutes to 24 hours for electricity generators. This suggests that there might be some practical advantage to combining prescriptive and market-based regulatory approaches to limit the potential for temporal hotspots.

The OTR program demonstrates promise for more decentralized multijurisdictional emissions trading programs. The final allocation for 1999 for the states that participated in the trading program was 218,738 tons, including 24,635 early reduction credits (U.S. EPA 2000). This budget reflects a 48% reduction from these states' five-month summer 1990 emissions of 417,444 tons.[25] Total emissions in 1999 amounted to 174,843 tons, and the remaining allowances were banked.[26]

Costs of NO_x Control under Cap-and-Trade

The cap-and-trade approach is expected to result in substantial cost savings compared with a prescriptive approach that would achieve the same level of emissions overall. Farrell et al. (1999) use a deterministic, dynamic, mixed integer-programming model of the proposed allowance market in the OTR to predict a total Phase II (1999–2002) compliance cost of about $900 million (1996$). Unfortunately, no *ex post* estimates of the total abatement cost of the program are available. However, we surmise that the total compliance cost was likely lower than this estimate. The Farrell et al. model does not allow for changes in dispatch and the option of reducing NO_x through trimming. These compliance measures were evident and, in the case of trimming, surprisingly ubiquitous in the OTR trading program.

The Farrell et al. model also predicts a NO_x allowance price of $1,331 in the first year of the trading program, rising to $1,718 by 2002 (1996$). Significant program uncertainty kept the allowance price above this range until the end of the first trading season. The subsequent downward movement in prices can be attributed to the unexpectedly large number of allowances banked in the first year of the program. As mentioned above, sources also realized that more low-cost abatement methods were available than initially anticipated (indeed, more allowances were banked the following season). In later years, the limitation on the number of banked allowances that could be converted into NO_x SIP Call allowances also kept downward pressure on allowance prices in the market for the OTR program.

Krupnick et al. (2000) examine the costs of NO_x emissions reductions in an annual trading program in a larger region of 12 states and the District of Columbia—jurisdictions that represent the major sources of emissions in the eastern United States, and one that is intended to capture most of the emissions within the broader SIP Call region. They find that emissions reduction targets can be met at roughly 50% cost savings under a trading program when there are no transaction costs, compared with a command-and-control approach. This provides a rare explicit analysis of the cost savings of trading for NO_x.

Burtraw et al. (2001) find that the seasonal NO_x trading program for the 19-state SIP Call region is expected to reduce annual emissions within the region from 3.45 million to 2.43 million tons in 2008. The annual cost of post-combustion controls to achieve these reductions is expected to be $2,146 million (1997$). They predict that the cost of emissions allowances—presumed equal to the marginal cost of adopting post-combustion controls per ton of NO_x reduction—will be $3,401 (1997$). This compares with a marginal cost of $1,356 for a policy applied just within the OTR.

Although the SIP Call policy will incur important costs, it is not expected to have a large impact on electricity prices. Burtraw et al. (2001) forecast that the average price in the SIP Call region in 2008 will increase from $64.4 per MWh to $65.1 per MWh (1997$). The reason that the effect on price may be negligible has to do with how the cost of capacity is reflected in electricity price. These forecasts assume that the current pattern of electricity regulation continues into the future. The authors assume that slightly more than half the generation in the SIP Call region remains in areas characterized by regulated pricing, under which capital and variable costs are annualized and spread over total sales to calculate the price of electricity. In these areas, introducing a new environmental policy that increases the costs of electricity supply leads directly to an increase in the electricity price.

However, the other part of generation in the SIP Call region is in the market pricing areas, where the electricity price is determined by the variable cost of the marginal generator plus the incremental capital cost of the marginal reserve unit. Policies that change the relative costs of facilities affect which facility is at the margin, which in turn affects prices and revenues earned by each facility, and thereby capacity investment and retirement. In these areas, introducing a new environmental policy that increases the costs of electricity supply may lead to an increase or decrease in the electricity price.

The primary analysis of the expected cost of NO_x reductions for EPA (1998) finds that the average cost per ton of NO_x emissions reduction achieved by the NO_x SIP Call is $1,807; the study does not report an estimate of marginal cost.[27]

The change in prices in the EPA study (1998) is about 1.6% assuming marginal cost pricing throughout the electricity sector, and about 1.2% assuming average cost pricing. The EPA study does not clarify whether this applies to the SIP region or the nation. Burtraw et al. (2001) forecast a rise in prices in the SIP region, which combines average and marginal cost pricing, of about 1.1%.

The estimates above are limited to compliance costs. The economic literature recognizes a broader definition of cost from environmental programs. The interaction of regulations with preexisting regulations or taxes is expected to cause the social cost of new regulations to be higher than the measure of compliance cost in a partial equilibrium analysis. In a general equilibrium framework, the costs of new regulations act like additional taxes in the economy, and they magnify the distortions away from economic efficiency associated with preexisting regulations or taxes.[28]

Goulder et al. (1999) find that the social costs under a cap-and-trade program actually may be higher than under command-and-control.[29] This can occur because the opportunity cost reflected in permit costs might increase product prices to a greater degree than would a CAC policy. However, an important distinction between these policies is that a cap-and-trade program has the potential to generate revenues while a CAC policy does not. An auction of emissions allowances can generate revenue that might be used to reduce preexisting distortionary taxes, thereby offsetting much of the hidden costs. In neither the cap-and-trade program in the OTR nor the SIP Call program are allowance auctions widely used.[30] Rather, permits are almost always distributed at zero cost, so no revenue is available to reduce taxes. In any case, the approach to distributing emissions allowances is one of the important issues in the design of future emissions trading programs.

Conclusion

Policy affecting NO_x emissions from coal-fired power plants in the United States has evolved from prescriptive approaches to more flexible cap-and-trade approaches. Both approaches remain important today, but the future is likely to show an expanded role for the use of cap-and-trade.

Prescriptive approaches include standards on new sources, which have led to emissions reductions; however, the perverse bias against new sources associated with NSPS is likely to have undermined the accomplishments of the program to some degree. A prescriptive approach also characterizes the first phase of the OTR program and the Title IV NO_x provisions. These programs also can claim substantial emissions reductions, and their costs appear to be lower than anticipated. Although the Title IV program is prescriptive in nature, it nonetheless shows considerable flexibility by allowing emissions averaging at commonly owned plants, alternative emissions limits, and so forth. To some degree, this may have undermined the effectiveness of the program because there is no cap on aggregate emissions.

The cap-and-trade program that began in 1999 in the OTR provides the maximum flexibility, but it also provides the maximum effectiveness by capping emissions. That program also can claim significant emissions reductions, at costs that are low compared with a CAC approach.

Both prescriptive policies and the cap-and-trade programs provide incentives to find low-cost ways to achieve reductions. Unlike the prescriptive policies, however, the cap-and-trade program provides an incentive to harvest low-cost emissions reductions at specific facilities even after the performance standard is achieved because those reductions can avoid investments at other facilities.

Future NO_x Policy for Power Plants

Two issues pertaining to the future of NO_x policy for power plants are in the forefront: extending NO_x trading spatially and throughout the year, and reevaluating how emissions allowances are distributed in cap-and-trade programs.

In the near term, NO_x controls, currently limited to the five-month summer season, should be extended year-round. The ozone problem is limited to summertime, and this has been the regulatory handle for the cap-and-trade programs we have discussed. However, other problems associated with NO_x occur throughout the year and, based on the public health and economics literature, are more serious than ozone. Particulate formation and its effect on health is especially important from an economic perspective and occurs year-round. Recent studies have shown hundreds of millions of dollars in annual net benefits (incremental benefits in excess of incremental costs) from extending the summertime NO_x trading programs to a year-round program.[31] The adoption of annual trading programs has already begun at the state level. New Hampshire adopted legislation that caps annual NO_x emissions from its three largest plants starting in 2006. In New York, proposed regulations would create a cap-and-trade program during the portion of the year outside the NO_x SIP Call trading season.[32]

In the long term, a critical issue in measuring the economic success of cap-and-trade programs is the way in which emissions allowances are distributed. Although there is evidence of cost savings from the cap-and-trade approach, this is limited to the measure of compliance cost. In a general equilibrium framework, the measure of the cost (and benefits) of environmental regulations will be greater than the measure of compliance cost, potentially to a significant degree. Unfortunately, the free distribution of emissions allowances precludes the possibility of capturing the value of allowances and using it to reduce preexisting tax distortions in the economy and thereby reduce the costs of the program. How emissions allowances are distributed has large efficiency and distributional implications that were not widely appreciated in the design of previous programs, but they will be central in the design of programs in the future.

Meanwhile, the evolution to the next stage for NO_x cap-and-trade policies is already under way. The trading program in the OTR, coupled with the success of the SO_2 program, provides momentum for expansion of the cap-and-trade approach under the NO_x SIP Call. The NO_x SIP Call subsumed the OTR trading program by significantly expanding the summertime trading program to 19 eastern states, thereby providing the second major experiment in emissions trading in the U.S. electricity industry.

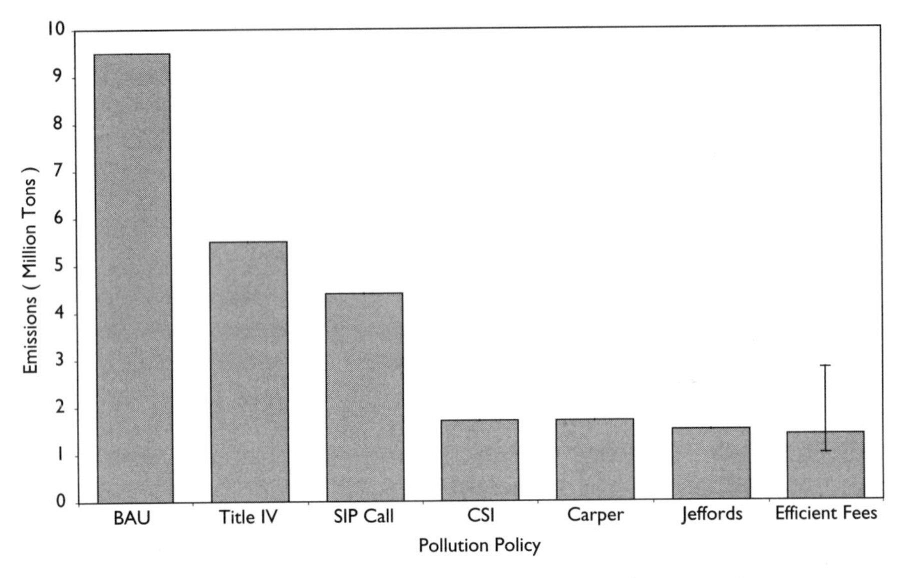

Figure 6-6. *Electricity Sector NO_x Emissions in 2020, by Scenario*

The long-run constraints on NO_x emissions from the power sector may be tightened much further. New legislative proposals are aimed at further reducing NO_x and other emissions from power plants and would expand the use of a cap-and-trade approach to achieve these emissions reductions by extending it to the national level on an annual basis.

The potential long-run annual emissions levels under the SIP Call and the multipollutant proposals are illustrated in Figure 6-6, which shows expected emissions from the electricity sector in 2020 under various scenarios. The first bar is a business-as-usual scenario, reflecting an emissions forecast that assumes the absence of Title IV and any further emission regulations. The second bar shows annual emissions under Title IV and the OTR programs as of 2000. The third bar shows annual emissions when the SIP Call program is implemented.

The next three bars represent forecasts for annual emissions under the leading three multipollutant proposals currently before Congress. The proximity of the long-run targets in these proposals is striking. Finally, at the right is an estimate of the efficient level for NO_x emissions under a cost-effective regulatory policy, such as emissions fees or a cap-and-trade program coupled with an allowance auction. This estimate is the result of an integrated assessment linking the Tracking and Analysis Framework model of atmospheric transport and health benefits from reduction in exposures to NO_x and nitrates, but excluding ozone from reductions in NO_x emissions, with a market model of the electricity sector. The bar indicates the 90% confidence interval around the mortality, morbidity, and valuation estimates in the benefits model (Banzhaf et al. 2004).

Figure 6-6 illustrates that if any of the multipollutant proposals on the table were to pass, emissions would fall within the range identified as efficient by benefit-cost analysis. All of these proposals would use a cap-and-trade approach, which is an important consideration. If the program is more expensive than modeled—perhaps because of hidden social costs associated with existing taxes, or technology standards implemented simultaneously, or state utility regulator interference in compliance planning—then benefit-cost analysis would suggest a greater level of emissions as the efficient target.[33] On the other hand, the exclusion of benefits from reducing ozone in the analysis suggests that emissions should be less than illustrated.

Both prescriptive and flexible approaches are likely to play important roles for NO_x controls at power plants in the future. However, the sun is rising for flexible cap-and-trade policies. At one time, a cap-and-trade approach was viewed by many as a sell-out to business. Today, this approach is embraced across the political spectrum as an important strategy for achieving emissions reductions at less cost than prescriptive regulation.

Notes

1. Eight-hour averaging is more consistent with the health information that prompted EPA to propose revisions to the standard. Also, by averaging over eight hours, the standard helps protect people who spend a significant amount of time working or playing outdoors and are thus particularly vulnerable to the effects of ozone.

2. Area classifications (status designations) under the new ozone standard were released in April 2004. States must plan for compliance with the new standard beginning in 2007. To attain the eight-hour standard, the three-year average of the fourth-highest daily maximum eight-hour average of continuous ambient air monitoring data over each year must not exceed 0.08 parts per million. Areas are expected to achieve the standard by 2009, although it is possible to extend the attainment deadline until 2014.

3. An important caveat to the 15% rate-of-progress rule was that a state did not need to reduce NO_x in the nonattainment region if it demonstrated that those reductions would result in an increase in ozone concentrations. In some circumstances where VOCs are the limiting precursor to ozone formation, additional NO_x will lead to a reduction in ozone. This phenomenon, "NO_x scavenging of ozone," typically happens only within 50 miles of an emissions source. At least 12 waivers were granted based on this provision. Occasionally, where NO_x abatement from sources in nonattainment regions was seen as undesirable, the burden for ozone control would fall upon VOC regulation under the 3% rate-of-progress. However, a strict focus on VOC control is an extremely burdensome way to achieve compliance.

4. The OTR comprises 11 northeastern and mid-Atlantic states stretching from Maryland to Maine, plus the District of Columbia and the northern counties of Virginia.

5. The program includes all fossil fuel–fired boilers with a maximum rated heat input capacity of 250 million Btus per hour or greater and all electricity-generating facilities with a rated output of 15 MW or more.

6. In New England states, RACT is defined as category-wide emissions rate limitations or control technology requirements. In Pennsylvania, RACT is explicitly defined as the implementation of low-NO_x burners. Throughout New England this technology was a common compliance strategy.

7. The NO_x budget for the entire OTR is calculated by applying emissions reduction factors to each source. Each state has discretion in allocating its share of the NO_x budget to its emission sources. Note that sources that do not fit these criteria may voluntarily participate in the program.

8. Environmental Science Services (1996).

9. While Section 110 of the 1990 CAAA allows the EPA administrator to determine a need for controls on upwind sources to ensure attainment of the ozone NAAQS in downwind states, Section 126 allows states to petition the administrator to require controls on these sources. In August 1997 eight northeastern states (joined by three other states in 1999) filed petitions under Section 126 requesting that EPA address emissions from these upwind sources. On April 30, 1999, EPA accepted the validity of these requests and initiated the Section 126 Federal NO_x Trading Program, which affects 13 states (a subset of the states affected by the NO_x SIP Call). The Section 126 program remains outside the bounds of the SIP process and is thus fully implemented by EPA. However, it was agreed by the petitioners and EPA that because the goals of the two programs are the same, the Section 126 program would be subsumed by the NO_x SIP Call, provided the latter survived legal challenge.

10. The EPA program for the SIP region, when fully implemented, would subsume the smaller OTR program in the Northeast.

11. The NO_x SIP Call is designed for attainment with the new eight-hour ozone NAAQS promulgated by EPA in July 1997. However, on May 14, 1999, the U.S. Court of Appeals for the District of Columbia Circuit (D.C. Circuit) ruled that the eight-hour standard for ozone was unenforceable. The court also granted a motion to stay the NO_x SIP Call deadlines for state SIP submissions on May 25, 1999. Without a schedule for NO_x SIP Call in place, EPA separated action on the Section 126 petitions from the SIP Call on June 15, 1999. The Section 126 program thus became the basis for proceeding with the reductions required by the SIP Call program. However, as it became more certain through subsequent court action that the NO_x SIP Call would proceed in generally the same form originally envisioned by EPA, the agency essentially amended the 126 program so that it could be subsumed by the SIP Call program.

12. After 2004, the trading season will begin May 1 and continue through September 31. Although the first year will have a shortened trading season, seasonal budgets (and thus NO_x allowances) are unaffected.

13. In March 2000, the Court of Appeals excluded Wisconsin from the program and raised questions about the inclusion of Georgia and Missouri. In August 2000, the court ruled that the compliance date would be May 2004.

14. U.S. EPA 1998, Figure 2-4 and Table 2-4. The EPA baseline includes the 1990 CAAA but only RACT controls in the OTR. There are two reasons these numbers are approximate: the reductions pertain to EPA's original program that targeted 22 states and the District of Columbia, and states have some latitude to change the portion of their emissions budget that is covered under the cap-and-trade program.

15. The Jeffords (I-VT) bill (S. 556) caps annual allocations of NO_x emissions allowances at 25% of 1997 levels (about 1.5 million tons). The Bush administration's Clear Skies Initiative, sponsored by Sen. Smith (R-NH) as S. 2815, caps annual emissions of NO_x at 2.1 million tons in 2008 and 1.7 million tons in 2018. The Carper (D-DE) bill (S. 3135) caps annual emissions of NO_x at 1.87 million tons in 2008 and 1.7 million tons in 2012.

16. However, it has become easier for firms to demonstrate the effectiveness of alternative control strategies because of the continuous emissions monitoring required by Title IV of the 1990 CAAA.

17. Annual demand growth in the early and mid-1990s was forecast by ICF (1990) to be 0.005% greater than what actually occurred. Adjusting the forecast of emissions in the baseline to account for lower emissions would lead to a lower measure of emissions reductions. Later estimates adjusted for this lower growth rate. ICF (1996) estimates emissions reductions of 2.06 million tons in 2000 with a lower projected growth rate. However, a comparison is difficult as it also assumes more stringent controls than ICF (1990). Similarly, Burtraw et al. (1998) estimate emissions reductions of 1.97 million tons in 2000 (Phase II), based on the lower estimate of future emissions.

18. The heat rate is the amount of energy, usually measured in million Btus, required as an input to obtain a unit of electricity, usually measured in kilowatt-hours. The lower the heat rate, generally the more efficient is the facility.

19. The sample is further limited to plants where generation from coal is at least 90% of total generation. The sample is about two-thirds of total coal-fired capacity in each case—the OTR sample and the Title IV sample not in the OTR. Some of the units in the Title IV sample are also subject to RACT rules given their location in nonattainment regions, but this is likely a small number of facilities. The two groups also differ because most of the OTR sample was affected by deregulation of the power sector in the 1990s. None of the plants in the Title IV sample were subject to deregulation during this period. Many observers argued that deregulation would lead to improved heat rates at power plants.

20. The costs are reported in current dollar terms, but the exact year is not indicated.

21. The realization that Maryland would not participate in the first season eventually resulted in downward pressure on the allowance price, because Maryland sources were expected to be net demanders of allowances (Farrell 2000). Maryland's participation was held up by a legal challenge from two utilities. The state signed a consent decree with the utilities that outlined their participation, which started in 2000.

22. Maine, Vermont, and the District of Columbia did not participate in the program. Maine and Vermont were already in compliance with the ozone standard and, along with District of Columbia, have few sources that would have been subject to it. Virginia was never expected to participate in the program because it did not sign the agreement that established it.

23. The OTR trading program also presented unique questions about the division of powers between executive administrators and state legislatures (Farrell 2000). However, the resolution of these debates will likely be applicable to future trading programs.

24. Similarly, the EPA had to demonstrate that abatement controls, as well as their applicability to U.S. facilities and coal-types, were available to achieve the cap under the NO_x SIP Call. Modeling of the electricity market was also required to show program feasibility.

25. This value is simply illustrative of the reductions that occurred over this period. A considerable amount of the abatement that occurred in the intervening years is attributable to the RACT and Title IV standards. The affect of the standards is significant. The total allocation including early reduction credits in 1999 was only about 16% lower than total emissions in 1998 (Farrell 2000). While informative, this comparison actually overstates the influence of these performance standards, as emissions in 1998 were controlled further to create early reduction credits, narrowing the gap between 1998 emissions and the 1999 allocation. It also ignores the fact that emissions in 1999 were actually 40% below 1998 due to allowance banking.

26. The trading program has a unique constraint on banking, called progressive flow control, which limits the number of allowances that can be withdrawn from a source's allowance bank on a one-to-one basis when the aggregate bank totals more than 10% of annual allocations. The NO_x SIP Call program also has this rule.

27. The EPA baseline includes only RACT controls in the OTR; Burtraw et al. (2001) include the OTR trading policy and its lower average emissions rates. Hence, there exist relatively low-cost abatement options in the EPA analysis that have already been included in the Burtraw et al. baseline. Also, Burtraw et al. assume 8.7% greater generation in 2008 than does the EPA baseline (for 2007). In addition, the emissions cap is 544,000 tons in the EPA study but 444,300 in Burtraw et al. Consequently, total reductions are 958,000 tons in the EPA study and 1,090,000 in Burtraw et al. The EPA study includes the original 22 eastern states plus the District of Columbia; the Burtraw et al. SIP Call region varies slightly.

28. Williams (2002) demonstrates an important corollary that there also may be important hidden benefits within a general equilibrium framework because of the effect of pollution on the productivity of labor.

29. Although this is an important possibility, the analysis does not account for the heterogeneity in generation technology (Burtraw and Cannon 2000) and the way that electricity price is determined at the margin (Palmer et al. 2002).

30. Few states have considered an auction under these programs, and even then only for a small portion of allowances.

31. Burtraw et al. (2001) find net benefits to be about $400 million (1997$). Burtraw et al. (2003) conduct a sensitivity analysis and find that in 18 of 18 scenarios, net benefits are positive, even in the western (upwind) coal-producing states in the SIP Call region.

32. Other states have adopted policies to reduce NO_x emissions outside the ozone season, but they have not used a pure cap-and-trade program. Massachusetts has adopted stricter annual output-based performance standards for its largest generating units. In Connecticut, sources are subject to a stringent input-based performance standard. Both of these states allow emissions averaging across all sources in a facility similar to the averaging provision in the Title IV program. North Carolina has adopted legislation that, starting in 2007, places annual NO_x emissions caps on its two largest utilities. The plants affected by this policy will not participate in the NO_x SIP Call starting in 2007. The utilities agreed to these caps in exchange for a five-year rate freeze and accelerated depreciation of pollution control equipment.

33. The Jeffords bill would distribute a large fraction of allowances through an auction, but it would not generate revenues that the federal government could use to reduce taxes. The Clear Skies Initiative would distribute a small but growing share of allowances through a revenue-raising auction.

References

Banzhaf, S., D. Burtraw, and K. Palmer. 2004. Efficient Emissions Fees in the U.S. Electricity Sector. *Resource and Energy Economics,* forthcoming.

Burtraw, D., R. Bharvirkar, and M. McGuinness. 2003. Uncertainty and the Net Benefits of NO_x Emissions Reductions from Electricity Generation. *Land Economics* 79(3): 382–401.

Burtraw, D., and M. Cannon. 2000. Heterogeneity in Costs and Second-Best Policies for Environmental Protection. Discussion Paper 00-20. Washington, DC: Resources for the Future.

Burtraw, D., A.J. Krupnick, E. Mansur, D. Austin, and D. Farrell. 1998. The Costs and Benefits of Reducing Air Pollutants Related to Acid Rain. *Contemporary Economic Policy* 16(October): 379–400.

Burtraw, D., K. Palmer, R. Bharvirkar, and A. Paul. 2001. Cost-Effective Reduction of NO_x Emissions from Electricity Generation. *Journal of Air & Waste Management* 51: 1476–89.

Chameides, W.L., R.R. Lindsay, J.L. Richardson, and C.S. Kiang. 1988. The Role of Biogenic Hydrocarbons in Urban Photochemical Smog: Atlanta as a Case Study. *Science* 241: 1473–75.

Ellerman, A.D. 1998. Note on the Seemingly Indefinite Extension of Power Plant Lives, A Panel Contribution. *Energy Journal* 19(2): 129–32.

Energy Information Administration (EIA). 1998. *Electric Power Monthly, May 1998.* Washington, DC: Department of Energy.

———. 2002. *Annual Energy Review 2001.* DOE/EIA-0384(2001). Washington, DC: Department of Energy.

Environmental Science Services (ESS). 1996. *NESCAUM/MARAMA NO_x Budget Model Rule.* Prepared for: the NESCAUM/MARAMA NO_x Budget Task Force, NESCAUM/MARAMA NO_x Budget Ad Hoc Committee and the Ozone Transport Commissions Stationary and Area Source Committee.

Farrell, A., R. Carter, and R. Raufer. 1999. The NO_x Budget: Market-Based Control of Tropospheric Ozone in the Northeastern United States. *Resource and Energy Economics* 21(1): 103–24

Farrell, A. 2000. The NO_x Budget: A Look at the First Year. *The Electricity Journal* 13(2): 83–93.

———. 2003. Temporal Hotspots in Emission Trading Programs: Evidence From The Ozone

Transport Commission's NO_x Budget. Paper presented at Market Mechanisms and Incentives: Applications to Environmental Policy, May 1-2, Washington, D.C.

Goulder, L.H., I.W.H. Parry, R.C. Williams III, and D. Burtraw. 1999. The Cost-Effectiveness of Alternative Instruments for Environmental Protection in a Second-Best Setting. *Journal of Public Economics* 72(3)(June): 329–60.

ICF. 1990. Comparison of the Economic Impacts of the Acid Rain Provisions of the Senate Bill (S.1630) and the House Bill (S.1630). Prepared for the U.S. Environmental Protection Agency (July).

———. 1995. Economic Analysis of Title IV Requirements of the 1990 Clean Air Act Amendments. Prepared for the U.S. Environmental Protection Agency (September).

———. 1996. Regulatory Impact Analysis of NO_x Regulations. Prepared for the U.S. Environmental Protection Agency (October).

Keating, T.J., and A. Farrell. 1999. *Transboundary Environmental Assessment: Lessons from the Ozone Transport Assessment Group.* Technical Report NCEDR/99-02. Knoxville, TN: National Center for Environmental Decision Making Research.

Krolewski, M.J., and A.S. Mingst. 2000. Recent NO_x Reduction Efforts: An Overview. ICAC Forum 2000.

Krupnick, A., and V. McConnell, with M. Cannon, T. Stoessell, and M. Batz. 2000. Cost-Effective NO_x Control in the Eastern United States. Discussion Paper 00-18. Washington, DC: Resources for the Future.

Mauzerall, D.L., and X. Wang. 2001. Protecting Agricultural Crops from the Effects of Tropospheric Ozone Exposure: Reconciling Science and Standard Setting in the United States, Europe and Asia. *Annual Review of Energy and Environment* 26: 237–68.

Palmer, K., D. Burtraw, A. Paul, and R. Bharvirkar. 2002. Electricity Restructuring, Environmental Policy and Emissions: Insights from a Policy Analysis. Annapolis: Maryland Department of Natural Resources, Power Plant Research Program. Also see RFF Reports, December 2002.

Swift, B. 2000. Grandfathering, New Source Review, and NO_x—Making Sense of a Flawed System. *Environment Reporter* 31(29): 1538–46.

———. 2001. How Environmental Laws Work: An Analysis of the Utility Sector's Response to Regulation of Nitrogen Oxides and Sulfur Dioxide under the Clean Air Act. *Tulane Environmental Law Journal* 14(2): 309–425.

———. 2004. Emissions Trading and Hot Spots: A Review of the Major Programs. *Environmental Reporter* 35(19): 1020–35.

Trainer, M., E.J. Williams, D.D. Parrish, M.P. Buhr, E.J. Allwine, H.H. Westberg, F.C. Fehsenfeld, and S.C. Liu. 1987. Models and Observations of the Impact of Natural Hydrocarbons on Rural Ozone. *Nature* 329: 705–07.

U.S. Congress, Office of Technology Assessment. 1989. *Catching Our Breath: Next Steps for Reducing Urban Ozone.* OTA-O-412. Washington, DC: U.S. Government Printing Office (July).

U.S. Environmental Protection Agency (EPA). 1997. *Ozone Transport Commission NO_x Budget Program: An Overview.* Washington, DC: Office of Air and Radiation, Acid Rain Division.

———. 1998. *Supplemental Ozone Transport Rulemaking Regulatory Analysis.* Washington, DC: Office of Air and Radiation (April 7).

———. 2000. *1999 OTC NO_x Budget Program Compliance Report.* Washington, DC: Office of Air and Radiation, Clean Air Markets Division.

———. 2002a. *Nitrogen: Multiple and Regional Impacts.* Washington, DC: Office of Air and Radiation, Clean Air Markets Division.

———. 2002b. *Acid Rain Program: 2001 Progress Report.* Washington, DC: Office of Air and Radiation, Office of Air and Radiation.

Williams, R.C. III. 2002. Environmental Tax Interactions When Pollution Affects Health or Productivity. *Journal of Environmental Economics and Management* 44: 261–70.

CHAPTER 7

CFCs: A Look Across Two Continents

James K. Hammitt

THE FIRST COMMERCIALLY important chlorofluorocarbons (CFCs) were developed in the 1930s for use as the fluid in refrigeration systems. These compounds are chemically stable, nonflammable, and nontoxic. By the 1970s, CFCs were widely used in a variety of industrial and consumer applications. In addition to their use in building, home and automobile air-conditioning, and refrigeration systems, CFCs were used as aerosol propellants for personal care and other products, as blowing agents in manufacturing rigid foams (for insulation and packaging) and flexible foams (for cushioning), and as solvents in the manufacture of electronic and other components. Closely related compounds were also used as solvents (e.g., methyl chloroform) and in fire-extinguishing systems (e.g., halons). CFCs and the related compounds were released to the environment either during use (e.g., aerosols) or through leakage or product disposal (e.g., refrigeration, closed-cell foams).

In 1974, Molina and Rowland published a paper suggesting that environmental release of CFCs might deplete stratospheric ozone. Because the compounds are chemically stable, they would not break down in the environment until they wafted into the stratosphere, where they would be exposed to intense ultraviolet radiation. Upon photodissociation, CFCs would release their chlorine in the stratosphere. The chlorine catalyzes a reaction that converts ozone (O_3) to molecular oxygen (O_2). By accelerating the ozone-destruction reaction, the increased presence of stratospheric chlorine would lead to a reduction in the concentration of stratospheric ozone. This reduction could be important because ozone absorbs much of the harmful ultraviolet light from the sun. With less ozone, more ultraviolet light penetrates to ground level, where it causes damage to human health (e.g., skin cancer and cataracts), ecosystems (e.g., destruction of phytoplankton), certain crops, and some materials (e.g., plastics).

This chapter describes the American and European policy responses to the threat that CFCs would deplete stratospheric ozone. At the time the threat was identified, the industrialized countries of North America and Europe accounted for a large share of CFC consumption, and probably more than three-quarters of the world's production (Hammitt et al. 1986). These regions were the primary actors in developing a global system to control CFC emissions.

The issue of CFC and stratospheric ozone has several characteristics that are important from a policy perspective. First, at the time regulatory decisions were taken, there was significant uncertainty about whether CFCs would in fact reduce the concentrations of stratospheric ozone, by how much, and how serious the consequences of ozone depletion would be for human health and the environment. Second, because the stratosphere is well mixed, the effects of CFC releases are independent of the location of release. It is not possible for any country to protect itself solely by reducing its own CFC emissions. Third, because of their chemical stability, CFCs survive in the atmosphere for long periods—for many of the compounds, on the order of a century. With emissions having increased for several decades, even if CFC emissions were to be eliminated, the amount of ozone depletion would continue to increase for years and the protective effects of reducing CFC emissions would lag well behind the costs incurred by reducing emissions.

Policy Response

The policy response to possible depletion of stratospheric ozone can be divided into several periods.[1] Soon after the issue was identified, the United States, Canada, Norway, Sweden, Denmark, Germany, Switzerland, and the Netherlands imposed unilateral restrictions to reduce CFC consumption and emissions. By 1980, these measures were in place and policy action stalled: the initial controls had significantly reduced CFC emissions, an economic recession suppressed the rate of growth of CFC production, and scientific developments suggested that the effect of CFCs on stratospheric ozone might be less severe than initially proposed. In the mid-1980s, however, economic growth and further scientific research shifted the balance toward greater concern, culminating in the signing of the Montreal Protocol in September 1987. The protocol provides an international framework to restrict both production and consumption of CFCs and other ozone-depleting substances (ODS). Since 1987, the protocol has been amended several times, in each case to tighten the controls or increase the scope of chemicals subject to its jurisdiction. Production of the primary CFCs (11, 12, and 113) was eliminated in the advanced countries by 1996, as was consumption, except for narrowly defined "essential uses." In recent years, attention has shifted to controlling emissions of some chemicals developed as substitutes for CFCs and other compounds that were not initially regulated, such as methyl bromide (a grain fumigant). Although the Montreal Protocol and its amendments provide a framework and limits on ODS production and consumption, implementation of these restrictions is the responsibility of the parties that have signed and ratified the protocol. Both the United States and the European Union (together with its member states) are signatories to the protocol and its amendments, and each has developed a set of regulations to ensure compliance with it. The details of these regulations are described below.

Early Regulations: 1974–1987

In the United States, there was a strong public reaction to the discovery that CFCs might deplete stratospheric ozone. Attention focused on the use of CFCs as a propellant in personal care aerosol products, such as deodorants and hair sprays, and included consumer boycotts and regulation in several states (e.g., Oregon prohibited CFCs in aerosols, and New York required that they be labeled). Industry responded quickly. For example, Johnson Wax announced in June 1975 that it would eliminate CFCs from its products. Removal of CFCs from aerosol products was accomplished relatively easily, in large part by substituting hydrocarbon propellants, which are flammable and less expensive, or by replacing aerosol dispensers with pump sprays that require no propellant. The federal government determined that under existing legislation it had authority to regulate only aerosol uses of CFCs, with different agencies having authority over different products. In 1978, the relevant agencies—the Consumer Product Safety Commission, the Environmental Protection Agency, and the Food and Drug Administration—issued a rule prohibiting use of CFCs as an aerosol propellant effective in 1979.

The reaction in Europe differed significantly among states. Sweden and Norway, which had little or no production, adopted aerosol bans at about the same time as the United States. Germany, which was a major producer, concluded a voluntary agreement with industry to reduce CFC use in aerosols by 30% from its 1976 level by 1979, and Denmark and Switzerland achieved similar voluntary reductions. The Netherlands imposed a labeling requirement. In contrast, there was little response in the United Kingdom, France, and Italy, which were also major producers.

The difference in response between the United Kingdom and France on the one hand, and the United States on the other may have been colored by the earlier debate about the effect of supersonic transport (SST) aircraft on stratospheric ozone. The first concerns about ozone depletion arose in the 1960s and involved the possibility that water vapor in SST exhaust, released in the stratosphere, would catalyze and accelerate the ozone-loss reaction. This concern was cited by the United States as one factor in its decision to halt an SST development program, although economic factors were seen as more important. In contrast, a consortium of the United Kingdom and France developed and eventually manufactured the Concorde supersonic passenger aircraft. The possibility of ozone depletion continued to be raised in disputes about landing rights for the Concorde in the United States. In addition, most of the scientists working on stratospheric ozone issues were in the United States, so the British and French governments did not have as much input from their own scientists on the CFC issue.

Because aerosol uses of CFCs accounted for about half of consumption, the aerosol bans and other responses in the United States and Europe led to significant reductions in CFC releases to the atmosphere. Growth in other uses was small in the early 1980s, due to slow economic growth associated with the oil price shock of 1979 and other factors. As scientific developments suggested that the effects of CFCs on ozone would not be as great as initially forecast, it appeared that the issue had been adequately addressed, at least for the moment.

By the mid-1980s, however, economic growth had resumed and CFC production began to increase. The science also began to point toward larger effects on stratospheric ozone, culminating in the 1988 Ozone Trends Panel Report (NASA 1988), which identified a statistically significant reduction in globally averaged stratospheric ozone. The 1985 discovery of the Antarctic "ozone hole" (Farman et al. 1985) attracted much scientific and policy attention. This hole is a large but geographically and temporally limited reduction in stratospheric ozone that occurs in the austral spring. It was discovered by observations at the British Antarctic research station and was neither predicted nor explained by current theories of ozone depletion. For this reason, its connection with CFCs was speculative.

In this climate, international negotiations led first to the 1985 Vienna Convention, an agreement to conduct research and monitor the situation, and subsequently to the 1987 Montreal Protocol. These negotiations were encouraged by Canada, some of the Nordic countries, and the United States. The CFC-producing and CFC-consuming industries were skeptical of the need for additional regulation, but U.S. industry eventually supported the international effort, in part because it feared that in the absence of international rules, additional domestic rules might be imposed that would weaken its competitive position.

The Montreal Protocol and its Amendments

The Montreal Protocol limits both production and consumption of CFCs, halons, and other ozone-depleting substances. Consumption is not measured directly but is defined as production plus imports minus exports. The limits apply at the national level, with the exception of the European Union, where production and consumption of the member states are not individually limited so long as the union as a whole complies with E.U.-wide limits.

Initially, the protocol limited CFC and halon production and consumption to their 1986 levels, with 20% and 50% reductions in CFCs scheduled for 1993 and 1998, respectively. These limits were strengthened by the 1990 London amendment and the 1992 Copenhagen amendment, which (with "essential use" exemptions) eliminated CFC production and consumption in 1996 and halon production and consumption in 1994. These and subsequent amendments have also enlarged the set of substances controlled under the Montreal Protocol, perhaps most significantly by including hydrochlorofluorocarbons (HCFCs), which have proved to be important substitutes for CFCs in refrigeration, foam-blowing, and other applications, and methyl chloroform (an industrial solvent).

The Montreal Protocol provides some flexibility in the particular ozone-depleting substances that are controlled. Production and consumption limits apply not to individual compounds but to the total quantities of ODS within a defined group, weighted by their relative ozone-depletion potentials. The 1987 protocol included the three primary CFCs (11, 12, and 113) and two potential substitutes (114 and 115) in one group, and the halons (1211, 1301, and 2402) in a second group. The HCFCs, added under the London amendment, form a third

Table 7-1. *Allowable Production and Consumption of Primary CFCs in United States and European Union*

	Percentage of 1986 consumption
Montreal Protocol, September 1987	1989: 100%
	1993: 80%
	1998: 50%
London amendment, June 1990	1995: 50%
	1997: 15%
	2000: 0%
Copenhagen amendment, November 1992	1994: 25%
	1996: 0%

Notes: Primary CFCs are 11, 12, 113, 114, and 115. Under the Montreal Protocol, restrictions apply to years beginning July 1. Under the London and Copenhagen amendments, restrictions apply to years beginning January 1.

group. Emission limitations of the Montreal Protocol and subsequent amendments are summarized in Table 7-1.

Implementation of the Montreal Protocol

Although the Montreal Protocol establishes limits on production and consumption of CFCs and other ozone-depleting substances, implementation of these limits is the responsibility of the parties. This section describes the rules implementing the protocol in the United States and in the European Union and its member states.

Implementation in the United States

Implementing rules for the Montreal Protocol in the United States were adopted in 1988 (53 FR 30566), and current authority is contained in Title VI of the 1990 Clean Air Act Amendments. U.S. implementation relies on a system of tradable production and consumption permits, supplemented by excise taxes and end-use controls.

The Environmental Protection Agency (EPA) issues annual permits for production (or import) and consumption to firms that manufacture ODS or import them to the United States. Allocation is based on historical production or import shares. The permits correspond to the ODS groups established under the Montreal Protocol and were initially denominated in ozone depletion potential–weighted pounds in a specified group, so firms may allocate their production and import among the ODS in a group in response to market conditions. Under the 1990 Clean Air Act Amendments, the permit system was altered so that permits are defined for each ODS, but intercompound trades based on relative ozone depletion potential are still allowed (Lee 1996). To introduce a quantity of an ODS into commerce, a firm must present the corresponding number of both production and consumption permits. The permits are tradable among firms without restriction and can also be banked (i.e., saved for later use). By controlling the quantities of ODS introduced into commerce, this system should ensure U.S. compliance with the Montreal Protocol.

Taxes were not included in the initial implementation policy, at least in part because EPA does not have authority to impose them (EPA was also uncertain about its authority to sell or auction the tradable permits). The Congress subsequently introduced ODS taxes, in part to capture some of the rents that producers and importers gain by receipt of tradable permits (Barthold 1994). Two types of taxes were imposed: an excise tax on new ODS produced or imported and a floor tax on ODS held in stock. The excise tax was initially set at $1.37 per ozone depletion–weighted pound for 1990 through 1992, and subsequently increased to $3.35 in 1993, $4.35 in 1994, and $5.35 in 1995. Some commentators credit the tax with holding U.S. consumption of CFCs below the limits required by the protocol (Barthold 1994; Hoerner 1995).

The economic incentive mechanisms were supplemented by a variety of command-and-control measures. These include specific prohibitions on ODS use in certain applications, such as nonpropellant aerosol uses not included in the 1978 rule (e.g., horns) and use as a blowing agent in flexible and packaging foams. Other rules require specific equipment and training for refrigeration and air-conditioning service personnel, and sales of small quantities of ODS were prohibited to prevent backyard mechanics without proper recycling equipment from recharging automobile air conditioners. Perhaps the most important of the command-and-control mechanisms is the significant new alternatives policy (SNAP), which prohibits the replacement of an ODS with certain substitutes if alternative choices would better reduce overall environmental or health risk. Under this policy, EPA provides lists of acceptable and unacceptable alternatives to ODS in a range of applications. In addition, the United States has adopted rules requiring labeling of products containing ODS and has revised federal purchasing guidelines to eliminate requirements for ODS and encourage substitutes. The influence of government purchasing specifications (e.g., MilSpec) extends beyond government purchasing, since these guidelines are often incorporated into private contracts as well (Wexler 1996).

Implementation in the European Union

In the European Union, implementation is organized through regulations imposed by the European Commission that are binding on, and enforced by, the member states. Some of the member states have supplemented the E.U. rules with additional regulations or voluntary agreements with industry. The negotiation and implementation of the Montreal Protocol coincided with the development of centralized authority for the environment at the E.U. level, which has also affected implementation.

The initial E.U. regulation (EC 3322/88) imposed a system of tradable production or import permits. Similar to the system in the United States, these permits apply to total quantities of ODS within a group, weighted by ozone depletion potential. These permits are tradable between firms within a single member state or between E.U. member states. EC regulation 3322/88 also required labeling CFCs and halons with a warning about their danger to stratospheric ozone.

Regulation EC 3093/94, adopted after the London amendments, extended the tradable permit system to the additional ozone-depleting substances added to the protocol. The permit limits were somewhat more stringent than required by the London amendments, creating a faster phaseout of CFCs in the European Union. This regulation also prohibited import of ODS and products containing ODS from countries that are not party to the Montreal Protocol. In addition, EC 3093/94 included a large number of end-use restrictions. It prohibited CFC uses except as solvents, refrigerants, or in rigid foam used for insulation or safety applications, with prohibition of CFC use in open solvent applications, domestic refrigeration, automobile and public bus air-conditioning after 1995; public rail air-conditioning after 1996; and large refrigerators (e.g., cold-storage warehouses) after 2000.

EC regulations are implemented and enforced by the member states. As described by Oberthür and Pfahl (2000), there is substantial variation across member states in implementation, with several states having programs that impose additional restrictions and other states that appear not to have the capacity or legal tools even to enforce the EC regulations. Member states can be divided into three classes. One group has adopted national legislation encompassing comprehensive use controls and often including more stringent reductions in ODS use. This group includes Austria, Denmark, Finland, Germany, Italy, Luxembourg, the Netherlands, and Sweden. A second group, including Belgium, France, Spain, and the United Kingdom, has no comprehensive legislation, although some of these countries have adopted some end-use controls (e.g., Belgium has controls on aerosol use and refrigeration, and France has controls on refrigeration). A third group, including Greece, Ireland, and Portugal, has not legislated sanctions and so may be unable to prosecute violations of the EC regulations.

Some E.U. member states have adopted specific provisions that supplement the EC regulations. For example, Germany has rules restricting use of any ODS in a variety of specified products, whereas other countries prohibit use of specified substances in certain applications. Germany also requires producers to take back ODS for destruction and requires labels on ODS-containing products. The Netherlands prohibits stocking of used refrigeration systems for commercial purposes. This rule is intended to discourage export of ODS-containing refrigeration systems, which might otherwise travel through Dutch ports to developing countries. The United Kingdom has adopted a system of information provision and codes of practice for maintenance and servicing of ODS-containing products.

A few E.U. member states have also adopted economic incentives. Austria created a deposit-refund system for refrigerants to encourage the recovery of ODS from systems at maintenance and disposal. In 1989, Denmark introduced a tax of DKK 30 (€3.7) per kilogram on ODS when produced or used in certain products (Oberthür and Pfahl 2000). From 1993 to 1997, Sweden imposed a fee on successful applications for exemptions from end-use controls. The fee was proportional to the quantity of ODS use exempted, at a rate of approximately $10 per kilogram in 1993 increasing to about $75 per kilogram in 1997.

Ex ante Analysis

Extensive prospective analysis of the effects of the Montreal Protocol and the implementing rules was conducted in the United States. Comparable analyses do not appear to have been conducted in the European Union.

In the United States, major rules are subjected to formal regulatory impact analysis under executive orders issued by Presidents Reagan and Clinton. Analysis of the rules implementing compliance with the Montreal Protocol focused on quantitative estimation of the benefits and costs of comprehensive limits on CFC production and use. Specifically, the regulatory impact analysis estimated the costs and partial benefits through 2075 of U.S. compliance with the protocol and with alternative levels of control (U.S. EPA 1988a, 1988b). Cost estimates were based on engineering cost studies of about 500 control options, including product substitution (e.g., substituting cardboard for foam egg cartons), recapture and reuse of ODS in manufacturing processes (e.g., foam blowing), and substitution of other chemicals (e.g., foam blowing using methylene chloride). The benefits analysis focused on reductions in human skin cancer, based on epidemiological studies of the effect of ultraviolet radiation on skin cancer combined with simulation modeling of the effects of ODS on stratospheric-ozone concentrations and ultraviolet radiation. Fatal skin cancers (which dominate the benefits estimate) were valued using standard estimates of the value of a statistical life and accounting for future increases in this value due to economic growth. The analysis concluded that the regulations easily passed a benefit-cost test, with estimated present-value benefits of approximately $6 trillion and costs of $30 billion.

The analysis also evaluated alternative methods for implementing ODS emissions reductions in the United States, including tradable production and import permits allocated to producers, tradable use permits auctioned by the government, fees on production or use, and command-and-control rules on manufacturing processes and end uses. The analysis of implementing mechanisms was less quantitative than the analysis of controlling CFC emissions. Tradable production and import permits were judged to be the best option because they could guarantee compliance with the quantity limits imposed by the Montreal Protocol with minimum administrative cost and complexity.

Ex post Analysis

In evaluating the measures imposed to control ODS emissions in United States and the European Union, it suffices to examine the two jurisdictions' compliance with their obligations under the Montreal Protocol and its amendments and the costs of compliance. The benefits of reducing ODS emissions depend only on the global reduction in emissions and are independent of the manner in which those reductions are achieved.[2]

Both the United States and the European Union have complied with the Montreal Protocol and the subsequent amendments that accelerated and strengthened CFC controls. As illustrated in Figure 7-1, both jurisdictions over-

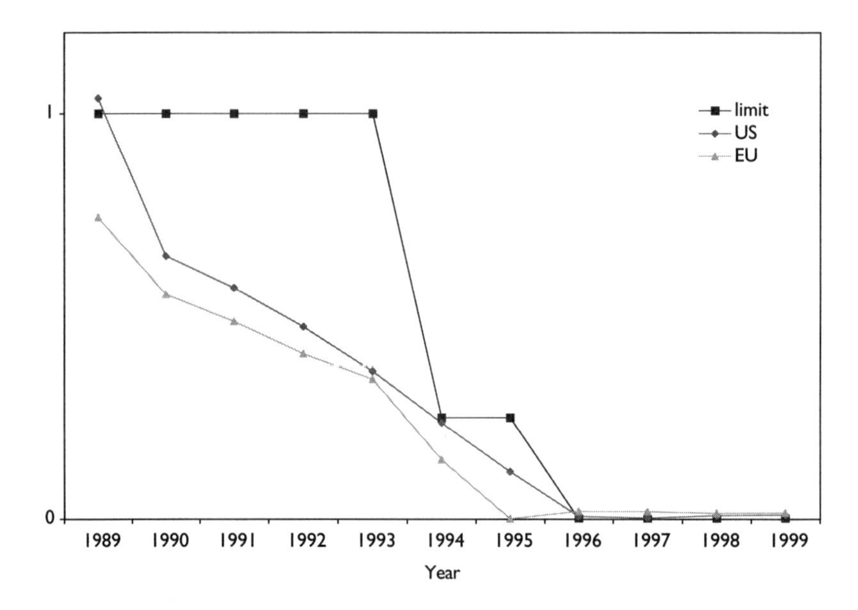

Figure 7-1. *Annual Consumption (Production Plus Imports Less Exports) of Primary CFCs*

Notes: Primary CFCs are 11, 12, 113, 114, and 115. Amounts are weighted by ozone depletion potential as reported to the Ozone Secretariat (UNEP). Consumption is graphed as a function of 1986 baseline consumption.

complied with their respective limits on consumption of the primary CFCs (11, 12, 113, 114, and 115). The European Union overcomplied somewhat more than the United States and eliminated CFC consumption (except for essential uses) in 1995, one year earlier than required by the protocol.

Some problems with illegal import and other black-market sale of CFCs have been reported in both the United States and the European Union, with several large cases prosecuted in the United States and in the Netherlands. Illegal imports are believed to have originated in Russia or China. Other illegal consumption is alleged to have resulted from the diversion to the domestic market of CFCs labeled for export to developing countries, mislabeling CFC-12 (used in automobile air-conditioning) as HCFC-22,[3] or mislabeling virgin material as recycled. Estimates of the quantities involved are necessarily uncertain, but the extent of black-market CFCs seems to have been comparable in the two jurisdictions, with estimates of 10,000–20,000 tons per year in the United States in 1993 through 1995[4] and something greater than 10,000 tons per year in the European Union in the mid-1990s (Oberthür and Pfahl 2000).

Those quantities are a small but nonnegligible share of permitted consumption, which in 1994 and 1995 was approximately 75,000 tons per year in each jurisdiction. In the European Union, black-market operations seem to have been limited to member states lacking comprehensive end-use restrictions, except for the

Netherlands, which was a site for illegal imports, presumably because of its status as a major port of entry to the European Union (Oberthür and Pfahl 2000).

With regard to compliance costs, the overall conclusion is that substantial reductions in CFC consumption were achieved with little economic disruption. In 1994, the Economic Options Committee of the UN Environment Programme reported that "ODS replacement has been more rapid, less expensive and more innovative than had been anticipated at the beginning of the substitution process. The alternative technologies already adopted have been effective and inexpensive enough that consumers have not yet felt any noticeable impacts (except for an increase in automobile air conditioning service costs)" (UNEP 1994).

For the United States, economic costs of compliance can be estimated using data on prices of CFCs, including the excise tax and permit price. These data suggest that the marginal cost of control increased to about $5 per ozone depletion potential–weighted kilogram by 1991, when consumption was about 50% smaller than it would have been in the absence of controls (other than the 1978 aerosol ban), and to more than $16 per ozone depletion potential–weighted kilogram in 1994, when consumption was about 85% percent smaller than it would have been (1986$; Hammitt 2000). Comparing these retrospective estimates with the prospective cost estimates reported in the regulatory impact analysis suggests that realized costs were somewhat higher than forecast for these levels of ODS consumption. This difference could reflect the fact that the London and Copenhagen amendments accelerated the reductions in CFC consumption relative to the scenarios analyzed in the regulatory impact analysis, and one would expect this more rapid reduction to have increased marginal control costs by limiting the time available for producer and user industries to develop and implement substitute technologies.

Little information on CFC prices in the European Union seems to be available, although prices apparently rose more slowly there than in the United States (Andersen and Sarma 2002). In the late 1990s, reported CFC prices in the United Kingdom were £10–12 per kilogram (Oberthür and Pfahl 2000), somewhat below the U.S. price, which reached about $24 per kilogram by 1994.[5] It is likely that marginal compliance costs were smaller in the European Union than in the United States because the Montreal Protocol required each jurisdiction to reduce its CFC consumption by the same fraction from its 1986 consumption baseline. By 1986, the United States had eliminated its use of CFCs as an aerosol propellant, but this use continued to account for a substantial share of E.U. consumption (perhaps 15%; Hammitt et al. 1986), and so the European Union should have been able to achieve part of its consumption reduction at low marginal cost by restricting aerosol use of CFCs. As discussed below, the lower price in the European Union could also reflect greater suppression of demand resulting from more stringent end-use controls.

In comparing compliance costs between the two jurisdictions, it is important to note that there has been extensive international cooperation in developing alternatives to CFCs, both alternative compounds and not-in-kind alternatives. Producer and user industries have collaborated across countries in safety and performance testing, and diffusion of alternatives has been encouraged through trade shows, sometimes with government sponsorship. Under the Montreal Protocol, technology and economic assessment panels were established to review and discuss alternatives, further helping to share knowledge and best practices among countries. As a result of these activities, country-by-country differences in information about technological options to reduce CFC consumption are probably small. For products that are manufactured for global markets, such as automobiles, one might also expect that producers would frequently use the same alternatives in production for both American and European markets. For other markets, there are differences in technology. For example, domestic refrigerators in Europe tend to use hydrocarbon refrigerants, whereas HFC 134a is used in the United States. Although hydrocarbons are flammable and present a safety risk, this risk is apparently more manageable for typical European units, which are smaller and less likely to include a heater for frost-free operation (Cook and Kimes 1996).

Implications for the Hypotheses

Both the United States and the European Union adopted mixed systems of tradable permits and command-and-control regulations to limit ODS consumption. In a mixed system, it is difficult to determine which elements are most consequential in influencing efficacy and costs. It is generally perceived that the U.S. system relied more heavily on economic incentive mechanisms—tradable permits and taxes—than the E.U. system. Consistent with this view, there appears to have been a greater degree of permit trading in the United States than in Europe, suggesting that the permits were the more important constraint. In the United States, there were about 750 trades (Solomon and Lee 2000), and 1992 trading volume accounted for 13% of allowable ODS consumption (Lee 1996). In contrast, there were very few trades in Europe (Andreas Kraemer, personal communication, Ecologic, June 2003).

Permit trades in the United States were predominantly between firms. They appear to have resulted in a substantial reallocation of consumption among the regulated compounds, suggesting that the permits were the more influential component and that significant inter-ODS differences in marginal control costs existed. Differences in control costs—across ODS or across uses of an ODS—are a prerequisite for trading to reduce compliance costs. In 1994, when ODS consumption was limited to 25% of baseline levels, actual consumption of the three major ozone-depleting substances was 10% for CFC-11, 39% for CFC-12, and 7% for CFC-113 (Lee 1996).

Despite the difficulty of determining the relative importance of the economic incentive and command-and-control elements of the ODS regulatory systems, comparing the U.S. and E.U. experience provides some insights into the differential effects of command-and-control and economic incentive regulatory instruments. This section addresses the implications of the ODS case for the project hypotheses.

Static Efficiency

The degree of environmental protection (i.e., the magnitude of reduction in ODS use) was specified by the Montreal Protocol and its amendments, not as part of the U.S. or E.U. implementing regulations. In comparing the two systems, the relevant question is thus one of cost-effectiveness in achieving the stated goals. Unfortunately, the available data are not sufficient for comparing compliance costs in the two jurisdictions. The price of ODS appear to have been somewhat lower in the European Union than in the United States (e.g., £10–12, or $16–19, per kilogram in the United Kingdom, versus about $24 per kilogram in the United States in the mid-1990s). This observation suggests that (marginal) compliance costs were smaller in the European Union than in the United States but is not conclusive. For example, greater reliance on end-use restrictions in Europe could have suppressed the demand for ODS, thereby reducing the price even if control costs were identical to those in the United States. If the marginal compliance costs were smaller in the European Union, this could reflect that fact that the United States had already eliminated a low-cost source of use reduction from its Montreal Protocol baseline by banning nonessential aerosol applications in 1979.

Information Requirements

The evidence in this case is consistent with the hypothesis that economic incentive instruments require less information than command-and-control instruments. The tradable permit systems that were established depended primarily on information already contained in the Montreal Protocol, such as the quantity to be permitted and the ozone depletion potential of each of the regulated substances. In the United States, permits were allocated to firms based on their historical production and import levels—information obtained from the firms themselves. In contrast, the command-and-control instruments required information about specific ODS applications and alternative technologies. Ensuring compliance with the production and consumption limitations of the Montreal Protocol through a pure system of end-use restrictions would have required information on the magnitude of ODS use in each application. In contrast, the use of tradable production and consumption permits provided a simple mechanism for ensuring compliance in each jurisdiction.

Dynamic Efficiency

This case provides little information about the dynamic efficiency of alternative regulatory mechanisms. Information about alternatives to use of ODS was widely shared through a variety of mechanisms. Many of the firms that produced and consumed ODS were multinational, often operating in both jurisdictions. The technology and economic assessment panels established under the protocol provided another mechanism for information exchange. The rapid and near-complete elimination of ODS under the Montreal Protocol amendments provided a strong incentive for firms to develop and commercialize substitute products and processes.

Effectiveness

Tradable permits appear to have been highly effective in ensuring compliance with the ODS production and consumption limits specified in the Montreal Protocol and its amendments, with one important caveat. In both jurisdictions, significant quantities of ODS were illegally imported, circumventing the consumption cap. Comprehensive end-use restrictions, as in some E.U. member states, appear to have reduced demand for illegal imports and thus proved more effective in reducing ODS consumption, or at least enhanced the efficacy of the tradable permit systems.

Regulatee Burden

In the United States, tradable permits were allocated without charge to producers and importers on the basis of their historical market shares. Since these permits were valuable assets, the allocation provided a direct benefit to the firms, at least partially offsetting their losses from restrictions on future ODS sales. Parts of these rents were captured by the federal government through excise taxes imposed later. In contrast, ODS user industries received no such transfer but faced higher ODS prices on both marginal and inframarginal quantities.

Administrative Burden

Tradable permit systems were adopted at least in part because of their greater administrative simplicity. Although firms producing and importing ODS in the United States and Europe numbered in the tens, firms consuming ODS numbered in the tens of thousands. EPA estimated that a traditional command-and-control approach would cost the agency \$23 million per year and require 32 staff to administer, and that it would impose on firms \$300 million per year in reporting and recordkeeping costs. The administrative costs of the tradable permit system are estimated to be much smaller, requiring only 4 agency staff and imposing only \$2.4 million annually in private reporting and recordkeeping costs (Lee 1996).

Hotspots and Spikes

Because ozone-depleting substances remain in the atmosphere for decades to centuries after release, they are well mixed in the atmosphere, and thus the spatial patterns of ODS use and emissions do not affect the environmental consequences. Similarly, environmental effects are insensitive to temporal variations in emissions for time scales of months and years. The tradable permit systems could therefore allow firms to bank production and consumption credits for use in future years, thereby providing operating flexibility and the potential for cost savings.

Monitoring Requirements

The design of the Montreal Protocol was strongly influenced by concern about monitoring requirements. This concern is evident in several aspects of the protocol. First, although ozone-depleting substances are dangerous to the environment only if they are released to the atmosphere, the protocol regulates production and consumption because these are much more easily observable than emissions. Second, rather than imposing a system of end-use restrictions, the protocol limits national and E.U. total production and consumption. Consumption is defined as production plus imports minus exports, so it can be measured by monitoring the activities of the small number of producing and importing firms rather than the numerous ODS-using firms. Moreover, although the protocol restricts import of bulk quantities of ODS, concern about monitoring requirements led to milder restrictions on import of products containing or produced with ODS. Restrictions on products containing ODS were postponed four years to allow for the development of a list of such products, and the protocol simply called for a process to determine the feasibility of restricting the import of products manufactured using ODS (e.g., electronic equipment cleaned using CFC 113). Implementing regulations in the United States and the European Union inherited this focus on total production and consumption and imposed relatively modest monitoring requirements because of the small number of ODS producers and importers in each jurisdiction. Illegal imports revealed limitations in the systems used to monitor and enforce compliance with import regulations.

Tax Interaction Effects

The tax interaction effects of ODS regulations have not been investigated. The use of excise taxes to capture some of the rents given to ODS producers and importers in the United States would be anticipated to reduce any adverse tax interaction effects.

Effects on Altruism

Both the United States and the European Union appear to have overcomplied with their Montreal Protocol obligations, holding ODS consumption substantially below allowed limits. The reasons for this overcompliance are not clear, especially in the United States, where there was an active tradable permit market with nonzero permit prices. In some ODS application sectors, consumption may have fallen because of consumer preferences and competition among firms to introduce substitute products and processes. These factors appear to explain the sharp reduction in U.S. use of ODS in personal care aerosol products in the 1970s, prior to the regulation banning such uses (Hammitt et al. 1986).

Adaptability

The tradable permit systems proved adept at incorporating changes in ODS production and consumption limits imposed by the London and Copenhagen amendments to the Montreal Protocol. The permitted quantities could be changed much more easily and quickly than end-use restrictions could be modified. In the United States, rulemakings to implement these amendments were completed within a year, substantially faster than most command-and-control rules. As a comparison, the regulations requiring recycling ODS from stationary air-conditioning and refrigeration systems took three years to develop (Lee 1996).

Cost Revelation

In principle, economic incentive instruments offer the ability to estimate marginal compliance costs by observing market prices. Under a pure approach, the gross market price (including the price of any tradable permits and taxes) of a regulated product identifies the demand for that product, which depends on the opportunity cost of the marginal unit (i.e., the incremental cost of using the best alternative to the regulated product). In systems that mix economic incentive instruments with end-use controls, as adopted by the United States and the European Union, the interpretation of market prices is less certain. The difficulty arises because the end-use restrictions may suppress demand for the regulated ODS, and so the gross market price of the ODS may be smaller than its social opportunity cost. As discussed above, the apparently lower ODS prices in the Europe than in the United States could reflect more stringent end-use restrictions in the European Union.

Conclusions

Formally, the regulations to control production and consumption of CFCs and other ozone-depleting substances in the United States and the European Union are quite similar. Both jurisdictions relied on a tradable permit system to enforce the quantity limits imposed by the Montreal Protocol and subsequent amendments, and supplemented these permits with a variety of command-and-control measures. In addition, excise taxes were imposed in the United States and some of the E.U. member states. Despite their formal similarity, the U.S. system appears to have relied substantially more heavily on tradable permits than the E.U. system, as evidenced by the much greater extent of trading in the United States.

In terms of effectiveness, there appears to be little difference in performance between jurisdictions. Both overcomplied with their obligations under the Montreal Protocol. The European Union overcomplied somewhat more than the United States in the early years, and complied with its own regulation to eliminate CFC consumption (except for "essential uses") one year earlier than required by the protocol. Both jurisdictions experienced illegal imports of roughly comparable magnitude, suggesting no great difference between jurisdictions in their susceptibility to evasion. There is some evidence suggesting that the E.U. member states with more comprehensive end-use restrictions were less likely to be a destination for illegal imports, suggesting that command-and-control instruments may prove more effective in preventing illegal imports.

There is limited evidence to suggest that the market price of CFCs was lower in the European Union than in the United States. If the two jurisdictions had relied on pure tradable permit systems to limit CFC consumption, this price differential would suggest that the marginal cost of compliance was smaller in the European Union. However, the price differential could also be explained by more stringent command-and-control regulations that suppressed demand for CFCs to a greater extent in the European Union. Moreover, the marginal control cost may have been smaller in the European Union because of differences in the structure of ODS applications. The Montreal Protocol required equal percentage reductions from a 1986 baseline. Because the United States had already eliminated most of its aerosol use before 1986 but the European countries had not, the European Union could achieve some part of its compliance at relatively low cost by eliminating remaining aerosol uses.

In summary, control of CFCs and other ozone-depleting substances in the United States and the European Union was coordinated through an international agreement that imposed quantity limits on each jurisdiction. Implementing regulations in each jurisdiction maintained a permit mechanism to ensure compliance with these quantity limits, and there was a high degree of information exchange between various industry sectors, governments, and international organizations in an effort to minimize compliance costs and disruption. In this context, it is perhaps not surprising that there do not appear to have been large differences in outcomes between the two jurisdictions.

Notes

1. The history of CFC regulation is described by Andersen and Sarma (2002), Benedick (1998), Cagin and Dray (1993), Cogan (1988), Dotto and Schiff (1978), Hammitt and Thompson (1997), Maxwell and Weiner (1993), Parson (2003), and Roan (1989).

2. Because the Montreal Protocol restricts only production and consumption of ODS, policies that comply with it could have different environmental effects if they lead to differences in the total quantities of ODS released to the atmosphere, or the timing of releases. Comparing U.S. and E.U. emissions of ODS would require information on the quantities of ODS consumed in various applications, which is not readily available.

3. HCFC-22 is a refrigerant that was first regulated under the 1992 Copenhagen amendment.

4. Parson and Greene (1995) report an estimate of the global black market as 10,000 tons of CFCs in 1994, Cook (1996) reports that 10,000 to 22,000 tons of CFCs were illegally imported to the United States in 1994 and 1995, Milmo (1996) reports 9,000 to 18,000 tons imported to the United States in 1995, and Hoerner (1996) reports annual illegal imports of CFC-12 to the United States of at least 16 million pounds (7,300 tons) in 1993 and 1994.

5. At the 1999 average exchange rate of £1 = $1.62, the U.K. prices were $16–19 per kg. The U.S. price was about $16 per kg in 1986 dollars (Hammitt 2000).

References

Andersen, S.O., and K.M. Sarma. 2002. *Protecting the Ozone Layer: The United Nations History*. London: Earthscan Publications.

Barthold, T.A. 1994. Issues in the Design of Environmental Excise Taxes. *Journal of Economic Perspectives* 8: 133–51.
Benedick, R.E. 1998. *Ozone Diplomacy: New Directions in Safeguarding the Planet*, enlarged edition. Cambridge, MA: Harvard University Press.
Cagin, S., and P. Dray. 1993. *Between Earth and Sky: How CFCs Changed Our World and Endangered the Ozone Layer.* New York: Pantheon Books.
Cogan, D.G. 1988. *Stones in a Glass House.* Washington, DC: Investor Responsibility Research Center.
Cook, E. 1996. Marking a Milestone in Ozone Protection: Learning from the CFC Phase-Out. WRI Issues and Ideas. Washington, DC: World Resources Institute.
Cook, E., and J. Kimes. 1996. Dangling the Carrot. In *Ozone Protection in the United States: Elements of Success,* edited by E. Cook. Washington, DC: World Resources Institute, 55–65.
Dotto, L., and H. Schiff. 1978. *The Ozone War.* Garden City, NY: Doubleday and Company.
Farman, J.C., B.G. Gardiner, and J.D. Shanklin. 1985. Large Losses of Total Ozone in Antarctica Reveal Seasonal ClO_x/NO_x Interaction. *Nature* 315: 207–10.
Hammitt, J.K. 2000. Are the Costs of Proposed Environmental Regulations Overestimated? Evidence from the CFC Phaseout. *Environmental and Resource Economics* 16(3): 281–301.
Hammitt, J.K., and K.M. Thompson. 1997. Protecting the Ozone Layer. In *The Greening of Industry: A Risk Management Approach*, edited by J.D. Graham and J.K. Hartwell. Cambridge, MA: Harvard University Press, 43–92.
Hammitt, J.K., K.A. Wolf, F. Camm, W.E. Mooz, T.H. Quinn, and A. Bamezai. 1986. *Product Uses and Market Trends for Potential Ozone-Depleting Substances: 1985-2000*, R-3386-EPA.
Hoerner, J.A. 1995. Tax Tools for Protecting the Atmosphere: The U.S. Ozone-Depleting Chemicals Tax. In *Green Budget Reform,* edited by R. Gale, S. Barg, and A. Gillies, London: EarthScan Publications, 185–99.
Hoerner, J.A. 1996. Taxing Pollution. In *Ozone Protection in the United States: Elements of Success,* edited by E. Cook. Washington, DC: World Resources Institute, 39–53.
Lee, D. 1996. Trading Pollution. *Ozone Protection in the United States: Elements of Success*, edited by Elizabeth Cook. Washington, DC: World Resources Institute, 31–38.
Maxwell, J.H., and S.L. Weiner. 1993. Green Conciousness or Dollar Diplomacy? The British Response to the Threat of Ozone Depletion. *International Environmental Affairs* 5: 19–41.
Milmo, S. 1996. Russia Exporting CFCs in Quantity. *Chemical Marketing Reporter* 249(20): 9, 12.
Molina, M.J., and F.S. Rowland. 1974. Stratospheric Sink for Chlorofluoromethanes: Chlorine Atom-Catalysed Destruction of Ozone. *Nature* 249: 810–12.
National Aeronautics and Space Administration (NASA). 1988. *Executive Summary: Ozone Trends Panel*, Washington, DC.
Oberthür, S., and S. Pfahl. 2000. *The Implementation of the Montreal Protocol on Substances that Deplete the Ozone Layer in the European Union.* Final Report to the European Commission, Environment DG and the UK Department of the Environment, Transport and the Regions. Berlin: Ecologic.
Parson, E.A., and O. Greene. 1995. The Complex Chemistry of the International Ozone Agreements. *Environment* 37(2): 16–20, 35–43.
Parson, E.A. 2003. *Protecting the Ozone Layer: Science and Strategy.* New York: Oxford University Press.
Roan, S. 1989. *Ozone Crisis: The 15-year Evolution of a Sudden Global Emergency.* New York: John Wiley and Sons.
Solomon, B.D., and R. Lee. 2000. Emissions Trading Systems and Environmental Justice. *Environment* 42(8): 32–45.
U.S. Environmental Protection Agency (EPA). 1988a. Protection of Stratospheric Ozone. *Federal Register* 53: 30566, August 12.
———. 1988b. *Regulatory Impact Analysis: Protection of Stratospheric Ozone.* Stratospheric Protection Program, Office of Program Development, EPA Office of Air and Radiation. Washington, DC.
UN Environment Programme (UNEP). 1994. *1994 Report of the Economics Options Committee for the 1995 Assessment of the Montreal Protocol on Substances That Deplete the Ozone Layer.* Nairobi.
Wexler, P. 1996. New Marching Orders. In *Ozone Protection in the United States: Elements of Success,* edited by E. Cook. Washington, DC: World Resources Institute, 77–86.

CHAPTER 8

Leaded Gasoline in the United States

The Breakthrough of Permit Trading

Richard G. Newell and Kristian Rogers

REFINERS IN THE UNITED STATES started adding lead compounds to gasoline in the 1920s to boost octane levels and improve engine performance by reducing engine "knock" and allowing higher engine compression.[1] Lead was used for boosting octane because it was less expensive than other fuel additives (e.g., ethanol and other alcohol-based additives), and because people were ignorant of the dangers of lead emissions, which include mental retardation and hypertension. The reduction in lead in gasoline in the United States came in response to two main factors: (1) the mandatory use of unleaded gasoline to protect catalytic converters in all cars starting with the 1975 model year, and (2) increased awareness of the negative human health effects of lead, leading to the phasedown of lead in leaded gasoline in the 1980s.

Initial Phasedown

As summarized in Table 8-1, the phasedown of lead in gasoline began in 1974 when the U.S. Environmental Protection Agency (EPA), under the authority of the Clean Air Act Amendments of 1970, introduced rules requiring new cars to be equipped with catalytic converters and requiring the use of unleaded gasoline in those models. The introduction of catalytic converters for control of hydrocarbons (HCs), nitrogen oxides (NO_x), and carbon monoxide (CO) emissions required that motorists use unleaded gasoline, since lead destroys the emissions control capacity of catalytic converters. A large proportion of the eventual phasedown of lead in gasoline is in fact attributable to the decreasing share of leaded gasoline that resulted from the transition to a new car fleet (Figure 8-1). To help increase the supply of unleaded gasoline and avoid misfueling, EPA also mandated that gasoline retailers offer unleaded gasoline and that car manufacturers of

Table 8-1. *Federal Standards for Lead Phasedown*

Deadline	*Standard*	*Exceptions*
July 4, 1974	Gasoline retailers must offer unleaded gasoline and must design fuel nozzles so that cars with catalytic converters can accept only unleaded gasoline.	Small retailers that sell less than 200,000 gallons annually and have fewer than six retail outlets are exempt.
July 4, 1974	Car manufacturers must design tank filler inlets to accept only unleaded gasoline and must apply "Unleaded Gasoline Only" labels.	The standard applies only to cars with catalytic converters (which became mandatory beginning with model year 1975).
October 1, 1979	Refineries must not produce gasoline averaging more than 0.5 gpg lead per quarter, pooled (leaded and unleaded).	The standard is relaxed to 0.8 gpg lead until October 1, 1980, if a refinery increases unleaded gasoline production by 6% over the prior year's quarter. Small refineries are subject to a less stringent standard. See Table 8-2.
November 1, 1982	Refineries must meet a leaded gas standard of 1.1 gpg. Interrefinery averaging of lead rights is permitted among large refineries and among small refineries, but not between refineries of different sizes.	Very small refineries are subject to a less stringent pooled standard. See Table 8-2.
July 1, 1983	Very small refineries are also subject to a standard of 1.1 gplg. Averaging is permitted among all refineries.	—
January 1, 1985	During 1985 only, refineries are permitted to "bank" excess lead rights for use in a subsequent quarter.	—
July 1, 1985	The standard is reduced to 0.5 gplg.	—
January 1, 1986	The standard is reduced to 0.1 gplg.	—
January 1, 1988	Interrefinery averaging and withdrawal of banked lead usage rights are no longer permitted. Each refinery must comply with the 0.1 gplg standard.	—
January 1, 1996	Lead additives in motor vehicle gasoline are prohibited.	—

Notes: gpg = grams (of lead) per gallon; gplg = grams per leaded gallon.
Source: U.S. Code of Federal Regulations (1996).

cars with catalytic converters design fuel inlets that fit only unleaded gasoline nozzles.

To further promote the production of unleaded gasoline, EPA scheduled performance standards requiring refineries to decrease the average lead content of all gasoline beginning in 1975. The standards were postponed until 1979 through

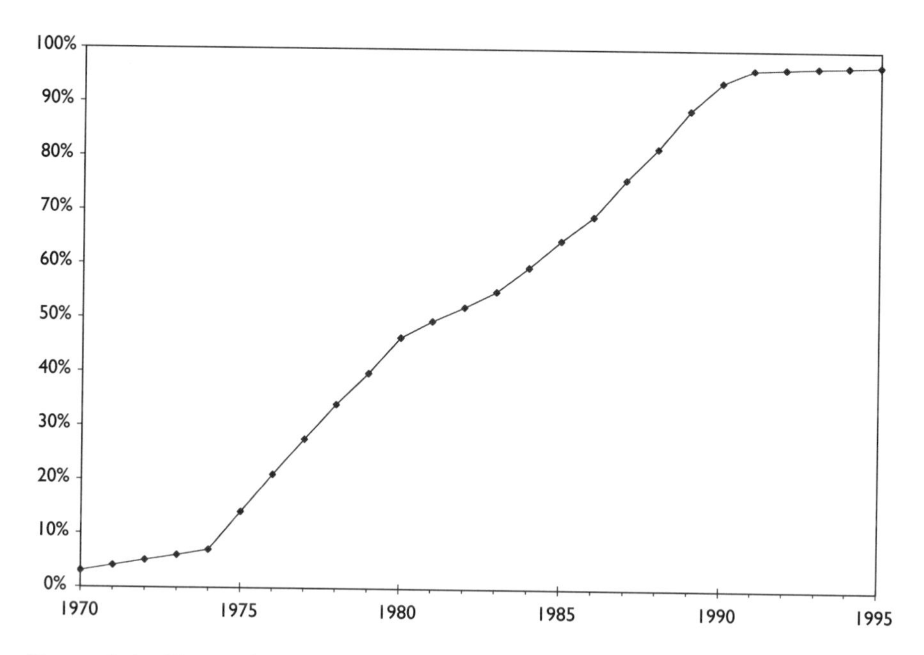

Figure 8-1. *Share of Unleaded Gasoline in Total U.S. Production*

a series of regulatory adjustments. By then, however, new studies provided increasing evidence of adverse effects of atmospheric lead on the I.Q. of children and on hypertension in adults (U.S. EPA 1985a).[2] Although lead use would have eventually dwindled as the last pre-1975 cars were retired, the additional evidence of health impacts led to a desire to accelerate the phaseout of lead in gasoline.

Under the individual facility performance standards in effect from October 1979 to November 1982, EPA set several standards, whose strictness increased with the size of the refinery. Production capacity of refiners ranged from 50 to 640,000 barrels per day (bpd), with an average of about 67,000 bpd.

Table 8-1 shows that large refineries —those with production capacity of more than 50,000 bpd and/or those owned or controlled by a refiner having total capacity greater than 137,500 bpd—were allowed to produce a quarterly average of no more than 0.8 grams per gallon (gpg) of lead for the first year and 0.5 gpg for the next two years. About 40% of refineries fell into this category.

Table 8-2 demonstrates that small refiners—those with capacity of less than 50,000 bpd—faced a scale of five standards, from 2.65 gpg for the smallest refiners to 0.8 gpg for the largest of the small refiners. It was up to the individual refiner to match these standards in the time allotted.

Notably, the early regulations set an average lead concentration for total gasoline output, both leaded and unleaded. This averaging method deliberately provided refiners with the incentive to increase unleaded production while not necessarily removing lead from their leaded gasoline—in fact, the regulation actually allowed refiners to increase overall lead concentration levels, provided they sufficiently raised unleaded output. Nonetheless, these regulations still led to a

Table 8-2. *Small Refinery Standards for Lead Phasedown*

Deadline	Standard (gpg)	Gasoline production in prior year (bpd)	Definition of small refinery
October 1, 1979	2.65 (pooled)	Up to 5,000	50,000 bpd or less crude oil throughput capacity and owned by a company with 137,500 bpd or less total capacity
	2.15 (pooled)	5,001 to 10,000	
	1.65 (pooled)	10,001 to 15,000	
	1.30 (pooled)	15,001 to 20,000	
	0.80 (pooled)	20,001 and over	
November 1, 1982	2.65 (pooled)	Up to 5,000	10,000 bpd or less gasoline production and owned by a company with 70,000 bpd or less total gasoline production
	2.15 (pooled)	5,001 to 10,000	
July 1, 1983 and after	Same as other refineries	—	—

Notes: gpg = grams of lead per gallon; bpd = barrels per day.
Source: U.S. Code of Federal Regulations (1996).

decrease in total lead usage because car owners were retiring their precatalyst automobiles and replacing them with new cars that required unleaded fuel.

By the early 1980s, gasoline lead levels had declined by about 80% as a result of both the regulations and fleet turnover (Nichols 1997). EPA was considering deferring the deadlines and relaxing the standards in response to growing complaints that the small refiners were having difficulty complying within the regulatory timeframe. However, regulatory relaxation met very strong opposition, both from within the agency, and from environmental groups and public health officials. The agency subsequently withdrew its consideration and instead decided to tighten the standards. It narrowed the definition of a small refinery, phased out special provisions for such refineries by mid-1983, and recalculated lead limits to be an average of lead in leaded gas only (unleaded fuel was by then a well-established product). Small refineries challenged the new regulations but gained only a slight extension in some of their compliance deadlines.

The new rules specifically limited the allowable content of lead in leaded gasoline to a quarterly average of 1.1 grams per leaded gallon (gplg). Very small refineries faced less stringent standards until 1983. From 1983 to 1985, EPA conducted an extensive cost-benefit analysis of a scenario under which the lead standard would be dramatically reduced to 0.1 gplg by 1988. The analysis suggested that not only was this goal feasible, but that an even tighter standard might be achieved, partly because large refiners had already acquired the technology to reduce lead to levels below the standard (Nichols 1997). In August 1984, the agency proposed a reduction of lead to 0.1 gplg by January 1, 1986. EPA understood, however, that some refineries might not be able to achieve this reduction

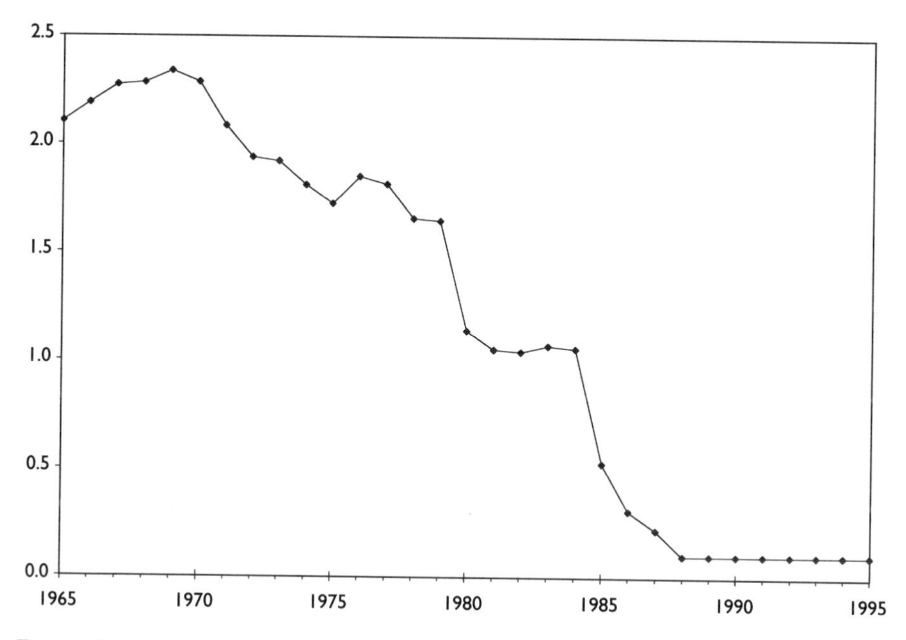

Figure 8-2. *Lead Content in Leaded Gasoline (U.S. Average)*

that quickly, so the agency also considered a more gradual phasedown involving banking, which would reach 0.1 gplg by January 1, 1988. The proposal also hinted that the agency was considering a total ban on lead, but only in the long run. Thus, during 1985 the standard was reduced to 0.5 gplg, and beginning in 1986 the allowable content of lead in leaded gasoline was reduced to 0.1 gplg.[3]

With the exception of the Lead Industries Association, support for the phasedown was generally widespread, encompassing environmentalists, the medical community, the general public, and the Office of Management and Budget, which reviewed the regulations before their enactment. By this time, even most refiners accepted the reasons for removing lead from gasoline, although some expressed reservations about the timeline.

To ease the transition for refineries, the regulations permitted both trading and banking of lead permits through a system of interrefinery averaging. Trading of lead credits among refineries was allowed from late 1982 through the end of 1987. Banking was allowed from 1985 to 1987. Beginning in 1988, EPA reimposed a performance standard of 0.1 gplg applicable to individual refineries. Lead was then banned as a fuel additive in the United States beginning in 1996. Figure 8-2 shows the decline over time in the lead content of gasoline in the United States.

The phasedown and ultimate ban on lead did not negate refiners' goals of boosting octane in gasoline. A constraint on the amount of lead led to the demand for a substitute to boost the octane in gasoline. Two basic approaches emerged. One was the use of other octane-enhancing additives, such as methyl tertiary-butyl ether (MTBE). These additives are more expensive than lead and constitute only one part of the long-term solution. Additives including MTBE provided about 30% of the octane lost due to the removal of lead in the final phasedown. The other approach is to increase refineries' abilities to produce high-octane gasoline components through process changes, primarily isomerization and reforming. In the short run, existing equipment can be run more intensively to increase octane production, but eventually new investment is required. Isomerization provided another 40% of additional octane requirements, and alkylation, catalytic cracking, and reforming together provided most of the remaining 30% of lost octane. A refinery could also adjust somewhat by altering the type of crude oil it purchased, by buying intermediate products with higher octane content, or by changing its output mix to one requiring less octane.

Lead Trading and Banking

Until 1982, EPA took a command-and-control approach to regulating lead, based on technology standards and individually binding refinery performance standards for lead content. However, the agency realized by the early 1980s that this approach was causing small refineries substantial difficulty in meeting the standards by the regulatory deadline. Small refineries faced higher compliance costs with the lead phasedown because they typically lacked the more sophisticated processing equipment needed to replace lost octane (e.g., reformers and alkylation). The lack of such equipment also increased the costs of installing new technologies such as isomerization. At the same time, many large refineries had already succeeded in implementing technology that could remove more lead from their gasoline than required by the regulations, and at a cost lower than that faced by small refineries. The reconciliation of these two situations became the basis of the tradable lead credit program.

EPA's plan to continue lowering the standards over time compounded the small refineries' problem of high abatement costs because the cost of removing an increment of lead from gasoline increased as more lead was removed. The problem also raised issues of optimal timing of abatement investments. The solution to this issue was the banking program. The banking option was introduced at the beginning of 1985 and ended with the tradable lead credit program at the end of 1986, although refiners could use their banked rights through 1987.

The new marketable permit system allowed for interrefinery lead averaging, whereby some refiners could produce higher concentrations of lead than others as long as the average across refineries met the agency's standard. This system alleviated a portion of the financial burden faced by many small refineries, and it allowed the entire refining industry a measure of flexibility in allocating the reduction among firms and in allocating investments over time, resulting in a more cost-effective reduction.

The regulations of interrefinery averaging left the logistics of trading up to the refiners. Interrefinery averaging allowed all gasoline refiners and importers, whether owned by the same refiner or not, to average lead usage over a calendar quarter through "constructive allocation." This process allowed refiners to comply with the applicable lead content standard by allocating actual lead usage "in any manner agreed upon by the refiners"—so long as average lead usage over the quarter did not surpass the applicable standard (e.g., 1.1, 0.5, or 0.1 gplg). Refiners or importers engaging in interrefinery averaging were free to carry out constructive allocation through whatever means they saw fit, including trades and negotiations, both monetary and otherwise. Because interrefinery averaging was offered as an alternative to individual refinery compliance, only those refiners who found this alternative beneficial used it.

Under the basic lead content regulations, refiners were required to report quarterly on the quantity of leaded and unleaded gasoline produced and quantities of lead used. Specifically, refineries engaging in interrefinery averaging needed to provide the following information:

- total grams of lead that the reporting refinery allocated (sold) to other refiners, and the names and addresses of such other refiners (A);
- total grams of lead that the reporting refinery was allocated (bought) from other refiners, and the names and addresses of such other refiners (B);
- total grams of lead "constructively used" by reporting refinery (C = actual lead usage – A + B);
- "constructive average" lead content of each gallon of leaded gasoline produced by the reporting refinery during the compliance period (C/total gallons produced); and
- if compliance was demonstrated through averaging with more than one other refiner, supporting documentation showing that all parties agreed to the constructive allocation.

The second market-based component of the lead phasedown was a banking scheme. Introduced in 1985, it was intended to offer a buffer for refineries facing the significant lead content decreases slated for 1986. Banking provided temporal flexibility to refiners in addition to the interrefinery trading flexibility established in 1982. Under the banking mechanism, refiners who used less than 0.5 gplg but more than 0.1 gplg of lead in leaded gas in 1985 were permitted to use this same amount of lead in gasoline between 1985 and 1988, in addition to the lead permits issued and bought during that time period. Production of leaded gasoline with less than 0.1 gplg did not generate additional credits. Thus the banking regulations extended a refinery's timeframe for compliance with the 0.1 gplg standard.

The 1985 regulations also eliminated the interrefinery averaging provisions of the 1982 regulations as of January 1, 1986, although refiners were permitted to buy credits from other refiners' banks until the end of 1987. EPA was concerned that the interrefinery trading provisions encouraged the production of leaded gasoline that still contained trace amounts of lead. The agency believed that engines designed to use leaded gasoline required at least 0.1 gplg to operate properly, and it wanted to eliminate any incentive to generate lead credits by producing leaded gasoline with concentrations below this threshold. Thus, with the end of the banking regulation in 1988, the lead trading program was completed.

Table 8-3. *Physical Measures of Estimated Benefits of Final Lead Phasedown Rule*

Estimated effects	*1985*	*1986*	*1987*	*1988*
Reduced cases of children with blood lead levels above 25 micrograms/dl (1,000s)	64	171	156	149
Reduced emissions of conventional pollutants (1,000 tons)				
HC	0	244	242	242
NO_x	0	75	95	95
CO	0	1,692	1,691	1,698
Reduced blood pressure effects in males aged 40–59				
Hypertension (1,000s)	547	1,796	1,718	1,641
Myocardial infarctions	1,550	5,323	5,126	4,926
Strokes	324	1,109	1,068	1,026
Deaths	1,497	5,134	4,492	4,750

Sources: Nichols (1997, Table 1); U.S. EPA (1985a).

Table 8-4. *Estimated Monetized Costs and Benefits of Final Lead Phasedown Rule (in millions 1983$)*

Estimated effects	*1985*	*1986*	*1987*	*1988*
Monetized benefits				
Lead-related effects in children	223	600	547	502
Blood pressure–related effects (males, 40–59)	1,725	5,897	5,675	5,447
Conventional pollutants	0	222	222	224
Maintenance and fuel economy	137	1,101	1,029	931
Total monetized benefits	2,084	7,821	7,474	7,105
Costs				
Increased refining costs	96	608	558	532
Net benefits				
Including blood pressure	1,988	7,213	6,916	6,573
Excluding blood pressure	264	1,316	1,241	1,125

Source: U.S. EPA (1985a, Table VIII-7c).

Ex ante Analysis

Ex ante estimates of the effects of the lead trading program were derived primarily from an EPA regulatory impact analysis (RIA) performed between 1984 and 1985, which predicted the costs and benefits of bringing the lead standard down to 0.1 gplg by the beginning of 1986.[4] Table 8-3 describes physical measures of the benefits of the phasedown program, and Table 8-4 depicts monetized costs and benefits.

Projected Benefits

The benefits associated with EPA's proposed rule to more quickly tighten the lead standard fall into four categories: children's health effects related to lead, health and environmental effects of nonlead pollutants, vehicle maintenance and fuel economy effects, and blood pressure effects (Nichols 1997), all expressed here in 1983 dollars. The first benefit, children's health effects related to lead, was quantified in monetary terms as the avoided costs of medical treatment and remedial education that would be incurred if existing (1982) standards (1.1 gplg) remained in effect. The avoided medical costs were estimated at $900 per child with blood lead levels above 25 micrograms per deciliter (μg/dl). The estimates for compensatory education averaged about $2,600 per child with blood lead levels above the same threshold. The total benefits in this category ranged from about $600 million in 1986 to $350 million in 1992 (U.S. EPA 1985a).

The second category of benefits, health and environmental effects of nonlead pollutants, were quantified in two ways. EPA first used a direct valuation method to estimate the physiological responses to various doses and to estimate and assign dollar values to health and welfare endpoints. However, these values were deemed highly uncertain, and they did not include values for some potentially important impacts (Nichols 1997). For example, the study considered only the effects of reductions in HC and NO_x, and they omitted CO as a factor. Internal EPA offices had argued that CO effects were too uncertain to include in the analysis. The second quantification method was an implicit valuation of the reductions, in which EPA used the forgone expenses of repairing damaged catalytic converters to indicate a minimum value of preventing the pollution. Catalytic converters were damaged when cars were misfueled with leaded gasoline. The final estimates were based on an average of the two methods, and they totaled $222 million in 1986 (U.S. EPA 1985a).

To calculate vehicle maintenance and fuel economy effects, EPA estimated the maintenance benefits to be about $.0017 per vehicle mile, or an aggregate of about $900 million in 1986. The agency also identified additional fuel economy benefits of about $200 million per year (U.S. EPA 1985a).

Finally, EPA included limited estimates of the proposal's effects on blood pressure. The RIA predicted that the policy would reduce the incidence of hypertension in middle-aged men by about 1.8 million cases in 1986, at a value of $220 per case per year of hypertension avoided (U.S. EPA 1985a). Also, the agency projected that reduced hypertensive blood pressure would mitigate the likelihood of other cardiovascular afflictions. Based on epidemiological studies, the estimates yielded benefits of $60,000 per heart attack avoided and $40,000 per stroke avoided. Added to the benefits of reduced mortality rates, these figures result in total blood pressure–related benefits of more than $5 billion per year from 1986 to 1988.

Projected Costs

The estimated costs of EPA's proposed rule include the cost to refiners of additional processing or use of other additives to replace the fuel octane previously supplied by lead, plus the cost of lost consumer surplus due to higher gasoline prices. The estimates also took into account the costs saved through the banking program. The additional processing costs (primarily from reforming or isomerization) under the 0.5 gplg standard totaled less than $100 million for the second half of 1985 (U.S. EPA 1985a). Under the 0.1 gplg rule, the projected costs fell over time, from $608 million in 1986 to $441 million in 1992, because of projected declines in the demand for leaded gasoline even in the absence of the new standard (Nichols 1997).

The RIA further predicted that refiners would achieve substantial cost savings through the innovative banking program. It estimated that refiners would collectively bank between 7.0 billion and 9.1 billion grams of lead in 1985, which would reduce the present value costs of the 0.1 gplg rule by $173 million to $226 million, or about 16% to 20%, depending on when refiners began banking (U.S. EPA 1985a). In actuality, refiners began banking immediately upon being permitted to do so, in line with the higher cost-saving estimate. The RIA did not estimate the cost savings from allowing trading relative to a command-and-control alternative.

At the time of the RIA, the average retail price of unleaded gasoline was about $.07 per gallon greater than that of leaded gasoline. However, all other measures of the marginal value of lead in gasoline (e.g., wholesale prices, lead permit prices, and lead shadow prices) indicated a significantly more narrow differential, less than $.02 per gallon. EPA believed the $.07 figure was mainly a result of marketing strategies and that the $.02 figure was more representative of real resource costs (Nichols 1997).

As an addendum to the RIA, EPA also estimated the benefits and costs of a complete ban on lead by 1988—that is, moving from the 0.1 gplg limit to zero. A ban on leaded gasoline, EPA reported, would further reduce the number of cases of children with toxic blood lead levels by about 7,000 in 1988, prevent as many as 100,000 more cases of hypertension among middle-aged men, and reduce heart-related fatalities by about 400 (U.S. EPA 1985b). The incremental cost to refiners of a complete ban was predicted to be $149 million, and the incremental benefits were placed between $193 million and $635 million (U.S. EPA 1985b). Although these results provided justification for a ban on lead in gasoline, EPA chose to wait in order to minimize the risk of valve seat recession damage in older engines (Nichols 1997). A ban was enacted in 1996, but by then virtually all lead had already been eliminated from gasoline.

Ex post Analysis

Probably the most useful measure of the phasedown's overall effectiveness is the extent to which the regulations accelerated the reductions in lead consumption that were already being made through fleet turnover. The phasedown program along with the fleet turnover achieved in 1981 what the fleet turnover alone would not have achieved until around 1987. From the start of the phasedown in 1979 until the completion of the marketable permit program in 1988, regulations on the refineries accounted for about 36% of the total gasoline lead reduction, amounting to more than 500,000 tons of lead that would otherwise have been emitted (Holley and Anderson 1989). The use of the banking program further accelerated lead reductions relative to what they would have been in the absence of banking.

Static Efficiency

The static efficiency of the marketable lead permit program can be measured by the cost savings it achieved—that is, the difference in the cost to abate lead under uniform standards versus the cost to abate the same amount of lead under the tradable permit program. EPA, however, collected no comprehensive data on permit prices, so the cost savings can only be estimated. Anecdotal evidence suggests that prebanking permit prices (i.e., under the 1.1 gplg standard) were typically less than $.01 per gram of lead, and that they were $.02 to $.05 per gram after the banking feature began (Hahn and Hester 1989). Based on these figures, Hahn and Hester estimate that the marketable permit program saved hundreds of millions of dollars in abatement costs.

The fact that permits were traded at all suggests that the tradable permit program allowed for lower costs than comparable uniform standards. Assuming that refiners were not systematically shooting themselves in the foot, it follows that they traded permits because doing so saved money. Low-cost refineries were able to abate a portion of their lead and sell the corresponding permits to high-cost refineries, realizing a net gain in revenues in the process. The high-cost refineries that bought the permits did so because the permit price was less than the cost for them to reduce the corresponding amount of lead, allowing them to save money. Indeed, the lead rights market was very active in terms permits traded, and activity increased as the trading program matured. The volume of lead rights traded (as a percentage of all lead produced) increased from about 7% in the third quarter of 1983 to more than 50% in the second quarter of 1987 (Hahn and Hester 1989).

In addition, the mechanics of the marketable permit program were such that transaction costs did little to inhibit permit trading (Kerr and Maré 1997). Refineries incurred these costs from establishing their marginal value of lead, collecting information on permit prices, finding trading partners, collecting information on the validity of the permits to be traded, negotiating permit quantities or prices (or both), and releasing potentially sensitive business information in the process of trading. Selling permits also meant parting with their option value, which would be important in the event of abatement cost shocks (Kerr and Maré 1997). This may imply that transaction costs are likely to be more burdensome for small refiners, which lack the scale and resources that would keep these costs relatively low. Using econometric methods, Kerr and Maré (1997) estimate that about 10% of the value of all trades that would have occurred did not transpire because of transaction costs, all else being equal.

Dynamic Efficiency

The banking program offered additional cost savings to participating refiners. It allowed refiners to lower their overall costs of abatement by "smoothing out" their emissions over time. This was important for many firms because their marginal cost schedules increased rapidly with increasing lead restrictions. This situation is evidenced by the fact that both large and small refiners produced lead in concentrations below the standards early in 1985, the year banking was introduced, implying that they were banking the difference. Both groups then exceeded the tighter standards in 1986 and 1987, when they used the saved permits to ease their transition to even tighter standards. EPA's *ex ante* projection that banking would save more than $226 million probably turned out to be an underestimate—the agency's figures assumed 9.1 billion grams of lead would be banked, whereas 10.6 billion grams were actually banked, starting at the earliest possible date (Hahn and Hester 1989).

Kerr and Newell (2003) address dynamic efficiency in the context of the U.S. lead phasedown through analysis of the adoption of octane-enhancing technology (e.g., isomerization) to replace lead. They investigated the influence of refinery characteristics (e.g., size of refineries or firms, technological sophistication), technology costs, and most importantly, regulatory variables, including regulatory stringency and form (e.g., tradable permits versus individually binding performance standards). They found a large positive response in lead-reducing technology adoption to increased regulatory stringency, indicating that the regulations effectively provided incentives for dynamic changes in technology. In addition, they found a pattern of technology adoption across firms that is consistent with an economic response to market incentives, plant characteristics, and alternative policies.

Economic theory suggests that tradable permit programs create an incentive for more efficient technology adoption—that is, they provide incentives for reducing abatement costs as much as possible, including dynamically over time. Taking the price of permits as given, permit sellers (i.e., refiners with relatively low abatement costs) would want to invest in technology that would lower their marginal abatement cost, which would allow them to capture a greater surplus in selling the permits. However, permit buyers (i.e., refiners with relatively high abatement costs), whose technology adoption would still be insufficient to lower abatement costs below permit prices, would have a disincentive to adopt that technology. Doing so would merely reduce their cost savings. The incentives to adopt technology would thus be lower for buyers under the permit system than under uniform standards, because they could buy permits rather than being forced to self-comply with relatively expensive reductions.

The tradable permit system, therefore, provides incentives for more efficient adoption, but it can lower adoption incentives for some plants with high compliance costs.[5] Under a nontradable performance standard, such opportunities for flexibility do not exist to the same degree. If plants face individually binding standards, they will be forced to take individual action—such as technology adoption—regardless of the cost, with the resultant inefficiency reflected in a divergence across plants in the marginal costs of pollution control.

As suggested by theory, Kerr and Newell (2003) found a significant divergence in the adoption behavior between refineries with low versus high compliance costs under the tradable permit program. Namely, the positive differential in the adoption propensity of expected permit sellers relative to expected permit buyers was significantly greater under market-based lead regulation than under individually binding performance standards. Overall, Kerr and Newell's results corroborate findings that the tradable permit system provided more efficient incentives for technology adoption decisions, although the level of adoption is in fact lower for certain types of firms.

Revelation of Costs

In theory, market-based instruments tend to equate marginal abatement costs across firms, with the market price of permits converging upon marginal abatement costs. Hahn and Hester (1989) estimate from anecdotal evidence the convergence price to be under $.01 per gram prior to banking and from $.02 to $.05 during the banking phase, when standards were becoming increasingly stringent. If the lead program had been designed more in the spirit of the sulfur dioxide (SO_2) trading program for acid rain, with clearly specified lead allowances rather than the lead averaging scheme, an even clearer market price would likely have emerged—as it has in the SO_2 market.

Distributional Effects

Empirical evidence suggests that many very small refineries, those with the highest cost structures, were inevitably eliminated from the market by the phasedown, and the ones that did survive were more likely to become permit buyers than sellers. Hahn and Hester (1989) report that net transfers of lead rights tended to be from large refiners to small ones (large refiners tending to have lower marginal abatement cost structures than small ones). Small refiners had to purchase permits from large ones, effecting a transfer of private revenue from small refiners to large ones. Nevertheless, relative to a uniform performance standard, small refineries were better off under the tradable permit policy.

In the case of lead emissions from automobile exhaust, environmental hotspots and spikes were not a significant concern. Pollution is created through gasoline consumption, not production, and there is little relationship between the location of refineries and automobile exhaust across the country. Even if a local region were predominantly served by small refineries producing gasoline with relatively high lead content, a comparable command-and-control policy would not have granted exemptions to those small refineries, as they had in the past. Thus, command-and-control instruments had no clear advantage over market-based incentives with respect to hotspots and spikes of atmospheric lead from gasoline.

Administrative Burden

EPA took on considerable burden with the lead permit program—probably more than if the agency had employed a tradable permit program akin to that used for SO_2. The administrative burden, however, was not necessarily higher than what one might have expected from command-and-control regulation. The actual design of the lead phasedown rule was fairly simple: the agency had only to establish the desired lead concentration and review refiners' reports regarding their lead usage, gasoline production, and averaging. But the output-based averaging allowed by the marketable permit system created substantial monitoring and enforcement problems for EPA (Holley and Anderson 1989).

The most significant problem was related to the unexpected creation of a quasi-industry of "alcohol blenders," which were mainly large service stations that added alcohol to leaded fuel. In doing so, these blenders lowered the average lead content of the aggregate volume of fuel, thereby generating lead credits that could be sold in the permit market to other refineries. This approach to compliance was made possible by the fact that the lead performance standard was measured as a ratio of output, and there were few restrictions on who could participate in lead trading.[6] EPA's rules effectively considered the blenders to be refiners, and the agency's enforcement and monitoring mechanisms treated them as such. By the beginning of 1985, 300 blenders reported permit trades, and within a year the figure had doubled. EPA had expected to receive reports from only about 250 traditional refineries, and the unexpected inflow of 600 additional lead production and trading reports significantly slowed the agency's monitoring and enforcement processes. To make matters worse, the blenders' reports were relatively disorganized and replete with errors, causing problems with the agency's report-processing computer system. While the reports were being manually processed, invalid permits might have been sold or even resold, and financially unstable market participants might have "disappeared" before their violations were ever detected (Holley and Anderson 1989).

Independent of the blender problem, the lead permit program gave rise to a number of other administrative and enforcement issues. Violations fell into a few common categories:

- self-reporting excess lead usage;
- failing to report regulated activities as required;
- incorrectly reporting compliance in cases where the average lead usage was actually above the standard, because of either use of more lead than reported or production of less leaded gasoline;
- failing to include shipments of imported gasoline in reports;
- using in blend stock materials that already had been reported in the previous quarter as leaded gasoline;
- falsifying banked rights;
- changing accounting systems that resulted in the "disappearance" of lead that should have been accounted for; and
- claiming lead rights based on fictitious production.

Because lead credits were fully fungible and because false credits could be traded several times before their discovery, EPA had difficulty tracing invalid rights to their sources (Holley and Anderson 1989). The agency had expected most of the violations to be committed by a small number of large refiners and planned its enforcement policies accordingly. But in fact, most of the violations were committed by a fairly large number of small refiners with small amounts of lead rights, to which the existing enforcement mechanics were less easily applied.

EPA began to perform audits of suspect refineries simultaneous with the introduction of the banking program in 1985. Until that point, the agency had detected violations through inconsistencies and inaccuracies in refinery reports, resulting in 71 notices of violation with proposed penalties totaling $17.8 million (Holley and Anderson 1989). After the agency started auditing, 1987 alone saw 17 notices issued, with proposed penalties topping $54 million. In some settlement cases, refiners were presented with the option of paying direct financial penalties or retiring a portion of their lead rights. Approximately 150 million grams of lead pollution were forgone in this manner (assuming those permits would have been used), representing an estimated value of about $40 million (1983$; Holley and Anderson 1989).

Holley and Anderson suggest that the relatively high level of enforcement activity through audits reduced the noncompliance rate—as EPA devoted an increasing amount of resources to audits and as the number of audits performed increased, the number of noncompliance cases decreased. But despite EPA's success in detecting violations through audits, the flexible nature of the agency's marketable permit approach itself increased the likelihood of administrative difficulties and violations. Much of difficulty could have been avoided, however, by limiting the universe of market participants to traditional refiners. On the other hand, such restrictions could have limited the potential for unforeseen opportunities for low-cost mitigation.

Monitoring Requirements

Although the marketable permit program may have required monitoring a greater quantity and variety of information than under a command-and-control policy, the collection of this information was fairly straightforward and inexpensive. EPA delegated the responsibilities of data collection and assimilation to the refiners themselves, who then reported their figures to the agency. The agency set up a computer system that processed refinery reports to detect inconsistencies and probable inaccuracies. Participating refiners were required to report their quarterly lead rights transactions, including trade volumes and the names of trading partners; refiners who used the banking option were also required to report deposits and withdrawals. All the information required by the reports was readily available to the refiners, so the added costs of monitoring were relatively low (Holley and Anderson 1989). Figures on lead usage were easily checked against sales figures of additive suppliers. Gasoline volume was less easily monitored, however, and more enforcement cases involved misreporting output than misreporting lead use.

Conclusions

One can draw several conclusions from the U.S. experience with phasing the lead out of gasoline. The program not only effectively met its environmental objectives but did so more quickly than it would have without the banking provisions. The phasedown from 1979 to 1988 accelerated the virtual elimination of lead in gasoline by at least a few years and an additional 500,000 tons, compared with what fleet turnover alone would have achieved.

The marketable lead permit system was highly cost-effective, saving hundreds of millions of dollars compared with command-and-control policies. The banking program itself saved more than $225 million because it allowed for a more cost-effective allocation of technology investment within the refining industry. Estimates also suggest that transaction costs brought about only a modest 10% reduction in the overall efficiency of the market-based program.

The market-based nature of the lead permit program also provided incentives for more efficient adoption of new lead-removing technology, relative to a uniform standard. The pattern of technology adoption under this program was consistent with an economic response to market incentives and plant characteristics. As theory suggests, the adoption behavior of refineries significantly diverged according to their compliance costs. Namely, expected permit sellers (low-cost refineries) demonstrated a higher likelihood of adoption than did expected permit buyers (high-cost refineries) under market-based lead regulation.

Although distributional issues are always valid concerns, the lead permit program was likely more responsive to the high costs faced by small refineries than a similar command-and-control policy would have been. Moreover, environmental hotspots, which can be an issue with some localized pollutants, were not a significant concern in this case.

The flexibility of the lead trading program, however, increased the likelihood of both intentional and unintentional violations, especially on the part of fuel blenders and small refiners. The profile of violators added an unexpected administrative burden to EPA's monitoring and enforcement costs. On the other hand, there were likely to have been efficiency advantages to the participation of blenders in lead compliance, since they apparently offered a cost-effective means to reducing lead content. This serves as a reminder that one of the advantages of flexible, incentive-based programs is that they provide opportunities and incentives for unanticipated means of cost-effective compliance.

Overall, the benefits of the U.S. lead phasedown likely outweighed its costs by 10 to 1, with lead trading and banking significantly lowering those costs. The phasedown was effective in meeting its environmental objectives, and did so more quickly with the allowance of permit banking.

Notes

1. Octane is a characteristic of fuel components that improves the performance of engines by preventing fuel from combusting prematurely in the engine. The availability of high-octane fuel allows more powerful engines to be built. Cars will not operate efficiently with a lower-octane fuel than that for which they were designed. In addition, some older cars need more than a minimum level of lead (less than 0.1 grams of lead per gallon) to prevent a problem called valve seat recession.

2. As described in Nichols (1997), lead emissions from gasoline are linked to elevated blood lead levels, which are associated with significant health effects, especially in the case of young children. In sufficiently high doses lead can cause severe retardation and sometimes even death. Moderate to high blood lead levels are sufficient to damage cognitive performance in children, though the magnitude of cognitive effects due to low-level lead exposure are still disputed. In addition, studies suggest that elevated blood lead levels are associated with

increased blood pressure and hypertension rates in middle-aged adults. Lead in gasoline can also raise maintenance costs by causing salt corrosion in an automobile's engine and exhaust system and by damaging the muffler, spark plugs, and other components.

3. The decision to tighten the lead standard so dramatically came in light of new scientific studies that linked two sorts of health problems *directly* to the ingestion of lead from fuel emissions. The first negative effect associated with lead, identified by the Centers for Disease Control and other health agencies, was mental retardation and in some cases death, especially in the case of young children. The second negative effect linked lead to elevated blood pressure, at least in middle-aged adults. Even without factoring in the blood pressure effects of lead, cost-benefit analysis unambiguously suggested the desirability of a substantial tightening of the standards. The particulars of this analysis are described in subsequent sections of this chapter.

4. After an initial analysis of achieving the standard by 1988, in which the benefits easily outweighed the costs, EPA proposed the even closer deadline of January 1, 1986. The following presents estimates from the RIA for the later proposal.

5. Whether any of these policies provide incentives for fully efficient technology adoption depends on a comparison with the social benefits of technology adoption and the usual weighing of marginal social costs and benefits.

6. See Helfand (1991) for an assessment of the incentives given by alternative design of regulatory standards.

References

Hahn, R.W., and G.L. Hester. 1989. Marketable Permits: Lessons for Theory and Practice. *Ecology Law Quarterly* 16: 380–91.

Helfand, G.E. 1991. Standards versus Standards: The Effects of Different Pollution Restrictions. *American Economic Review* 81(3): 622–34.

Holley, J., and P. Anderson. 1989. Lead Phasedown—Managing Compliance. Draft Internal Report, Appendix A. Field Operations and Support Division, Office of Mobile Sources, U.S. EPA.

Kerr, S., and D. Maré. 1997. Transaction Costs and Tradable Permit Markets: The United States Lead Phasedown. Manuscript. College Park: University of Maryland.

Kerr, S., and R.G. Newell. 2003. Policy-Induced Technology Adoption: Evidence from the U.S. Lead Phasedown. *Journal of Industrial Economics* 51(3): 317–43.

Nichols, A.L. 1997. Lead in Gasoline. In *Economic Analyses at EPA: Assessing Regulatory Impact,* edited by R.D. Morgenstern. Washington DC: Resources for the Future, 49–86.

U.S. Environmental Protection Agency (EPA). 1985a. Costs and Benefits of Reducing Lead in Gasoline: Final Regulatory Impact Analysis. EPA-230-05-85-006. February. Washington, DC: EPA Office of Policy Analysis.

———. 1985b. Supplementary Preliminary Regulatory Impact Analysis of a Ban on Lead in Gasoline. February. Washington, DC: EPA Office of Policy, Planning, and Evaluation.

CHAPTER 9

Leaded Gasoline in Europe

Differences in Timing and Taxes

Henrik Hammar and Åsa Löfgren

THE PHASEOUT OF leaded gasoline in the European Union began in the 1970s, when concerns about the negative health effects of lead started to draw increased public attention to leaded gasoline. Lead has been linked to environmental and health toxicity, including neurological dysfunction, renal damage, and death (Lovei 1996). Children are especially sensitive to lead exposure, and the effects on children include learning difficulties, lowered I.Q., behavioral problems, and hyperactivity (Thomas 1995).[1]

At the same time that lead was gaining notoriety, catalytic converter technology reached Europe. Catalytic converters were introduced to help curb emissions of carbon monoxide (CO), nitrogen oxides (NO_x), hydrocarbons (HCs), and volatile organic carbons (VOCs). Attention given to this group of emissions engendered a vehicle-based policy requiring new cars to be equipped with catalytic converters. Catalytic converters require the use of unleaded gasoline because lead destroys the emissions-reducing function of the converter. The requirement that new cars be equipped with catalytic converters for solving other emissions problems also made it possible to devote attention to the lead problem. The phaseout of lead from gasoline was at first merely a by-product of the more stringent emissions requirements for CO, NO_x, HCs, and VOCs, which caused a positive regulatory externality in the form of lead reductions.

Europe's vehicle-based policy requiring catalytic converters was complemented by a fuel-based policy in the form of a tax differential favoring unleaded gasoline. Compared with the United States, European gasoline taxes were (and are) high. The tax incentive made unleaded gasoline more attractive for the refineries to produce, and it encouraged them to produce greater amounts of unleaded gasoline than might have been produced otherwise. As is shown by Löfgren and Hammar (2000), the tax differential between leaded and unleaded gasoline contributed significantly to the phaseout process in the European

Union (the tax incentive was not, however, directed toward lead content, which was regulated at the national level and at the E.U. level). Moreover, the lower tax on unleaded gasoline—which translated to a lower price at the pump—also worked as an incentive to prevent misfueling of leaded gasoline in cars with catalytic converters. The lower price thereby secured the performance of the emissions-reducing technology of the converter.

Although the phaseout of leaded gasoline in the European Union got off to a late start compared with the United States (with the exception of Germany, the E.U. phaseout did not start until the mid-1980s), the reduction in lead emissions has been achieved despite a high and steady increase in transportation volumes. In fact, in combination with the specific policies addressing lead, the high purchase rates of new cars equipped with catalytic converters (driven by growth in the gross domestic product) has greatly contributed to the reduction of lead emissions from automobiles. The significance of domestic growth and share of catalytic converters in explaining the speed of the phaseout of leaded gasoline is supported by Löfgren and Hammar (2000).

In this chapter, we provide an overview of E.U. policy instruments used to decrease lead emissions in the transportation sector, and we further discuss the effect of the instruments used. Although the member states of the European Union are heterogeneous, we discuss the critical features of various countries' policies, and we also present detailed case studies for Germany and Sweden.

E.U. Policies for Lead Phaseout

The phaseout process in the European Union has differed among countries, but all have incorporated the same basic features: (1) the regulation of lead content, (2) the use of a tax differential, (3) the requirement of new cars to be quipped with catalytic converters, and (4) for countries further along in the phaseout process, a ban on leaded gasoline when the phaseout process neared completion. The policy instruments adopted, therefore, have been a combination of command-and-control regulation and economic incentive.

As a response to the adverse health effects of lead (and possibly following developments in the United States), the European Union fixed the maximum limit of lead content in gasoline at 0.4 grams per liter (g Pb/l) in 1981. But at the same time, no individual member state was allowed to reduce its gasoline lead content below 0.15 g Pb/l because of E.U. laws pertaining to trade restrictions (Council Directive 78/611/EEC).[2] The law also emphasized that no other pollutant could be increased because of the decrease in lead content, and that the quality of gasoline could not decline with the removal of lead.

In 1985, the European Union mandated that by 1989 unleaded gasoline containing a maximum level of 0.013 g Pb/l should be available for sale in all member states (Council Directive 85/210/EEC), and unleaded gasoline was thereafter defined as gasoline containing no more than 0.013 g Pb/l. Member states were also encouraged to adopt a maximum limit of 0.15 g Pb/l for leaded gasoline, but this limit could be adopted when the individual country found it appropriate.

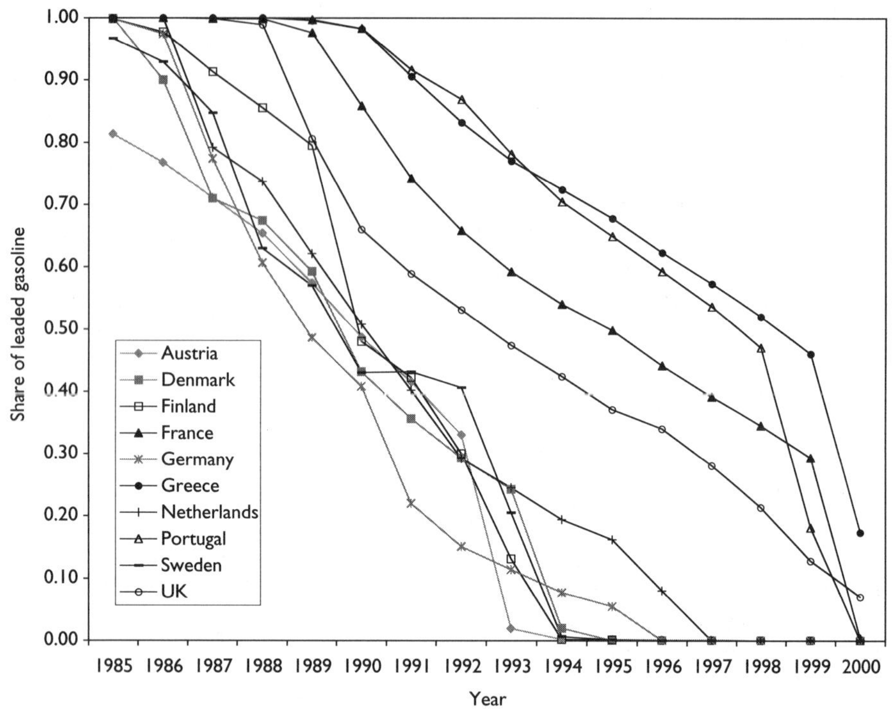

Figure 9-1. *Leaded Gasoline Consumption, by Country*

Later E.U. policy (Council Directive 87/416/EEC) further emphasized that unleaded gasoline should be available in all member states. Moreover, the European Union had also lifted the trade restrictions, making it possible for member states to ban leaded gasoline entirely. In 1998, all Western European countries and nearly all Eastern European countries had signed the Aarhus Treaty, which stated that only unleaded gasoline should be used in Europe by 2005.

The introduction of unleaded gasoline differed. Austria, Denmark, Germany, the Netherlands, and Sweden promoted the introduction of unleaded gasoline, especially by introducing cars with catalytic converters. France, on the other hand, protected its small car-export industry (which did not want to install catalytic converters because of the additional production costs) by resisting catalytic converter technology and the introduction of unleaded gasoline. The share of leaded gasoline consumption over time for each country is shown in Figure 9-1. The phaseout was not completed by 2000 for Greece and the United Kingdom.

Member states have shown comparatively greater uniformity with regard to their tax policies. Each country's tax policies have favored unleaded gasoline since 1986. In general, countries have provided tax exemptions for unleaded gasoline based on its higher production cost, compared with leaded gasoline. Hence, the tax has been based on the cost of production, rather than on the environmental and health costs to society. The instruments used in Europe have been estimated to correspond to an incremental cost of lead reduction of US$200–500 per kilogram of lead (Sterner 2003). This is about 10 times the

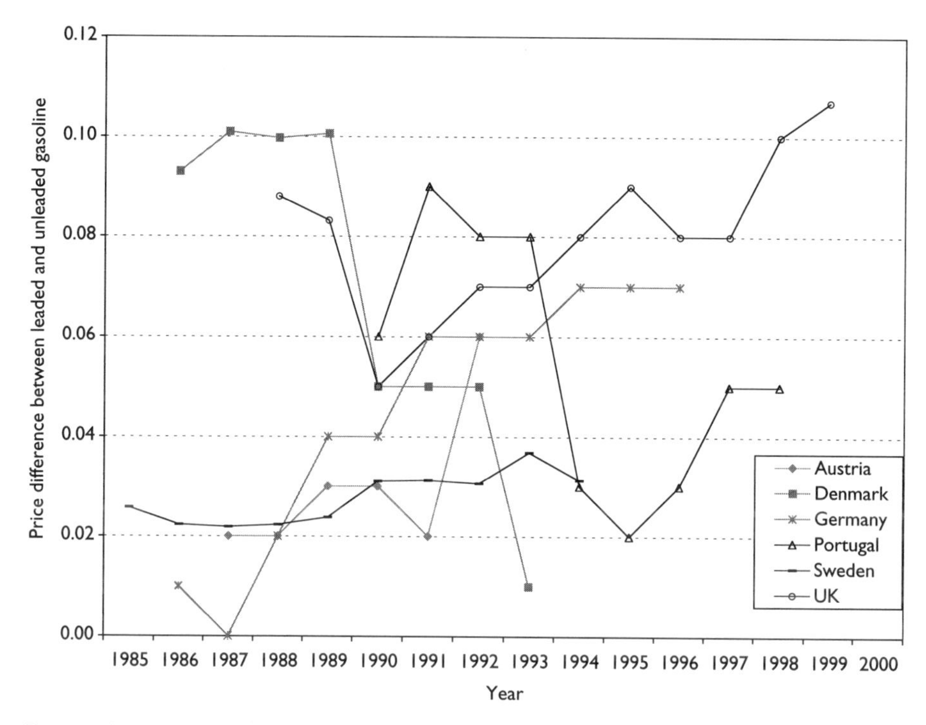

Figure 9-2. *Reported Price Difference between Leaded and Unleaded Gasoline (1990$ PPP adjusted)*

marginal cost estimated from data gathered on the tradable permit program in the United States. Although the evidence is not strong, it indicates that oil refineries overstated the production cost of unleaded gasoline when reporting their costs to regulators.

The price of leaded gasoline was always higher than unleaded during the phaseout period (Figure 9-2).[3] Aside from this general trend, however, prices have differed between countries and over time. For some countries—Finland, Germany, the Netherlands, and Sweden—the price differential between leaded and unleaded gasoline increased as the phaseout approached completion. Denmark, on the other hand, illustrates the reverse pattern. There is no clear trend for the other countries.

Regulations requiring cars to be equipped with catalytic converters were introduced in the European countries during the mid- to late 1980s (and even during the 1990s). Because using leaded gasoline in cars with catalytic converters destroys the catalytic converter, the introduction of catalytic converters increased consumption of unleaded gasoline. Figure 9-3 shows an increasing trend for catalytic converters for all countries over time. The introduction of catalytic converters was a major factor in the increased production of unleaded gasoline (although high-octane unleaded gasoline had been available earlier as a "green" product). For the countries that had completed the phaseout by 1996, the share of catalytic converters ranged between 30% and 60% by the end of the phaseout

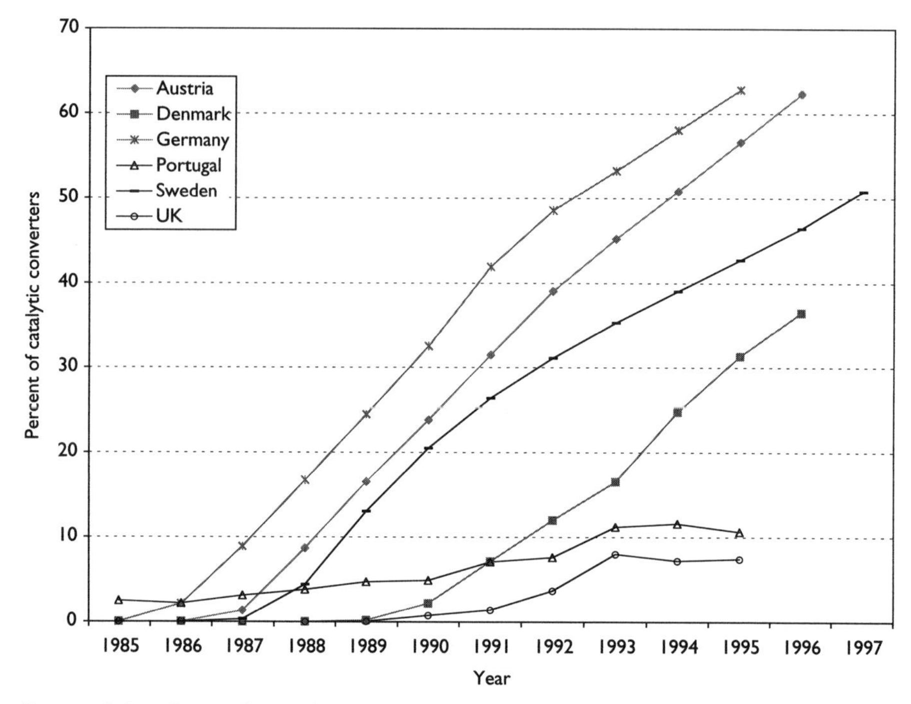

Figure 9-3. *Share of Catalytic Converters*

period. In 1995, at least 80% of all gasoline sold in Austria, Denmark, Germany, the Netherlands, and Sweden was unleaded, but only 30% of all gasoline sold in Italy, Greece, and Portugal was unleaded.

The policies used to combat lead originating from gasoline consumption in the European Union have had two components: to secure the supply of unleaded gasoline, and to speed up the phaseout process. Central to both is an understanding of the degree of substitutability between leaded and unleaded gasoline for cars not equipped with catalytic converters. In the mid-1980s, the degree of substitutability was largely unknown, and this lack of knowledge should have affected the measures taken.

Case Study: Sweden

All gasoline sold in Sweden before 1986 had been subject to an equal level of taxation.[4] In January 1986, the Swedish government differentiated the gasoline tax between leaded and unleaded gasoline to favor the production of unleaded gasoline. The tax change followed on the heels of the government's decision to impose stricter rules for emissions from new cars. Cars produced in 1989 and thereafter faced these stricter emissions rules, which could be achieved only if the cars were equipped with catalytic converters. A lower sales tax on cars equipped with catalytic converters was imposed to stimulate sales of such cars. The tax reduction compensated car manufacturers for the increased production costs associated with catalytic converters.

Before 1986, the supply of unleaded gasoline was limited, and to phase out leaded gasoline, a lower tax on unleaded gasoline was introduced. The tax compensated oil refineries for the higher cost of producing unleaded gasoline. This cost was estimated to be approximately $.01 per liter of gasoline, and the tax difference was set even higher, because the government wanted to see a rapid phaseout of leaded gasoline. Furthermore, the sale of unleaded gasoline was very low in 1986, and the reform was assumed to be tax neutral in the first year.

The initial phaseout of leaded gasoline went more quickly than expected, and in 1987 the government decided to increase the tax on leaded gasoline.

The problem of valve seat recession hampered the phaseout, however: people who owned older cars were reluctant to switch to unleaded gasoline. This issue was resolved in 1994, when one oil company introduced a new unleaded gasoline in which lead was substituted with sodium. Sodium's lubricating properties negated the problem of valve seat recession. After that, the phaseout progressed quickly.

Case Study: Germany

Germany was the first country in Europe to impose restrictions on leaded gasoline.[5] In 1972, Germany set a limit of 0.4 g Pb/l (the earlier limit had been 0.6 g Pb/l) for both production and importation of gasoline. This restriction was tightened in 1976 to 0.15 g Pb/l. By comparison, the E.U. limit of 0.4 g Pb/l came into force in 1981 (with a lower-bound limit of 0.15 g Pb/l). Germany introduced unleaded gasoline (containing no more than 0.013 g Pb/l) in 1984. At that time, a ban on leaded gasoline by an E.U. member state was not possible because of trade restrictions set by the European Union. Instead of using a ban, Germany introduced a tax differential between leaded and unleaded gasoline, a pattern later followed by most of the other European countries. German refineries were concerned about increased costs following the 1972 regulation of lead content. As it turned out, however, costs actually decreased after the regulation because of savings from the reduction of lead additives. Costs did not rise until the tightening of restrictions in 1976 because new additives were needed to increase octane. Between 1972 and 1976, lead emissions decreased significantly in Germany, but lead emissions decreased more slowly than expected after unleaded gasoline was introduced the mid-1980s. The explanation is that at first, unleaded gasoline was more expensive than leaded gasoline (although by 1986 the prices were essentially equal, and from 1986 onward the tax difference increased the price of leaded gasoline).

Many gas station owners did not support the introduction of unleaded gasoline because they expected to incur large investment costs to accommodate the new fuel. This fear was not completely unfounded, especially for the independent station owners, who had to build new gasoline pumps and containers. Over time, independent stations lost market share despite targeted investment subsidies.

The German automobile industry was not directly affected by the lead regulations, but the introduction of catalytic converters in the 1970s imposed additional costs. Car manufacturers, such as Daimler-Benz and Volkswagen, which had experience with catalytic converters from the United States, were favored (in competitive terms) by the increased use of catalytic converters.

Technical Feasibility of the E.U. Lead Phaseout

When unleaded gasoline was first introduced to European markets, its use was limited for a number of reasons: unleaded gasoline was more expensive than leaded, the quality of unleaded gasoline was lower, fears had increased regarding other pollutants associated with unleaded gasoline, gasoline station owners faced high investment costs (Hagner 2000), and cars were still being manufactured with soft valve seats, which many consumers thought required the use of lead or other lubricants for proper engine maintenance. Failure to lubricate would cause valve seat recession (VSR). [6,7]

Cars can be divided in three categories regarding their gasoline use flexibility and their sensitivity to VSR: cars with hard valve seats that have catalytic converters, cars with hard valve seats that do not have catalytic converters, and cars with soft valve seats. Cars with hard valve seats and catalytic converters are not sensitive to VSR, but they must use unleaded gasoline because leaded gasoline destroys the catalytic converter. This raises the concern that misfueling can occur. Cars with hard valve seats and no catalytic converter are not sensitive to VSR and can use either leaded or unleaded gasoline. Because unleaded and leaded gasoline are more or less perfect substitutes for these cars, the use of an appropriate tax differential can be sufficient to guide these car owners to choose unleaded gasoline. The problem arises in cars with soft valve seats—basically old cars—which are sensitive to VSR. These cars still constitute an important part of the total car fleet and constitute a significant portion of the gasoline users on the market, although the portion decreases as these cars are retired.

The crucial question is whether cars with soft valve seats can use unleaded gasoline. This matter has been under heavy investigation, but conclusions differ, leaving much of the matter open to interpretation. All studies show that under extreme driving conditions, cars with soft valve seats run the risk of VSR. On the other hand, several studies show that cars with soft valve seats can use unleaded gasoline. These studies suggest that for cars with soft valve seats, the maintenance cost of using only unleaded gasoline compared with leaded gasoline does not differ significantly. These studies, however, have been mostly ignored in Europe, and drivers of cars with soft valve seats have been advised to use leaded gasoline every third to sixth fill (if not more often). Hence, Europe has been one of the biggest markets for lead additives (Thomas 1995).

One aspect that researchers, oil companies, and vehicle salespeople agree on is that only 0.05g Pb/l is sufficient to prevent VSR (McArragher et al. 1994; the limit of 0.05g Pb/l for lead was tested under severe driving conditions). Therefore *ex post* analysis suggests that the E.U. directive that set the leaded gasoline maximum limit at 0.15g Pb/l leaded gasoline could have been adjusted downward.

Adding nonlead compounds such as methanol, ethanol, and methyl tertiary-butyl ether (MTBE) can increase octane. Gasoline additives such as sodium, potassium, and phosphorus can also be used in gasoline as substitutes for lead. Nonetheless, the use of nonlead additives and the production of unleaded gasoline increases production costs. Refinery production of unleaded gasoline requires more crude oil and more energy to run the refinery components than would be needed to produce leaded gasoline of the same octane rating.[8] But although more energy is needed to produce unleaded gasoline, the unleaded gasoline will provide more energy because the lead additive itself has no energy value.

Ex ante Analysis

The lead phaseout policies adopted in all E.U. countries was a combination of regulation and tax. Because the projected benefit of reducing lead in the atmosphere was large, the regulation was a way of controlling actual emissions. The tax differential gave an incentive for the refinery industry to develop substitutes for lead, resulting in a choice not only between leaded and unleaded gasoline but also between leaded and unleaded gasoline with a lead substitute. In fact, in Sweden the lead substitute replaced lead entirely in 1994 (even though most consumers at that point still believed they were using leaded gasoline).

The combination of tax and regulation instead of a tradable permit program was most likely chosen because of the thin and heterogeneous market in Europe. Most European countries have only a limited number of refineries, making a national tradable permit market less workable. Also, the administrative burden of a tradable permit system in the European Union would be large because different currencies, different institutional arrangements, and more complicated monitoring requirements would create high transaction costs. Furthermore, the purpose of the tax differential between leaded and unleaded gasoline served not only to compensate oil refineries for increased production costs but also to prevent misfueling and to alleviate regulatory burdens on the refinery industry.

The administrative burden of a tax differential used together with the regulation was assumed to be fairly low, but to our knowledge, no *ex ante* cost-benefit analysis has been conducted.

Because catalytic converters require unleaded gasoline, gasoline stations must supply unleaded gasoline. The 1989 E.U. regulation requiring unleaded gasoline to be supplied in all member countries helped ensure the supply. As already mentioned, the tax differential was partly aimed at preventing misfueling, and since individuals did not want their catalytic converters to break down, little monitoring was necessary. In addition, the inlet in cars requiring unleaded gasoline was made too small to fit the nozzle for pumping leaded gasoline, and labels on gas pumps were color coded to differentiate gasoline types. Notably, during initial attempts to phase out lead (in Germany, for example), unleaded gasoline was more expensive, resulting in misunderstanding and misfueling. That experience led to the European Union's subsequent use of a higher tax on leaded gasoline.

The information that was needed to set the tax "right" was not considered a significant problem, because the tax was set at a level to compensate the refinery industry for additional production costs. As E.U. regulations have gradually imposed more stringent levels on lead content, the size of the tax differential has been determined via discussions with refiners about the additional cost of producing unleaded gasoline.

Cars are mobile, and drivers may need to fill gas in different countries. Heterogeneity in regulations between countries caused a coordination problem with the supply of unleaded gasoline. Sweden addressed this problem in the beginning of the phaseout period when unleaded gasoline was not available all over Europe by providing maps that identified gas stations selling unleaded gasoline in other countries.

Ex post Analysis

The European Union's decision to phase out lead from gasoline through regulation can be explained primarily by the large negative health effects of lead together with the potential risk of engine breakdown (which we now know was greatly overstated) and uncertain regulatory burden. Because new cars in the European Union have been equipped with catalytic converters since the end of the 1980s, the goal of a lead phaseout from gasoline will eventually be achieved through the natural turnover rate (which is dependent on economic growth). Whether the use of a tax differential to speed up the phaseout is efficient depends on the estimated costs and benefits of this policy. In the case of the European Union, the tax was not an efficient tax, because marginal damages were not part of the calculations. It is difficult to conclude that any tax could be efficient under such circumstances. The tax and the regulation together have resulted in a fairly rapid phaseout for some countries, but for other countries it has progressed more slowly. The development of new additives seems to indicate that the economic incentive to develop innovative technology has worked to some extent. The actual policy goal from the beginning was not a complete phaseout of leaded gasoline (at least not before all old cars had been scrapped), but over time policymakers adjusted to new information (such as information about lead additives and substitutability between leaded and unleaded gasoline), and the policies used have been dynamically effective, even if not dynamically efficient. Löfgren and Hammar (2000) find that the price difference between leaded and unleaded gasoline significantly affected the phaseout of leaded gasoline but at a magnitude much smaller than the effect of catalytic converters (and hence the turnover in the car stock). Although the tax differential has prevented misfueling and covered higher production costs, it has not provided sufficient incentive for owners of cars without catalytic converters to use unleaded gasoline. Apparently, drivers' concerns about VSR proved stronger than the incentive. Furthermore, Hagner (2000) refers to a study by German Shell asserting that only 15% of the potential consumption of unleaded gasoline was used in 1986. This indicates that the incentive provided by the tax differential might have been effective had it been coupled with reliable information on substitutability.

Another reason for the tax differential was hastening the introduction of unleaded gasoline. *Ex post,* we know that problems with VSR were overstated and that substitutes for lead in leaded gasoline could have been developed. Hence, in retrospect we see that policymakers did not account sufficiently for health effects and overstated the costs of regulation (regulatory burden for car owners and car and refinery industry).

Given the evidence for the health effects of lead, why didn't more European Union countries ban leaded gasoline? Many Latin American countries, in contrast, banned leaded gasoline successfully in a short time. The answer may lie in the complexity of the European market system. Most Latin American countries had only one state-run oil refinery and no oil production.

And why did European countries use a tax differential instead of another policy instrument? The challenge of a tax differential is that the regulator has to guess the marginal production costs of unleaded gasoline, or rely on refiners' information. This sets up a situation of uncertainty and asymmetric information between regulator and refiner. In a tradable permit system, trading within the refinery industry reveals these costs, but monitoring costs are significantly higher, as shown by the U.S. tradable permit program (Newell and Rogers 2003).

Another important aspect of the effectiveness of E.U. policies is the regulation of lead content. If the tax differential between leaded and unleaded gasoline had been combined with a tax on lead, the phaseout of lead would have been faster. The regulation of lead content at the E.U. level was not as stringent as it could have been. Even the lower lead limits in some countries (e.g., Sweden and Germany) probably could have been lower still if coupled with a tax on lead. Furthermore, agreement was growing that lead content could be much lower (0.05g Pb/l gasoline) and still not cause problems with VSR.

Regulatory Burden

The lower tax on unleaded gasoline in Europe has compensated refineries for the additional production cost of unleaded gasoline. The use of a tax differential reduced the tax on unleaded gasoline, corresponding with a lower or equal regulatory burden for refinery industries. This made the tax politically easier to implement. The tax compensation might not have been sufficient for all refineries to cover the extra production costs of unleaded gasoline, and therefore inefficient refineries may have been outcompeted. This is not necessarily a bad thing from an efficiency perspective, but it might have resulted in a more concentrated refinery market. This can be compared with the effects in the United States, where small refineries were more affected by the policy. Distributors also faced a regulatory burden by being compelled to outfit their facilities with separate tanks and pumps for different gasoline types.

Monitoring and Information Requirements

Monitoring the phaseout was not a problem because the tax differential discouraged misfueling and provided an incentive for the refinery industry to supply unleaded gasoline.

Ex post analysis suggests that an outright ban on leaded gasoline would have required less information dissemination in the market than was required by having both unleaded and leaded gasoline on the market. With both types of gasoline available, drivers needed better information about the different qualities of the fuels. Moreover, environmental quality objectives—that is, the goal of zero emissions of lead from the transportation sector—would have been reached much faster had information on substitutability been convincing.

Administrative Burden

The administrative burden of the lead phaseout, to our knowledge, has not been analyzed. No actual figures have been calculated *ex post.* Hence, the administrative cost of using a tax differential in combination with a regulation is unknown. However, the costs can be considered to have been low.

Adaptivity and Tax Interaction Effects

A tax differential between leaded and unleaded gasoline appears to be adaptable, as demonstrated by the many times the differential changed in most European countries. By comparison, the regulation of lead content ensued at a much slower rate.

In Sweden the tax differential policy was intended to be revenue neutral; there was no attempt to make gasoline more expensive relative to other goods, such as labor. Therefore, tax interaction effects following the use of a tax differential do not exist or are minor. However, because the phaseout progressed more quickly than expected, there was a loss of tax revenue that makes it possible to argue that there were tax interaction effects. These effects, however, involved very small amounts of revenue because adjustments in the tax rates compensated for the unexpected revenue losses.

Equity

Only new cars were required to be equipped with catalytic converters, and typically poorer people own old cars. One could therefore argue that the phaseout hurt the poor because the tax differential policy raised the price of leaded fuel. However, taking into account that poorer people typically live in densely populated areas and are therefore more exposed to emissions, their health benefits from the phaseout would also be higher.

The poorest countries in the European Union have not yet phased out leaded gasoline. Personal incomes are highly correlated with the turnover rate of the car fleet, and these countries' lower per capita incomes probably explain their slow phaseout trajectories. Given the delay by the poorer countries in phasing out lead, a stricter regulation or general ban in the European Union would have been more equitable.

Uncertainty

Uncertainty influenced the choice of policy instruments to a high degree. Policymakers possibly gave too much weight to refiners' arguments about difficulties and high costs associated with the phaseout and not enough to health advocates' arguments about the benefits of a total ban on lead emissions.

The risk-averse strategy of using a tax differential rather than a total ban was justified *ex ante* because policymakers did not have a sufficient understanding of the production of unleaded gasoline or the difference in maintenance costs between leaded and unleaded gasoline in cars with soft valve seats. *Ex post,* however, we know that unleaded gasoline can substitute for leaded gasoline at low cost. We also better understand the health effects of lead exposure. With this uncertainty now resolved, there is reason *ex post* for supporting a total ban on leaded gasoline, or at least for phasing lead out sooner than has been the case for most E.U. countries.

Conclusions

The European policy for phasing out leaded gasoline did not assume that people would make enlightened, "green" choices.[9] Rather, it assumed that consumers would make rational financial choices about the costs and benefits of driving.[10] Conversely, the U.S. tradable permit approach targeted the refiners. In the tradable permit scheme, the cost was borne by refiners, and possibly passed on to consumers.

In the United States, the automobile industry was required to develop catalytic converters by environmental standards imposed during the early 1970s. This resulted in the availability of new cars equipped with catalytic converters in 1974. Many Europeans, however, were opposed to legislation requiring installation of catalytic converters in new cars; such regulation was viewed as giving an advantage to U.S. car manufacturers exporting to Europe. The estimate of the cost of installing catalytic converters varied significantly between different automobile producers (ranging from $250 to $750). Because of the cost of installing catalytic converters, France, Italy, and the United Kingdom classified diesel engines as low-emissions cars to protect their exports (Hagner 2000). The resistance to laws concerning catalytic converters in Europe is probably one of the reasons for the delayed phaseout process in Europe, compared with the U.S. experience.

Germany was the first country to impose a regulation on the maximum allowable lead content in gasoline; it had limits in place by the early 1970s. The phaseout process in the majority of the European countries, however, did not start until the mid-1980s. Despite earlier resistance, during the late 1980s several countries passed laws that all new cars had to be equipped with catalytic converters.

Because of the administrative challenges associated with a tradable permit program, Europeans chose tax and command-and-control policies. The tax was primarily used to compensate refiners for higher production costs, but it also encouraged consumers to choose unleaded gasoline, both to prevent misfueling (in cars with catalytic converters) and for financial reasons (in cars without catalytic converters).

Gasoline station owners opposed the introduction of unleaded gasoline because they faced additional investment costs. The number of gasoline stations has decreased over time, indicating that station owners' concerns were not unfounded. The phaseout of leaded gasoline therefore favored larger owners, making an already concentrated market more concentrated.

European Union countries may have benefited from their delay in initiating the phaseout process. The lag gave them time to take advantage of new developments in nonlead additives and to learn that unleaded gasoline can serve as an almost perfect substitute for leaded gasoline in cars with soft valve seats.

Owners of cars without catalytic converters, who could choose between leaded and unleaded gasoline, were subject to imperfect information. Policymakers must take imperfect information into account to reach efficient and effective policies. The crucial question was how could behavior be changed most efficiently. We argue that car sellers and mechanics had incentives not to use unleaded gasoline[11] because damage to the valve seats was attributed to the use of unleaded gasoline, while increased maintenance costs for engines using leaded gasoline were more diffuse and not as easily assigned to fuel type.

Nevertheless, the policies chosen in the European Union—the combination of a tax differential and regulation—have been effective. Implementation of a tax differential alone (without the support of regulation) would have slowed the phaseout process significantly. The introduction and increased use of catalytic converters have been the driving force for the phaseout. The substitutability of unleaded for leaded gasoline points to the opportunity to consider a total ban on leaded gasoline, especially in poor countries with low fleet turnover rates.

Although a tradable permit program was not chosen for Europe during the phaseout period, it is worth mentioning that the analysis does change over time. Compared with the beginning of the 1980s, the current integration of markets and the implementation of the Euro make it more worthwhile to consider tradable permit schemes (e.g., for greenhouse gases) than in the past.

Notes

The authors would like to thank Winston Harrington, Richard Newell, Thomas Sterner, Jonathan Wiener, and participants at the RFF Workshop, "International Experience with Competing Regulatory Approaches: Six Paired Cases," for their constructive comments.

1. See Thomas (1995) for an overview of studies concerning the health effects associated with leaded gasoline.

2. 1 U.S. gallon = 3.785 liters.

3. This pattern was not true earlier, when unleaded was more expensive than leaded gasoline.

4. The information for the Swedish case study is based on SOU (1997, 11).

5. The information for the German case study is based on Hagner (2000) and Storch et al. (2002).

6. This section is based upon Thomas (1995) and Lovei (1996) unless otherwise stated.

7. Valve seat recession occurs when valve seats wear and the valve sink recedes into the cylinder head. Lead oxide formed by the combustion of lead alkyls prevents valve seat recession by forming thin layers of lead oxides on the valve and seat faces. These layers prevent metal-to-metal contact and so eliminate wear of the seat.

8. Each unit of octane number provides the potential of a 1% increase in fuel efficiency of the car.

9. The discussion is based mainly on Löfgren and Hammar (2000) and Hagner (2000).

10. In Sweden, for example, even years after leaded fuel was banned, many people still think in terms of leaded and unleaded gasoline instead of unleaded gasoline with and without lubricating additives.

11. Because old cars have no catalytic converters and initially ran on leaded gasoline, this might affect future fuel choices as well. Hence the advice from car sellers and/or car manufacturers is potentially important. Of course, these agents might be subject to imperfect information.

References

Hagner, C. 2000. European Regulations to Reduce Lead Emissions from Automobiles—Did They Have an Economic Impact on the German Gasoline and Automobile Markets? *Regional Environmental Change* 1: 135–51.

Löfgren, Å., and H. Hammar. 2000. The Phase Out of Leaded Gasoline in the EU Countries—A Successful Failure? Transportation Research Part D. *Transport and Environment* 5D(6): 419–31.

Lovei, M. 1996. Phasing Out Lead from Gasoline: World-Wide Experience and Policy Implications. Environment Department Papers, Pollution Management Series, Paper No. 040. Washington, DC: World Bank.

McArragher, S., L. Clarke, and R. Paesler. 1994. Protecting Engines with Unleaded Fuels. Shell Selected Paper. London: Shell International Petroleum Co.

Newell, R., and K. Rogers. 2003. The U.S. Experience with the Phasedown of Lead in Gasoline. Working Paper. Washington, DC: Resources for the Future.

Octel Ltd. 1996. Worldwide Gasoline and Diesel Fuel Survey. London: Associated Octel Company Ltd.

SOU. 1997. Skatter, miljö och sysselsättning. (Taxation, Environment and Employment). Slutbetänkande från Skatteväxlingskommittén, Fritzes, Stockholm (a report from the Swedish Green Tax Commission).

Sterner, T. 2003. Policy Instruments for Environmental and Natural Resource Management, Resources for the Future, Washington D.C.

Storch, H., M. Costa-Cabral, C. Hagner, F. Feser, J. Pacyna, and E. Pacyna. 2002. Four Decades of Gasoline Policies in Europe: A Retrospective Assessment. Mimeo. Institute for Coastal Research, GKSS Research Centre. Geesthacht, Germany.

Thomas, V.M. 1995. The Elimination of Lead in Gasoline. *Annual Review of Energy and Environment* 20: 301–24.

CHAPTER 10

Trichloroethylene in Europe

Ban versus Tax

Thomas Sterner

INDUSTRIAL SOLVENTS ARE economically important chemicals, and chlorinated (or more generally, halogenated) hydrocarbons have properties that make them useful as solvents. Some, however, can cause some severe environmental problems and health hazards. Many of these compounds are hazardous or toxic either directly or indirectly (after transformation). Among the problems cited are the formation of dioxins, which are extremely toxic and can form when waste containing chlorinated hydrocarbons is incinerated. Another example is the effect of chlorofluorocarbons (CFCs) on the ozone layer. Still other examples are persistent and bioaccumulating chemicals, such as DDT and PCB. Policymaking in this area is difficult because genuine environmental, chemical, and medical uncertainties are considerable. This paper compares policy responses in various European countries: trichloroethylene (TCE) and several similar solvents were banned in Sweden, heavily taxed in Norway and Denmark, and subjected to strict regulation in Germany.

Table 10-1 lists the main uses and environmental and health effects of some of the most commonly used solvents. One of the most important uses for chlorinated solvents is metal degreasing in the metal industry. As soon as a metal object is cut or processed, the fresh surface can become corroded. To prevent corrosion between process steps, the object is covered in oil or grease. Before the next step in production, the grease generally has to be removed. Similarly, for lacquering, assembling, and delivery, goods must be clean and dry. Surface fats, oils, wax, or soil must be removed. Chlorinated solvents have been popular in metal degreasing because they can be applied on different kinds of materials and effectively remove fats and oils. The high volatility of chlorinated solvents ensures that the goods dry fast after degreasing. In the workplace these solvents are superior to certain other volatile solvents, like benzene, which are not only toxic but highly flammable and explosive as well.

Table 10-1. *Some Chlorinated Solvents*

Solvent	*Main use*	*Environmental and health effects*
111-trichloroethane ($C_2H_3Cl_3$)	Metal cleaning and degreasing	Ozone depleting
CFC-113 ($C_2Cl_3F_3$)	Degreasing, dry-cleaning	Ozone depleting
Carbon tetrachloride (CCl_4)	Laboratory	Ozone depleting, carcinogenic
Trichloroethylene (C_2HCl_3)	Metal degreasing	Toxic, likely carcinogenic
Perchloroethylene (C_2Cl_4)	Dry-cleaning	Toxic, likely carcinogenic
Methylene chloride (CH_3Cl)	Laboratory	Toxic, likely carcinogenic

Health and environmental issues related to these chemicals are complex, and priorities change with knowledge. Before damage to the ozone layer were discovered, a number of ozone-depleting substances (ODS), such as the CFCs, were introduced as substitutes for other chlorinated solvents because they were less hazardous to human health. TCE, in fact, was once used as an anesthetic. It should be noted, however, that all these solvents are hazardous to human health because they pass easily through skin and membranes and dissolve fats, including those surrounding nerves and other vital organs. Some of these chemicals can be extremely hazardous, depending on how they are used. In the early days of degreasing, some workers died while welding metals that were still contaminated by TCE, which upon combustion can form phosgene ($COCl_2$), a lethal gas used in some chemical weapons. Following the debate on the ozone layer, ODS solvents were phased out, and some industries tried alternatives to chlorinated solvents in metal degreasing, such as water-alkaline processes and low-aromatic mineral oils. There was a lot of experimentation with, for instance, lemon peels, which are clearly "natural" but contain very strong and quite toxic aromatic compounds. Other users, however, reverted to TCE and other similar chemicals, such as methylene chloride. It was the risk of this type of substitution that was one of the driving forces behind the Swedish ban. Decisionmakers wanted to avoid creating new workplace health hazards as a by-product of addressing environmental problems, such as the ozone layer.

Chlorinated solvents are regulated by several international treaties. The Montreal Protocol of 1987 led to the phaseout of the worst ozone-depleting substances (a number of so-called HCFCs, which have some but less ozone-depletion effect, are still permitted). Following the Montreal Protocol, the Swedish parliament adopted a plan to abolish the use of ODS in Sweden. The main policy instruments used were import restrictions and total bans on ODS use. The ozone-depleting chlorinated solvents were prohibited in the early 1990s. Between 1988 and 1994 the use of ODS decreased by 93% (Östman et al. 1995). The Swedish strategy to regulate ODS was thus quite successful, but because of fear that it would lead to greater use of other hazardous solvents, TCE and several similar (but not ozone-depleting) chemicals were also prohibited.

Perchloroethylene (PER), chemically close to TCE, is used mainly in dry-cleaning but also to some extent in the printing and metals industries. The only regulation of PER in Sweden[1] is a maximum exposure limit in the working environment of 10 parts per million (ppm). Although there is no ban on the professional use of PER, its use decreased from 1,600 tons in 1988 to 250 tons in 1995. The decrease is due mainly to a modernization of the machinery used in dry-cleaning, from open to closed systems (Naturvårdsverket 1997). Despite that decrease and the stringent ambient standard, a number of old PER machines are still around. Some of these are open systems, and in a recent report by the Swedish environmental protection agency (Naturvårdsverket 1997), many were described as poorly maintained and having insufficient reporting routines, high emissions, and other problems. Only 20% of the open machines and 55% of the closed ones have carbon filters. According to the report, considerable reductions could be achieved by forbidding open systems and requiring other technical improvements at a cost of SEK 23 to 47 per kilogram.[2] It was earlier believed there were no technical alternatives, but it is now clear that alternatives are in fact becoming more common. Because many dry-cleaners are small enterprises located in the midst of apartment blocks and other housing and commercial areas, the risk of unintended exposure is relatively high. It might be considered a paradox that TCE should be forbidden but not PER.

Policy Responses

The choices available to policymakers are much more subtle than either command-and-control or economic policy instruments—that is, taxes. Market-based instruments range from taxes, charges, and deposit refunds to tradable permit schemes for fishery management and pollution control. Information provision, ecolabeling, liability, refunded emissions payments, subsidies, voluntary agreements, and many other schemes show that there is in fact a menu of policies (Sterner 2002). The selection and design of policy instruments depends on both ecological and economic conditions and the selection criteria. With several goals, such as efficiency, incentive compatibility, fairness in the distribution of costs, and political feasibility, one would expect to find different combinations of policy instruments for different tasks.

The choice of appropriate criteria depends on the conditions that characterize the issue. For environmental problems with moderate abatement costs in an economy with an even distribution of income, equity issues may be less important and efficiency paramount. Conversely, for issues that affect health and ultimately life in countries with large income disparities, distributional concerns and fairness may be more important than efficiency. In markets characterized by powerful monopolies or marked information asymmetries, the issues of incentive compatibility may well dominate. In other cases it may be the complexity of the ecosystem that determines the design of the instrument. If there is a risk of serious and irreversible damages, then caution may dictate the use of some very direct instruments, like outright bans. But if the prohibitions are not effective and lead to lobbying rather than research into new technologies, then market-based instruments that encourage such research may be more efficient.

In this chapter it is reasonable to question whether bans are good policy instruments in the case of TCE. Traditional environmental economic analysis (Weitzman 1974) suggests that if the marginal cost of abatement is very steep or close to the zero limit (and the environmental damage curve is not so steep), then taxes (or some other price instrument) would appear better. If technological progress is expected to be fast but unevenly distributed among specific application areas, this preference for taxes would be strengthened by considerations of dynamic efficiency.

The presence of asymmetric information makes the choice of instrument more complex. When an environmental protection agency imposes regulation, the regulated industry has an obvious incentive to overestimate abatement costs to get a generous emissions concession. When taxes are the instrument, industry has no incentive to overestimate abatement costs, since that would imply that the equilibrium tax necessary is high, too. One might think that a firm has an incentive to underestimate, but this incentive is likely to be weak because it amounts to acknowledging that abatement is easy. Hence the tax instrument is likely to lead to fairly truthful reporting. This is an example of the "revelation principle" used to deal with policymaking under asymmetric information. Furthermore, a tax promotes rapid technological change: if a company has a 10-year concession, it has little incentive to develop or even adopt new technology, but if taxes are used, it will be motivated to adopt new technology as soon as possible.

Competition and the political economics of firm behavior with respect to environmental policy are also important, but here, as pointed out by Albrecht (1998), the hypotheses are contradictory. The "industrial flight and pollution haven" hypothesis states that strict regulation will lead to industrial relocation. This is in principle what most economists would expect based on comparative advantage, although the effect is likely to be very small and the empirical evidence is not clear. We also have the Porter hypothesis, formulated by Harvard management guru Michael Porter, who suggests that environmental regulation will increase productivity through the secondary effects of innovation.

The Porter hypothesis has been criticized by economists who argue that if the productivity opportunities are real, they will be exploited irrespective of legislation (Oates et al. 1994). As the critics recognize, however, it is possible to construct models with some other market or regulatory imperfection that could lead to the existence of a Porter effect; see, for instance, Bonato and Schmutzler (2000). The model by Xepapadeas and de Zeeuw (1999) confirms the logical impossibility of the pure Porter effect but points to a number of mechanisms implying that the cost of compliance with regulation may be very low.

Compliance has also received considerable attention; it is sometimes thought to be a paradox that firms comply even if monitoring and enforcement are far from perfect. Various explanations have been offered for this so-called Harrington paradox, including the notion that compliance confers benefits, such as market advantages related to image (Arora and Gangopadhyay 1995; Harrington 1988; Arora and Cason 1996). Heyes has written extensively on monitoring and regulation, and Heyes and Liston-Heyes (1999) conclude that many business executives agree with Porter and then through lobbying build a model for regulation to show that the regulating agency should take account of Porter's views even if it considers them wrong and misguided.

One practical aspect when it comes to regulating chemicals is that there are so many compounds with rather stark differences in their effects. This implies that if price instruments are used, a detailed and complex vector of prices (taxes) must be set. To date this has really been tried (on a small experimental case) only in some of the former Soviet countries where environmental taxes had rather a different connotation anyway. On the one hand this is laborious and complex work, and on the other hand it may appear unethical in those cases where the main effects are on workplace environment and employee health. In the area of occupational health, economic policy instruments have not been frequently used and there appears to be unease about the mere suggestion that firms would make tradeoffs between cost savings and occupational risks (see Torén and Sterner 2002).

Sweden: A Ban on Trichloroethylene

The Swedish parliament passed a law in 1991 banning the use of TCE in consumer products starting in 1993, and prohibiting the professional use of TCE and methylene chloride, effective January 1, 1996. A complete prohibition on all use might seem a very strong policy instrument, but it has not been wholly effective. The reason appears to be that the very strength of the instrument is in some sense its weakness: the ban is so absolute that it creates strong opposition among some users, who either find it particularly difficult to replace TCE or simply disapprove of the timing or policy method. The Swedish experience has shown that some firms spend a great deal of effort and resources in appealing and lobbying against the ban and gather support from industry associations and others. They even tried to get support from E.U. institutions on the grounds that the ban would threaten the free mobility of goods.

The reason for the ban was TCE's detrimental health and environmental effects. The environmental consequences have been challenged, however, and in fact the environmental damage of TCE (particularly atmospheric releases) does not appear to be as serious as that of CFCs or even HCFCs. Abrahamsson et al. (1995) and Abrahamsson and Pedersén (2000) even report very substantial natural production of TCE, PER, and similar halogenated compounds in marine algae. If the health effects are no worse than for certain other solvents, perhaps restrictions on maximum exposure in the working environment would have been enough. Sweden does have a stringent exposure limit of 10 ppm (eight-hour time-weighted average), compared with 100 ppm in the United States and United Kingdom. Most European countries have intermediate levels—75 ppm in France, 50 ppm in Austria and Belgium, 35 ppm in the Netherlands[3]—but Germany also has a very strict standard. The German limit of 20 ppm, combined with additional technical and workplace requirements, including completely closed systems for operation and even storage and transport, effectively reduces ambient levels to even lower levels (Lerrach 2000), with the result that some degreasing units currently operating in Sweden would not be able to operate in Germany.

The comparisons are particularly relevant because TCE was brought in as a replacement when other solvents, such as highly aromatic hydrocarbons and CFCs, were phased out for health or environmental reasons. This was, for instance, the case at the SKF factory in Gothenburg, which in the early 1980s used about 3,000 to 4,000 tons of TCE per year (almost half Sweden's consump-

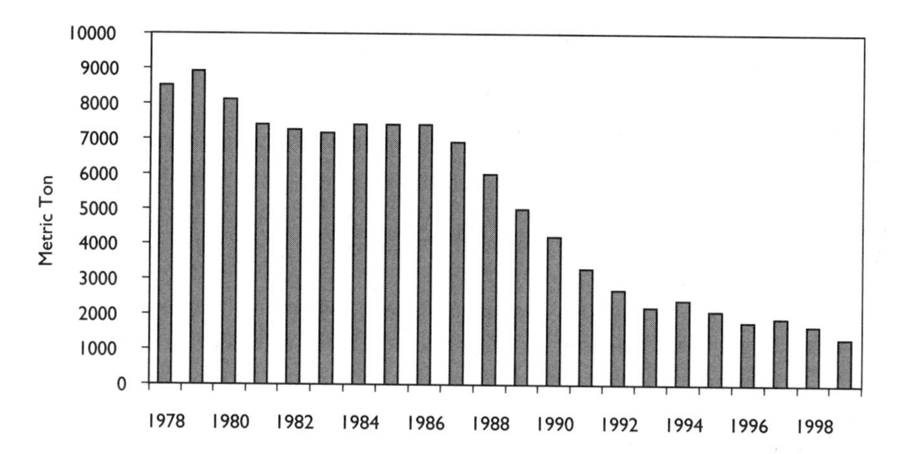

Figure 10-1. Use of TCE in Sweden, 1978–1999 (Decision on Ban 1991; Total Ban 1996)

Source: Naturvårdsverket (Swedish environmental protection agency).

tion at the time). The firm, a manufacturer of ball bearings, introduced TCE as a replacement for both CFCs and highly aromatic nafta products.

The use of TCE in Sweden is shown in Figure 10-1. In 1996 it was supposed to be completely phased out. Quite clearly, the ban did not cause the phaseout of TCE; perhaps it might be considered the logical last step in a phaseout, or the only instrument capable of stopping the residual applications after other policies had reduced its use. By the time the decision to ban TCE was made, consumption had already fallen from about 9,000 tons per year to 3,000 tons. This reduction was much faster than the overall reduction in European use of TCE and other halogenated solvents (a decrease of about 50% over 15 years). It is likely that this reduction was in response to a whole arsenal of strong policies pursued by the Swedish authorities, including the (threat of) prohibition. The actual decision to ban was nevertheless followed by a period of fairly stable use: it seems that industries did little or nothing between 1991 and 1995 to prepare for the ban, and in fact many executives have said in interviews that they did not think the authorities were serious, especially since TCE had a long history and no other countries were banning it. At the same time, Sweden was preparing for entry into the European Union, where it was not banned, and TCE became an almost disproportionately important item in the negotiations, where it was seen as a symbol of Swedish self-determination in the area of environmental policymaking.

In the second half of the 1980s large industries faced emissions standards and tighter exposure limits that forced them to reduce their use of TCE or phase it out completely. The standards and regulations led either to the use of other technologies and other solvents or, more commonly, to the adoption of closed systems with carbon filters, which allowed a drastic reduction. As an example, we may look at the case of SKF, which accounted for a large share of consumption.

SKF used TCE for three purposes: ordinary degreasing, as a solvent for fats used as antioxidants; and finally for dewatering of sensitive components.

SKF is an interesting case not only because of its size but also because the demands of ball bearing manufacture are unusually exacting. Ball bearing components must be dried within 30 seconds, after which corrosion becomes unacceptable. Moreover, the size and shape of the product make substitution more difficult than in many other types of manufacturing—for instance, those working with large sheets of metal. Thus, although one could argue that SKF had more resources and could more easily accommodate to environmental restrictions, at least technically one might argue the opposite: if SKF could eliminate TCE, then many other industries should have found it fairly easy.

The reduction in use of TCE at SKF was driven by two factors: the trade union's demands on the working environment and the environmental requirements for reduced emissions. When use was at its peak, SKF was emitting 250 tons per year into the air. In its 1983 permit SKF was ordered to reduce annual emissions to a maximum of 15 tons. Installing active carbon filters led to a two-thirds reduction, to about 80 tons per year, but further reductions required process changes. New degreasing processes that used water and (low-aromatic) oils, and new packaging and storage routines were combined with the use of lighter oils for conservation of ball bearings instead of wax, whose removal required TCE as a solvent.

After some time and further investments, only very small amounts of TCE were being used, and the costs of maintaining the handling, storage, and filter facilities became disproportionate. Furthermore, SKF discovered it could gain in goodwill and environmental image by exceeding the requirements instead of reacting passively to them. The company therefore decided to phase out all use of TCE. This decision applied to all SKF plants abroad, even if local authorities did not require it, for two reasons: environmental image, and uniformity of standards and processes to ensure homogeneous product quality.

Ensuing Legal Battles

The case of SKF is not typical; in some other companies the ban on TCE led to bitter opposition and protests. Many petitions and articles were written, and several companies decided to fight the legislation, threatened to close down or leave the country, and appealed to the courts. Although this is common in the United States, it is extremely rare for Swedish legislation to be challenged in this way. Leading points of contention were that the industries disapproved of the prohibition as a method, its timing, and its consequences, which we discuss below. Here we briefly recapitulate some of the more salient points.

In 1994, 39 industries published an open letter to Prime Minister Ingvar Carlsson as an advertisement in a leading Swedish newspaper (Dagens Industri 1994), saying that the prohibition was poorly motivated and prepared and should be withdrawn. They contended that their viability was threatened and that more than half of them would have to move abroad if the ban were enforced.

This letter was very strongly worded—even excessive, as several executives admitted later in interviews—but its tone was indicative of industry's strong resentment of the regulation. Although some companies did phase out TCE and as far as we know none moved abroad, many applied for waivers and tried to obstruct or circumvent the legislation. Originally the ban was to take effect on January 1, 1996. Because many companies had difficulties and there was considerable resistance, the Chemicals Inspectorate issued a general exemption for any firm that could report difficulties. In the first year (1996) some 500 companies were given waivers, effectively postponing the ban until January 1, 1997. After that date only companies that could show that they had made a serious effort to substitute for TCE and had a plan for doing so would be granted another, temporary exemption. These companies would also pay an exemption fee of SEK 150 per kilogram—a fee intended not so much as an environmental tax but as a way to remove any disadvantage that a complying company might suffer vis-à-vis noncomplying competitors.

Of the 283 companies that applied for exemptions for 1997, 137 were granted waivers, and 60 of the companies whose applications were rejected appealed the decisions. The Stockholm County Administrative Court revoked the decisions made by the Chemicals Inspectorate, and there have been several further rounds of appeal to higher courts. When Sweden joined the European Union in 1994, Holland, the United Kingdom, and the E.U. Commission were very critical of the ban, largely because Sweden did not produce TCE and was importing the chemical, and thus the ban would be a barrier to trade. However, Sweden fiercely defended the ban by citing politicians' pledge during the E.U. campaign that membership would not compromise the country's environmental goals. The ban thus became a symbol and the subject of a fight with high stakes. A step backward would have fed anti-E.U. opinion.

Meanwhile, Sweden's Chemicals Inspectorate modified its rules for exemption, dropping the requirement that firms present plans for the phaseout of TCE and modifying the fee structure. The new requirements were as follows:

- the company is actively researching other alternatives;
- no suitable alternative is readily accessible for the company's needs; and
- no harmful exposure results from the use of TCE.

The exemption fee was later withdrawn entirely, since the E.U. Commission considered it "out of proportion" to the environmental damage. All 220 companies that applied for waivers to continue using TCE after 1997 received exemptions, along with 121 waivers in 1998 and roughly 150 in 1999. The case of one company, Toolex Alpha, was referred to the European Court of Justice (by a Swedish court) to determine whether the Swedish prohibition was in accordance with the free movement of goods (case C-473/98).

Several interesting general principles of European law underlie the legal struggle. First, the prohibition must be in the public interest and not a hidden trade restriction. Second, it must be necessary and nondiscriminatory. Third, the law should be proportional—that is, not unreasonably harsh in comparison with its goal. On July 11, 2000, the European Court of Justice ruled that the Swedish prohibition did not run counter to E.U. legislation on the free movement of goods, reasoning that

- the basis for the prohibition was concern for health and the environment;
- the European Union had classified TCE as toxic and carcinogenic;
- member countries had the right to enact stricter environmental legislation;[4]
- there was no reason to assume the prohibition was motivated by an attempt to stop trade; and
- there were reasonable possibilities of getting a waiver.

The possibility of waivers was considered particularly important: it appears that an absolute phaseout with no exceptions would have been considered disproportionate. Hence, one might with some exaggeration say that the prohibition was accepted because it was watered down by the likelihood of obtaining a waiver. Considering the sensitive issue of national sovereignty in environmental decisionmaking and the relatively skeptical attitude of the Swedish electorate vis-à-vis the European Union, it seems wise that Sweden was allowed to keep its independent (and at least in some sense more radical) national policy on this issue. There remains, however, the broader question of whether the ban as such was a good policy instrument in this case. Recall that Sweden was alone in prohibiting TCE, and that TCE and methylene chloride were the only chlorinated solvents for which Sweden chose this rather drastic measure. Most notably, PER was not prohibited, even though it is chemically quite similar.

Ex ante Analysis

There was apparently no *ex ante* economic analysis, such as a cost-benefit study. Slunge (1997) and Slunge and Sterner (2001) did, however, study the marginal cost of abatement and looked at the applications for exemption from the 1996 ban, which contained information on the economic consequences of a substitution—that is, the marginal cost of substitution. Although these figures may reasonably be assumed to be biased estimates, since their very purpose was to justify an exemption, we also have data from other sources: detailed interviews with several executives on their firms' use of TCE and the actual costs of substitution, and other studies, including one by the Swedish Chemicals Inspectorate that contains figures on the cost of substituting for TCE. The data cover the whole range of companies, from SKF to some very small workshops, and represent different sectors of industry; they also include companies that both have and have not made the required investments.

The companies that had phased out TCE incurred an average and median cost of SKr 6 per kilogram of TCE. Those that had not phased out TCE calculated much higher costs, a median of 48 but an average of SKr 84 per kilogram; some of these very high figures may have been protest estimates. To shed light on which of these factors is most important, we looked in greater detail at those companies for which we have the best data and examined the marginal costs of abatement for different steps in phasing out TCE. As in the case of SKF, the first reductions in quantity were sometimes very easy, since they mainly required closing systems and installing carbon filters, which (at least in large-scale use) were inexpensive. However, the next step in abatement turned out to be much more difficult and expensive.

The evidence shows that in most cases, substitution of other chemicals for TCE is relatively if not very cheap. It might be very expensive in special cases, but it is difficult to judge whether this was tactical exaggeration or lack of knowledge by the individual firms, or a reflection of special needs or circumstances. For example, the residues might be very difficult to remove, or the demands on cleanliness very high, or the technical specifications in some other way exacting. More commonly, a company might have recently invested in expensive, closed-cycle TCE degreasing equipment and could not afford in the short run to throw out this equipment and install new degreasers. Some companies were small and lacked either space (the water-based equipment is typically larger) or time (if they were temporarily overloaded with work or understaffed). There might also be firms in special financial situations that could not borrow money for this type of investment. A small company whose owner was the only technical expert might find that he had to close the plant to install new equipment, and such a disruption to production could well be prohibitively costly. If a small company was working overtime to expand in a new direction or struggling to catch up with orders, a management distraction of even a week or two on "side issues" like degreasing might carry a very considerable cost: the opportunity cost of management. This type of cost is hard to quantify and not included in our material, but it is reflected in the strong statements of some executives from small companies.

In many instances companies facing tougher environmental restrictions initially reacted by saying that the new requirements were impossible to meet and would force plant closure. After a couple of years, however, they found it easier than anticipated, and there is even anecdotal evidence that new processes turned out to cost less or improve product quality. This is a version of the Porter hypothesis—that efficiency is enhanced by tough environmental standards. However, some such cases might be due to a coincidence of technological progress, and it is not easy to distinguish the companies that were bluffing from the ones that faced true difficulties.

Figure 10-2 shows the *reported marginal costs* of abatement for a fairly large group of companies. A very large share of emissions carries a low marginal cost of abatement, and an environmental tax or fee of SEK 50 per kilogram would probably have sufficed to effect most (around 90%) of the abatement. A higher fee would incidentally not appear to have much of an effect on this figure if the costs of abatement rise as dramatically at the tail end as they appear to in Figure 10-2. These cost estimates may be exaggerated, but neither the regulator nor the researcher can know for certain. Although there are environmentally reasonable and technologically viable solutions to degreasing without TCE, the compliance costs for an individual firm may still, for reasons already discussed, be high particularly if the firm is required to comply very fast and not given time to adapt.

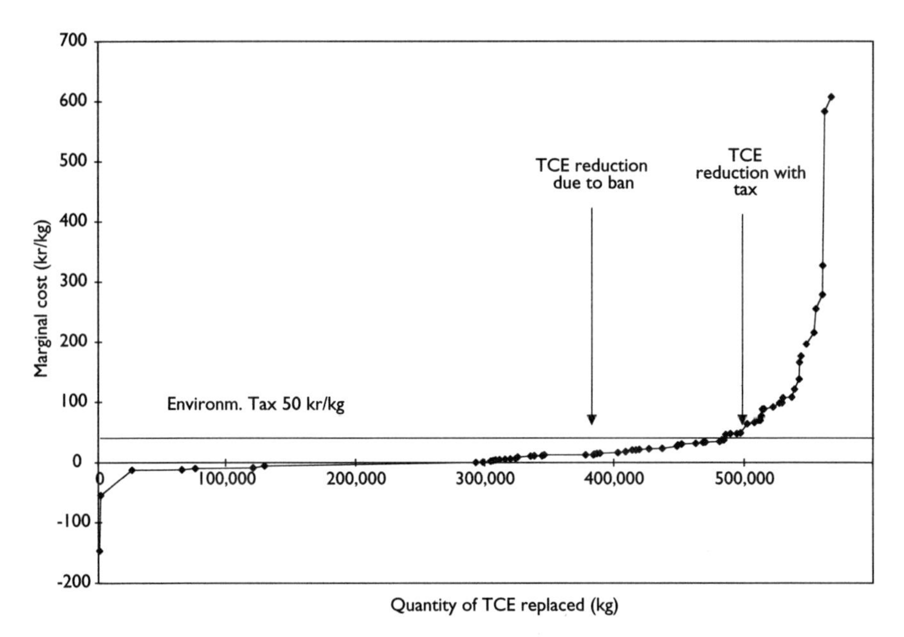

Figure 10-2. ***Marginal Abatement Cost and Effects of Tax Compared with Ban***

Source: Slunge (1997). Assumptions: 15-year equipment life, 4% real interest.

Ex post Analysis

There has not been any official cost-benefit evaluation of the ban in Sweden. Such evaluations are difficult without a counterfactual with which to compare. One promising possibility for an *ex post* analysis, however, is to compare the experiences in similar, neighboring countries, such as Denmark, Norway, and Germany, which have implemented very different policies.

Norway has a tax per kilo on both TCE and PER of 50 Norwegian crowns (worth some 10% more than Swedish crowns) that combines elements of both tax and deposit refund, since firms have the right to reclaim half the tax paid on delivery of TCE sludge delivered to special treatment plants or authorized recyclers. Considering that the market price of TCE is 10 to 15 crowns, a tax of 50 crowns may be expected to have a significant impact, and that in fact is what our firm-level inquiries suggest. The main alternative policy considered in Norway was a prohibition. Industry reactions to the tax have not been enthusiastic; nevertheless, there appears to be an appreciation of the fact that the tax allows firms much more flexibility than an outright ban.

The Norwegian tax took effect in 2000, and thus it is rather early to evaluate its effects. As Norwegian data show, there was hardly any decrease in Norwegian demand before 2000, and in fact a 25% increase during 1999 might be attributed to pretax hoarding. According to Eriksen (2000, 2001), the tax has been very effective, and preliminary figures show that purchases of TCE have fallen from more than 500 tons in 1999 to 82 tons in 2000 and 139 in 2001. For PER the figures show a reduction from 270 tons in 1999 to 26 tons in 2000 and 32 tons in 2001. Even if the 1999 purchases included some buildup of stockpiles, the reductions are close to 80%. All in all, we see that the Swedish policy was more effective than the pre-1999 Norwegian policy (which was fairly lax), but it seems to be less effective than the very high Norwegian tax.

Denmark also has an environmental tax. It is levied not only on TCE but also on PER and dichloromethane (CH_2Cl_2), but it is only 2 Danish crowns (roughly equivalent to the Norwegian crowns) and thus just 4% of the Norwegian tax but still enough to add some 25% to the price of these solvents in Denmark.[5] In Denmark, as in Sweden, the stated justification was concern about increased use as halons and CFCs were phased out (also by taxation, but at a rate of 30 Danish crowns per kilo). The Danish tax took effect on January 1, 1996, and use decreased very significantly by 1998 (DEPA 2000). Compared with an estimated 1992–1995 average, TCE declined from 1,000 to 356 tons, PER from 720 to 463, and dichloromethane from 483 to 0, giving an average reduction of over 60%—less than that achieved by the Swedish prohibition but still sizable, particularly considering that it was achieved with such a modest tax.

Germany has employed neither tax nor prohibition but very tough technical requirements concerning emissions that apply to both TCE and similar solvents, such as PER. When PER from dry-cleaners was found in the air and food of adjacent apartments, regulations were imposed, calling for completely closed systems for PER; that policy was then expanded to TCE. As a result, PER and TCE (as well as 111-trichloroethane and dichloromethane) have all been drastically reduced—roughly by a factor of 10 in 15 years (by almost two-thirds from 1986 to 1991, when the unification of Germany creates a break in the figures, and then another two-thirds from 1991 to 2000). These figures show a decline that is considerably faster than in the European Union as a whole. Comparing E.U. and German figures, it is clear that for 1993–1998, Germany alone accounts for half the reduction (and thus 10% of use in 1993 but only 5% in 1998). For PER, Germany accounts for 9 of the total 11 kilotons of E.U. reduction. It would seem that the German focus on health issues and thus the joint regulation of TCE and PER made good sense. At the same time the industrial policies, which were worked out in close cooperation with industry, also focused on workers' health and on technological improvement. This had the added advantage of being an incentive for the machine industry, and Germany is now the main exporter of high-quality closed-loop degreasing equipment.

Figure 10-3 compares the rates of phaseout in the four countries and compares these with the rest of Europe. Definitions vary between countries and even within them, depending, for instance, on the way TCE in imported products is treated. The sources also vary and the data are not complete. The data for the whole of Western Europe come from the manufacturers' association, and the figures for the rest of Europe are derived from the available national figures. To be as comparable as possible, the figures have been scaled in relation to population. Although the precision of the individual numbers varies, taken together, they appear to indicate two points. First, the regulation in Germany has been roughly

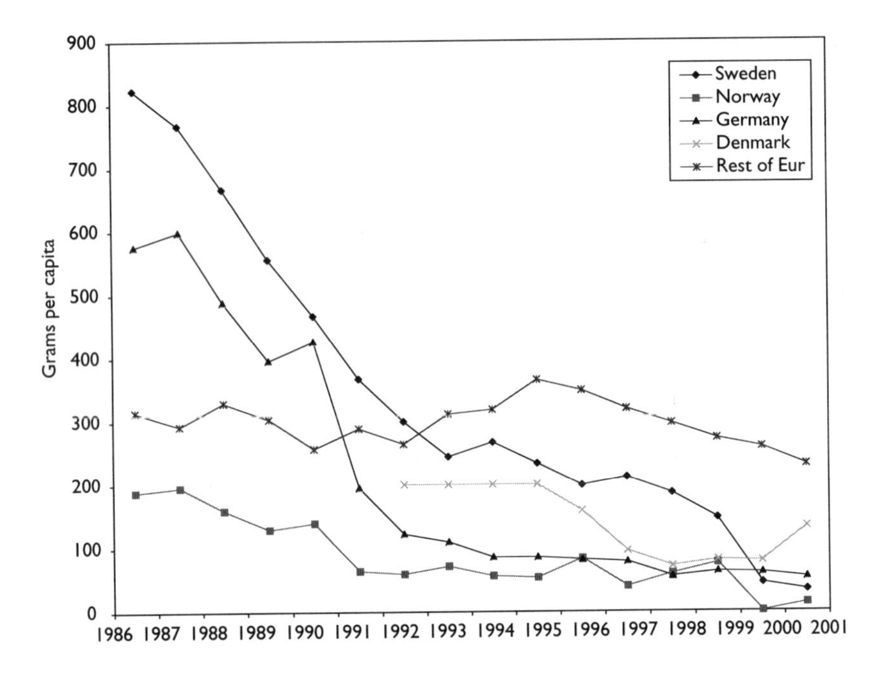

Figure 10-3. *Rates of Reduction of TCE*

Sources for data are the respective environmental agencies, statistical bureaus, or chemical inspectorates (see endnote 1). Data reflect sales, not production, and discrepancies between these two types of data are due to internal use as intermediary chemicals in some processes. Data for Germany cover the whole country from 1991. For 1986–1990 they are interpolated based on West Germany. Data for Norway are for imports of TCE only. Other figures for use (including TCE embodied in products) are somewhat higher. Data for Denmark are only averages for 1992–1995; and no earlier data are available. Data for the whole of Western Europe come from the European Chlorinated Solvents Association, http://www.eurochlor.org/chlorine/news/news25.htm. Our estimates for the "Rest of Europe" are the West European total sales minus the national figures used for Germany, Sweden, and Norway. All data are originally in kilotons and scaled by being divided by the (approximate) population in millions (9 million for Sweden, 5 million for Denmark, 4 million for Norway, 82 million for Germany, and 265 million for the remaining countries). This gives approximate use in grams per capita per year (with some underestimation of reduction since the increase in population is not taken into account).

as effective as the prohibition in Sweden. Second, the other Scandinavian countries—Norway and Denmark—had lower rates of use in the first place and then followed a somewhat slower rate of phaseout during the 1990s. At the end of that decade they then implemented taxes. The difference in response to the taxes is moderate, but the very high Norwegian tax had an immediate and drastic effect, at least as severe as the Swedish prohibition, while the low Danish tax has a much more modest effect. All of the policies appear to have been more effective than those of other European countries, whose policies were less stringent and where emissions have declined only gradually.

Conclusions

The Swedish ban reduced the use of TCE considerably but has not been entirely effective, and it appears likely that there were more cost-efficient options. Many companies with seemingly moderate abatement costs have chosen to fight the ban rather than abate. The threat of sanctions was apparently not credible. The risk of formal sanctions was reduced when a large number of firms coordinated their efforts, supposing that the Swedish authorities had a weak case. As an illustration of the power of informal market sanctions, it is noteworthy that larger companies, such as SKF, quickly eliminated their use of TCE. Small companies, in contrast, may have been less concerned about their image, perhaps because of their size or because they did not sell consumer products and did not think their corporate customers would be sensitive to this issue. In this case, the logic was clear: fighting the ban cost little more than the effort of writing the letters, and companies facing regulatory instruments had an incentive to exaggerate the costs of abatement.

Although the ban survived legal challenge (the opposite would have been extremely embarrassing to the Chemicals Inspectorate), implementation has been an uphill battle. Might this process have shown, once and for all, the legal powers of the national authorities and thus strengthened the likelihood that prohibitions would be used in similar future cases? We think that is an unlikely interpretation. In our view, the ban survived because it was watered down by the generous exemptions; the legal and media battles were exhausting but ultimately unproductive—or even counterproductive. Prohibition will no doubt be used again, but the authorities will restrict it to more clear-cut cases, as when health or environmental damage is more dramatic or international opinion more coordinated. Prohibitions that are, like this ban, not immediately successful portend problems for the policymaker:

- The planned environmental improvement is, after all, not achieved—at least not immediately.
- A greater improvement could perhaps have been achieved, at lower cost.
- The regulating agency suffered a loss of prestige (possibly one goal of the companies that fight the legislation), and the energy and confidence needed to enforce other regulations may be weakened. In Sweden, companies now have less incentive to comply with future regulations from the Chemicals Inspectorate.
- Companies that in good faith followed the regulations and invested in new equipment (in some cases at great expense) found that their efforts had been "unnecessary" and their competitors did not even have to pay environmental fines or compensatory fees. Complying firms' experience may therefore reduce the incentive for compliance with future regulations or initiatives.
- The regulating agency was obliged to allow polluters to apply for waivers. Besides wasting inspectors' time, this opens up the possibility of arbitrary decisions and, in theory, even corruption.

- The uncertainty caused by a proposed ban may discourage investment in new technology. For many years the Swedish industry was hesitant, and anecdotal evidence suggests that hazardous, temporary methods (such as manual degreasing in petrol in open air) were used because of the uncertainty.

It is perhaps an irony that half a dozen Swedish companies have now ordered German closed-loop degreasing machines, even though they do not know whether they will be allowed to run them on TCE in Sweden—but if not, they can always use PER instead. The fact that PER is not banned casts some doubt over the selective banning of TCE.

It is always risky to draw conclusions based on the partial experience of just a few countries. Any of various policies could decrease the use of chlorinated solvents. Observed results are due to a combination of policies and other factors, and it is hard to disentangle the effects and judge how beneficial they are from an environmental viewpoint. The Norwegian policy is so new that the data are insufficient. The German case is complicated and consists of many different elements but appears to have had the advantage of focusing on ambient conditions and promoting new technology. For Sweden, a tentative conclusion—drawn with the benefit of hindsight—is that the authorities would have been more successful implementing harsh technical requirements and a monetary instrument applied over a somewhat broader range of chemicals than banning just TCE. For instance, a fee on the use of a wide spectrum of chlorinated hydrocarbons would have provoked less resistance and probably achieved the same reduction—and maybe more. It would also have been more proportional.

Although TCE, PER and other similar solvents are dangerous to human health, there is fairly limited evidence of environmental damage. This suggests that the most important issue is ambient levels in the working environment. Exemptions could have been made for a few types of use where the costs of abatement were high and the health risk was low. Alternatively, since exemptions are undesirable, the fees could have been earmarked for environmental collaboration with the industry association to help those industries that could not easily find substitutes for TCE. In this case, the results might even have been enhanced at the same time as the industry lobby would have been considerably weakened.

Notes

1. In Germany, by comparison, regulators treat PER at least as stringently as TCE.

2. SEK 10 ≈ US$ 1.

3. Warner (2002).

4. The lack of harmonization of regulations and laws governing the use of TCE within the union was an important factor in favor of allowing Sweden's regulations. In areas of law where there is explicit harmonization, it would be much more difficult for one nation to have separate legislation.

5. In addition to being low, the tax is also refunded if the TCE is exported or incorporated in exported goods. The stated argument for this is to maintain Danish competitivity.

References

Abrahamsson, K., and M. Pedersén. 2000. Evidence of the Natural Production of Trichloroethylene. *Limnology and Oceanography* 45(2).

Abrahamsson, K., A. Ekdahl, J. Collén, and M. Pedersén. 1995. Marine Algae—A Source of Trichloroethylene. *Limnology and Oceanography* 40(7).

Albrecht, J. 1998. *Environmental Regulation, Comparative Advantage, and the Porter Hypothesis.* Fondazione Eni Enrico Mattei. Working Paper 59/98.

Arora S., and T. Cason. 1996. Why Do Firms Volunteer to Exceed Environmental Regulations? Understanding Participation in EPA's 33/50 Program. *Land Economics* 72(4): 413–32.

Arora, S., and S. Gangopadhyay. 1995. Toward a Theoretical Model of Voluntary Overcompliance. *Journal of Economic Behavior and Organization* 28(3): 289–309.

Bonato, Dario, and A. Schmutzler. 2000. When Do Firms Benefit from Environmental Regulations? A Simple Microeconomic Approach to the Porter Controversy. *Swiss Journal of Economics and Statistics* 136(4): 513–30.

Dagens Industri. 1994. Protestannons mot triförbudet (protest advertisement against the TCE ban), signed by 39 industrial companies. November 21.

Danish Environment Protection Agency (DEPA). 2000. *Ökonomiske styrningsmidler I dansk miljöpolitik.* Miljö og Energiministeriet, Mijöstyrelsen, Köbenhavn, Danmark.

Eriksen, H. 2000, 2001. Ministry of Environment, Oslo. Personal communication.

Harrington, W. 1988. Enforcement Leverage When Penalties Are Restricted. *Journal of Public Economics* 37(1): 29–53.

Heyes, A.G., and C. Liston-Heyes. 1999. Corporate Lobbying, Regulatory Conduct and the Porter Hypothesis. *Environmental and Resource Economics* 13(2): 209–18.

Lerrach, J. 2000. Use and Regulations of Trichloroethylene in Germany. Bundesministerium des Umwelt, Germany, prepared for the seminar on trichloroethylene, May 9, at the University of Gothenburg, Sweden.

Naturvårdsverket. 1997. Sveriges Kemtvättar, maskinpark och utsläpp. Rapport 4725. Stockholm.

Oates, W., K. Palmer, and P. Portney. 1994. Environmental Regulation and International Competitiveness: Thinking about the Porter Hypothesis. Discussion Paper 94-02. Washington, DC: Resources for the Future.

Östman, A., C. Nordin, and L. Hawerman. 1995. Utvärdering av ODS-avvecklingen (Evaluation of the phaseout of ODS). Rapport 4477. Naturvårdsverket, Stockholm.

Slunge, D. 1997. Förbudet mot trikloretylen. Memo. Gothenburg University.

Slunge, D., and T. Sterner. 2001. Implementation of Policy Instruments for Chlorinated Solvents. *European Environment* 11(5): 281–96.

Sterner, T. 2002. *Policy Instruments for Environmental and Natural Resource Management.* Washington, DC: Resources for the Future.

Torén, K., and T. Sterner. 2002. How to Promote Prevention—Economic Incentives or Legal Regulations or Both? Mimeo. Department of Economics, University of Gothenburg, Sweden.

Warner, B. 2002. Human Health Risk Reduction Strategy for Trichloroethylene. CAS No. 79-01-6. Mimeo, February draft. Available from bob.warner@hse.gsi.gov.uk.

Weitzman, M.L. 1974. Prices versus Quantities. *Review of Economic Studies* 41: 477–91.

Xepapadeas, A., and A. de Zeeuw. 1999. Environmental Policy and Competitiveness: The Porter Hypothesis and the Composition of Capital. *Journal of Environmental Economics and Management* 37(2): 165–82.

CHAPTER 11

Trichloroethylene in the United States

Embracing Market-Based Approaches?

Miranda Loh and Richard D. Morgenstern

UNLIKE THE REGULATION of sulfur dioxide and other so-called criteria pollutants, control of hazardous air pollutants (HAPs) in the United States is a relatively recent phenomenon. Apart from early steps to control toxics such as vinyl chloride, which the U.S. Environmental Protection Agency (EPA) regulated in 1976, most regulation of HAPs has been based on new provisions added in the 1990 Clean Air Act Amendments.

Claims about the effectiveness of toxics regulations or about the efficiency by which they achieve their objectives are typically based on studies carried out before the fact. *Ex post* studies of toxics regulations are truly rare. Although interest in the issue has been growing in recent years, remarkably little is known about how regulations of toxic substances actually perform.

Halogenated or chlorinated solvents—including methylene chloride, perchloroethylene (PCE), 111-trichloroethane (TCA), and trichloroethylene (TCE)—are widely used for metal cleaning and chemical manufacturing and as components of paints and other substances. Health impacts include skin irritation and neurological effects, as well as liver and kidney damage. There is also evidence for carcinogenicity.[1] Chlorinated solvents are regulated under at least four federal statutes plus numerous state laws.[2]

This case examines the U.S. regulation of TCE in degreaser cleaning, which accounted for about 90% of TCE use in 1992. A notable feature of the governing statute, Section 112 of the U.S. Clean Air Act, is the focus on design and performance-based national emissions standards for hazardous air pollutants (NESHAP), to be established for relevant source categories of HAPs based on the maximum achievable control technology (MACT) (U.S. EPA 1993).

The MACT standards are explicitly tied to technical feasibility and economic achievability rather than health criteria. On its face, the MACT for TCE is significantly less stringent than standards issued in Sweden and other northern

European nations around the same time, which virtually ban the use of TCE altogether. In general, design-based standards specify the method, and sometimes the actual equipment, that firms must employ to meet a particular regulation. Performance-based standards typically establish a uniform control target for firms while allowing some latitude in how this target can be met.

Over the years, as the emphasis on efficiency of regulation has increased, new federal rules have begun to embrace market-based, economic incentive (EI) approaches. In contrast to traditional command-and-control regulation, market-based rules allow greater flexibility in achieving environmental goals by creating financial incentives to firms to control pollution.

The approach adopted for regulating TCE under Section 112 of the Clean Air Act involves a combination of design and performance-based standards complemented by certain market-based elements. Although the regulation does not include a cap-and-trade system or other commonly used incentive instruments, it does allow limited within-facility averaging among specified HAPs—a type of emissions "bubble." An early reductions provision contained in the statute provides an additional economic incentive mechanism. Specifically, a firm is granted an extension of six years beyond the established compliance date to meet the standard if it can demonstrate that it has achieved at least a 90% reduction in emissions of HAPs below a baseline level prior to the date the MACT is proposed. Although a few additional restrictions apply, this early reductions provision introduces at least some financial incentives into the firm's decision on how to meet the stipulated environmental goal. Conceptually, the early reductions provision can be seen as a form of facility-specific inter-temporal emissions trading.

A particularly interesting aspect of the regulation is that in an *ex ante* analysis, the EPA estimated net *savings* rather than the more common net *costs* associated with most environmental regulation. That is, the analysis found that the emissions reductions stipulated in the regulation could be achieved while costs to the affected facilities also dropped. Such a finding, if supported by the *ex post* analysis, raises questions not only about the need for the regulation in the first instance but also about the suitability of adding an economic instrument to the regulatory mix. Are the market-based elements of the regulation—which are presented as an option for the facilities to adopt—sufficiently attractive to entice firms away from the strict command-and-control approach? Especially if the costs of reducing emissions are very low or even negative, what are the potential gains from the added flexibility provided by the market-based instruments? Issues considered in this case are whether the emission reductions occur in a timeframe and at a level stipulated by the regulation. Did firms actually take advantage of the flexibility mechanisms—that is, the early reductions and the within-facility averaging provisions? If so, what can be said about the economic savings, if any, associated with the economic incentives? Overall, what does the *ex post* analysis tell us about the costs of the regulation, especially the *ex ante* claim of net savings?

First we describe the regulatory framework adopted to control the risks associated with TCE and the *ex ante* analysis that indicated net cost savings. The following sections outline the methods and data used to conduct the *ex post* analysis and present the quantitative results drawn from analysis of a national data set as well as from supplemental interviews with selected firms. We then discuss and interpret the basic results. A final section offers concluding observations about the suitability of applying market-based instruments to regulations with low (or negative) compliance costs.

Regulatory Framework

In setting limits for individual source categories, EPA is required to take into account both technological and economic feasibility. The national emissions standards for hazardous air pollutants are not designed around health-based criteria or any type of benefit-cost analysis. Instead, the stringency level is established on a relatively pragmatic basis, reflecting the best-performing 12% of current industrial practice in the relevant industrial subcategory. Thus, the NESHAP regulations are designed to raise the laggards to the level of the best performers rather than force the adoption of exotic or unproven technologies. The fact that economic feasibility is so explicitly stated as a criterion requires EPA to avoid significant economic impacts or dislocations.

For major regulations (those with expected annual costs in excess of $100 million), determinations about costs, benefits, and economic feasibility are made via a regulatory impact analysis (RIA), also known as economic analysis (EA), developed in advance of the formal rulemaking. For regulations, such as the degreasing rule, that do not qualify as major rules, less formal economic impact analyses are typically conducted by the regulatory agency. From a technical perspective, the standards are based on the top-performing 12% of the industry in terms of emissions controls. States may adopt their own rules, which must be at least as stringent as the NESHAP.

Various industries use machines designed to clean parts with halogenated solvents. These degreasing machines are the principal targets of the NESHAP for TCE. Two specific classes of degreasers—batch vapor and in-line solvent (see Figure 11-1A and B)—are addressed by the regulation. The general mechanism of action of a degreaser involves either dipping the dirty parts in liquid solvent (cold cleaning) or passing them through solvent vapor (vapor cleaning), which removes oils, fats, and waxes and then allows the parts to dry. For certain applications, where the level of cleanliness of the parts is of critical importance, halogenated solvents are considered among the most effective cleaners. As its name implies, batch cleaners are used for a set of parts that are dipped in the solvent all at once. In contrast, in-line cleaners operate as a sequential system, where parts are placed on a conveyer and sent through the solvent. Automated in-line systems have the advantage of reducing worker exposure, as there is less need to keep opening and closing the machine. Degreasers can also be "open-top" (see Figure 11-1C) and operated without a lid, a process that allows significant evaporative losses. In closed systems, degreasing occurs in an encapsulated device, thereby limiting solvent evaporation (Michigan Department of Natural Resources 1995; University of Tennessee 1995).

The NESHAP allows a machine owner to choose either an equipment standard or an overall emissions limit. The former consists of specified work practice

A. Batch Vapor Cleaning Machine

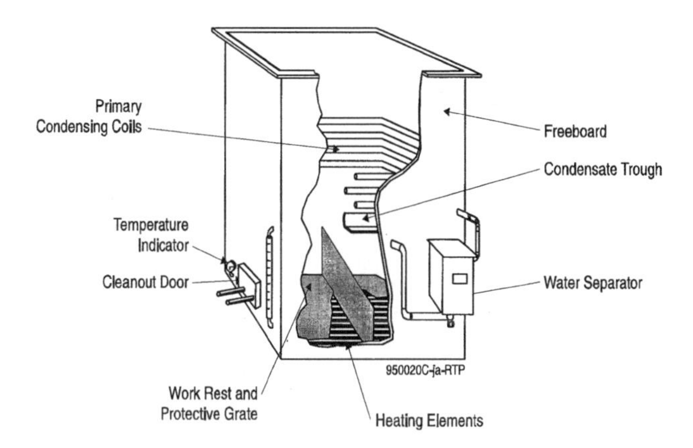

B. In-line Cleaning Machine

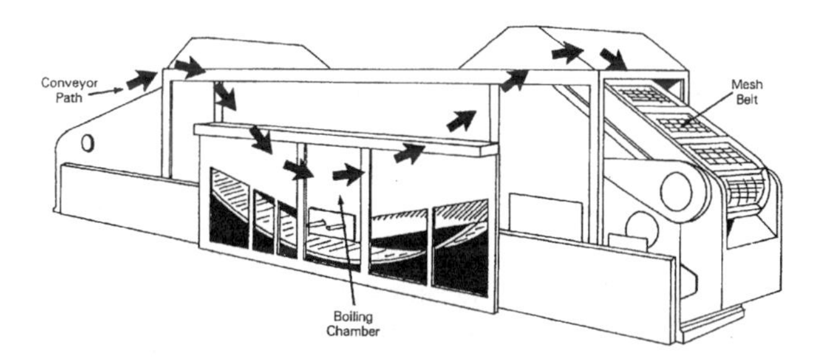

C. Immersion Cleaning Machine

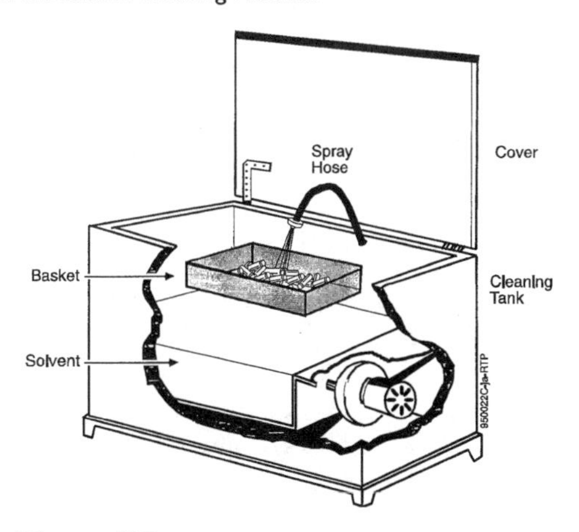

Figure 11-1. *Types of Degreasers*

Source: EPA. 1995. Guidance Document for the Halogenated Solvent Cleaner NESHAP. EPA-453/R-94-081, April.

Table 11-1. *Degreaser DESHAP Timeline*

Date	*Regulatory action*
November 29, 1993	Standard proposed
December 2, 1994	Standard promulgated
December 2, 1997	Compliance date for existing machines

requirements and either a combination of machine controls to reduce loss of vapor solvent or the adoption of an emissions limit while idling. The overall emissions limit does not require any specific work practices to be followed, but the firm must demonstrate that it will not exceed a specified three-month average emissions limit. On a case-by-case basis, facilities may also apply to EPA or the state agency to use some other means of compliance, as long as their methods are demonstrated to have an equivalent effect to the MACT. Different compliance timelines are adopted for new versus existing machines.[3]

The MACT is generally applicable to major sources,[4] but "area sources" (those emitting below the major source cutoff) are subject to a somewhat less stringent standard, the generally applicable control technology (GACT). In some cases the MACT and GACT are the same. In general, both major and nonmajor sources need to obtain an operating permit if the source is regulated under Section 112 of the Clean Air Act. Batch cold cleaning machines that are a nonmajor source are exempt from the permitting program. Certain nonmajor source degreasers—for example, those that previously installed so-called best available control technology (BACT) or technology designated as achieving the lowest achievable emissions rate (LAER)—are allowed to defer permitting for five years.[5]

HAPs cannot be traded, but with some restrictions, emissions of one HAP can be offset against another, thereby creating a facility-wide "bubble." Thus, if a facility makes a physical modification that causes it to exceed the specified limits for one pollutant, it can remain in compliance by demonstrating an equal or greater decrease in emissions of another HAP. However, the offsetting reductions must be made with a HAP that is considered by EPA to be at least as potent. The offset provision is not applicable to certain pollutants considered extremely toxic, especially above *de minimis* levels. Final determinations on the allowable emissions are made during the permitting process.

For added flexibility, Section 112 of the Clean Air Act also contains an early reductions provision. To qualify, a source must demonstrate that it has achieved at least a 90% reduction in emissions of HAPs in advance of the required compliance date with respect to a baseline year of 1987 or later. If so, it is allowed an additional six years to comply with the MACT. Not surprisingly, the statute enumerates specific safeguards designed to prevent abuse of the early reductions provisions. For example, the baseline year of 1987 must not be "artificially or substantially" greater than emissions in other years prior to implementation of emissions reductions measures.[6]

The relevant timeline for the degreaser NESHAP is shown in Table 11-1. Based on this schedule, one would expect to see the sharpest reduction in TCE use between 1993 and 1997.

Eligibility for the early reductions provision under Section 112 requires a regulated source to demonstrate that it achieved a 90% reduction before November 1993.

Ex ante Analysis

In November 1993 EPA published a draft economic impact analysis (EIA) of the degreasing NESHAP. The analysis is described by EPA as based largely on worst-case assumptions. EPA's justification for this approach was that since the preliminary estimates, which did not incorporate significant offsetting cost reductions within the facilities, indicated either zero or negligible economic impacts overall, the main concern was to identify those sectors or subsectors with potentially larger impacts, regardless of the likelihood of occurrence. In this instance, the worst-case scenario models effects on small facilities.[7] The draft EIA contains three principal parts: an industry profile, a methodology for estimating costs, and actual estimates of the economic impacts. The industry profile presents extensive technical information on degreasing equipment, a description of the alternative solvents available for degreasing, and an overview of the industries using degreasing equipment. In addition, there is a detailed analysis of the economic and financial characteristics of the small business sector most adversely impacted by the regulation—automotive repair.

The methodology section of the EIA presents an engineering model plant approach for calculating baseline and incremental costs associated with the rule. Both capital and operating costs, principally solvent use, are considered. Cost estimates are developed for several types of degreaser equipment as well as for the substitution of aqueous and nonhalogenated solvents for the halogenated products. Although the emphasis is on the solvent users, there is also some consideration of the impacts on both solvent producers and manufacturers of degreasing equipment.

Two cost measures were calculated in the EIA: the change in degreasing costs above baseline estimates, and the ratio of control costs to total costs. The increase in degreasing costs associated with the new rule was estimated to be slightly less than $2,000 per small model plant. Table 11-2 presents the EPA estimates of the *ex ante* control costs as a percentage of operating and total costs for 36 industries in the three-digit level of the Standard Industrial Classification (SIC). Perusal of this table indicates relatively small gross impacts of the proposed regulation: an increase above baseline cleaning costs of 6.4% for existing machines in a model plant and 6.0% for new machines in a model plant.[8] In terms of total production costs, impacts range from 0.02% to 0.61% for existing degreasers and 0.01% to 0.58% for new cleaners.[9]

The actual regulation, published in 1994, contained an updated analysis of various aspects of the regulation. EPA estimated that this NESHAP would reduce HAP emissions by 77,400 Mg (85,300 tons) per year by 1997, a 63% reduction below baseline (U.S. EPA 1994). The agency also estimated that the savings from decreased solvent use would exceed the additional costs of control equipment plus monitoring and recordkeeping, leading to a net reduction in

Table 11-2. *Cost Data from EPA's Economic Impact Analysis*

			Control costs as percentage of baseline cleaning costs		Control costs as percentage of total costs of production	
SIC code	Industry	Baseline cleaning costs as percentage of operating costs	Existing cleaners	New cleaners	Existing cleaners	New cleaners
254	Partitions and fixtures	6.2	6.4	6.0	0.39	0.37
259	Misc. furniture and fixtures	8.1	6.4	6.0	0.52	0.49
332	Iron and steel foundries	6.0	6.4	6.0	0.38	0.36
335	Nonferrous rolling and drawing	2.1	6.4	6.0	0.13	0.13
336	Nonferrous foundries (castings)	5.2	6.4	6.0	0.33	0.31
339	Misc. primary metal products	4.3	6.4	6.0	0.27	0.26
342	Cutlery, gandtools, and gardware	5.5	6.4	6.0	0.35	0.33
343	Plumbing and heating, except electric	3.7	6.4	6.0	0.30	0.28
344	Fabricated structural metal products	4.3	6.4	6.0	0.27	0.26
345	Screw machine products, bolts, etc.	5.4	6.4	6.0	0.34	0.32
346	Metal forgings and stampings	4.9	6.4	6.0	0.31	0.29
347	Metal services, n.e.c.	7.5	6.4	6.0	0.48	0.45
348	Ordinance and accessories, n.e.c.	8.6	6.4	6.0	0.54	0.51
349	Misc. fabricated metal products	5.4	6.4	6.0	0.34	0.32
351	Engines and turbines	4.2	6.4	6.0	0.27	0.25
352	Farm and garden machinery	5.2	6.4	6.0	0.33	0.31
353	Construction and related machinery	4.1	6.4	6.0	0.26	0.25
354	Metalworking machinery	6.5	6.4	6.0	0.41	0.39
355	Special industry machinery	5.3	6.4	6.0	0.34	0.32
356	General industrial machinery	4.3	6.4	6.0	0.27	0.26
357	Computer and office equipment	3.9	6.4	6.0	0.25	0.23
359	Industrial machinery, n.e.c.	9.7	6.4	6.0	0.61	0.58
361	Electric distribution equipment	4.2	6.4	6.0	0.27	0.25
362	Electrical industrial apparatus	5.6	6.4	6.0	0.36	0.34
364	Electric lighting and wiring equipment	4.8	6.4	6.0	0.31	0.29
366	Communications equipment	5.0	6.4	6.0	0.32	0.30
367	Electronic components and accessories	6.9	6.4	6.0	0.44	0.41
369	Misc. electrical equipment and supplies	5.9	6.4	6.0	0.38	0.35
371	Motor vehicles and equipment	4.1	6.4	6.0	0.26	0.24
372	Aircraft and parts	5.3	6.4	6.0	0.33	0.31
376	Guided missiles, space vehicles, parts	1.0	6.4	6.0	0.02	0.01
379	Misc. transportation equipment	5.7	6.4	6.0	0.36	0.34
382	Search and navigation equipment	6.2	6.4	6.0	0.39	0.37
382	Measuring and controlling devices	5.6	6.4	6.0	0.36	0.33
39	Misc. manufacturing industries	7.6	6.4	6.0	0.48	0.45

annual costs of $19 million—equivalent to finding the proverbial $50 bill lying on the ground. The EPA estimate includes a net annualized savings from installation of control devices of $30.5 million and total monitoring, reporting, and recordkeeping costs of $11.6 million. Based on these estimates, the EIA concluded that the incentive for solvent substitution was small. In addition, other considerations, such as level of necessary cleanliness, specifications from end users, and space and energy requirements, further reduce the incentives for substitution of nonchlorinated solvents.

In the final rule, EPA reported on the economic analysis as follows:

> The economic impact analysis done at proposal showed that the economic impacts from the proposed standard would be insignificant ... While the estimated annual costs for the regulation have increased since proposal, there are still cost savings for most affected entities (Federal Register Notice, 4).

There was no discussion in the final rule of the rationale for the chosen regulatory design, particularly the decision to adopt a hybrid rule that embodies elements of both command-and-control and economic incentive instruments. Even though economic incentives are mandated in the statute, the rule contains no analysis of the expected gains from this approach.

Methods and Data Collection

Ideally, one would use detailed source-specific data to examine the actual performance of the degreaser NESHAPS. Since many of the TCE sources are small and medium enterprises, it would be particularly important to have a data set with broad coverage of affected sources.[10] Such detailed data would enable us to develop a clear picture of the nature and types of sources that actually reduced their emissions in response to the regulations.

Unfortunately, emissions data for small and medium sources are not readily available. EPA does not maintain a national database of TCE users. Individual states keep their own records of facilities that release or use TCE, but these data are neither comprehensive nor uniform across states. Given the difficulties of developing a complete picture of the full range of TCE users in the United States, we decided to focus on the larger sources. Though relatively few in number, these sources are disproportionately affected by the regulation and likely represent the bulk of the TCE reductions (measured in pounds) actually achieved by the rule. For this purpose we rely on the Toxics Release Inventory (TRI), a well-known national data set.

TRI was mandated under the Emergency Planning and Community Right-to-Know Act (EPCRA) and consists of self-reported releases by major sources. If the chemical is used at the facility, the reporting threshold is 10,000 pounds per year. If the chemical is processed or manufactured at the facility, the minimum reporting level is 25,000 pounds per year. The TRI contains annual estimates of total air emissions of TCE since 1988. Reported releases of TCE account for about 40% of total TCE use in the United States.[11]

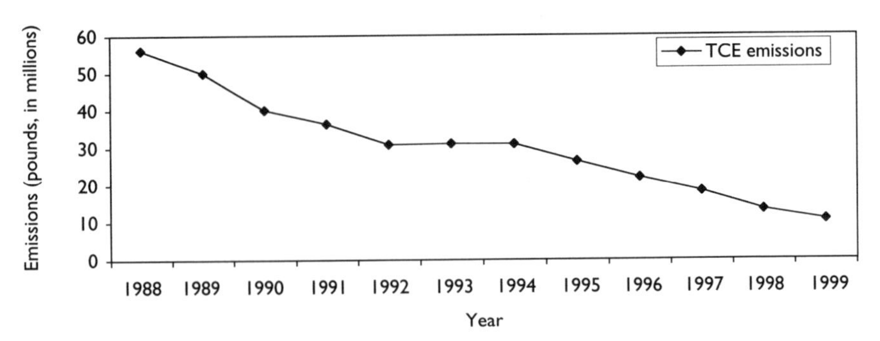

Figure 11-2. *Total TCE Emissions*

Source: Toxic Release Inventory (U.S. EPA).

To examine the industry-specific effects of the regulation, we disaggregated air releases data to the two-digit SIC level for the years 1988–1999. Specifically, we examine SICs 33 (primary metals), 34 (fabricated metals), 35 (machinery), and 36 (electrical and electronics equipment). These industrial categories had the largest number of reporting facilities in the TRI database. They were also identified as major users of degreaser machinery in the EIA.

Since we are particularly interested in assessing the economic incentive aspects of the early reductions provision, we selected a subsample of facilities for closer examination based on whether they qualified for the program, as determined by their reported TRI releases. Because a demonstration of at least a 90% reduction by the proposal date is required, facilities with reported TRI emissions in 1993 (or earlier) at least 90% below baseline were identified as potential candidates for the intertemporal trading provisions.[12] The first year of reporting (1988 or 1989) was adopted as a baseline.

Quantitative Information

Toxic Release Inventory Data

TCE releases reported in the TRI totaled almost 56 million pounds in 1988 and declined to about 10 million pounds in 1999 (see Figure 11-2). Over the 10-year period, this represents a reduction of more than 80% at the large (reporting) facilities. Trend analysis indicates a sharp decrease from 1988 to 1992 (45%), a slight increase from 1992 to 1995 (less than 1%), then a dramatic decline through 1999 (66%). Because of the minimum thresholds for reporting releases, these declines in emissions probably overstate the true reductions. Nonetheless, comparison with the limited production data publicly available indicates that the TRI data are generally consistent with production trends. Because of the discrepancies between TRI data and total releases, and because EPA's baseline scenario may have already incorporated some decrease in TCE use over time, the actual reductions observed in TRI are not totally comparable to the 63% decline in TCE releases predicted by EPA. However, assuming these factors introduce

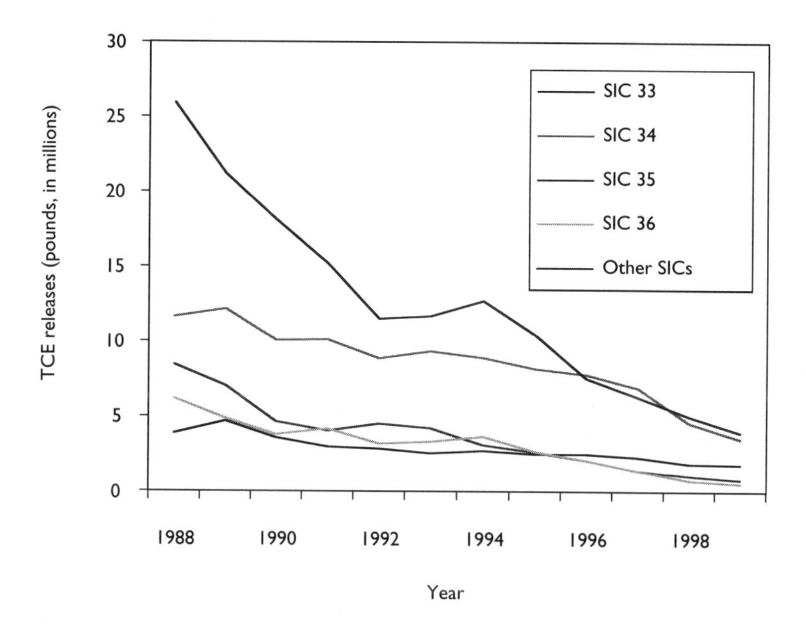

Figure 11-3. *TCE Trend by SIC Code*

Source: TRI.

only small biases, the observed 66% reduction in releases between 1995 and 1999 is roughly in line with the forecasts.

The downward trend in TCE releases from 1988 to 1992, before the NESHAP was proposed, is likely attributable to the stricter requirements for hazardous waste disposal as stipulated in the implementing regulations for the 1984 Hazardous and Solid Waste Amendments to the Resource Conservation and Recovery Act. The advent of required reporting to the Toxics Release Inventory may also, in part, explain the decreases. By the late 1990s, however, the observed declines are almost certainly attributable to the NESHAPS.

Figure 11-3 shows the trends in TCE air releases for the selected two-digit SIC industries. Of the four SIC codes examined, SIC 34 (fabricated metals) accounted for the largest total reduction in emissions. In both SIC 33 and SIC 34, emissions also decreased (62% and 72% from 1989 to 1999) but by a lesser extent than in SIC 35 and SIC 36 (about 90% each from 1988 to 1999). Interestingly, the industries with the highest growth rates—industrial machinery (SIC 35) and electronics (SIC 36)—also had the greatest emissions reductions. In these sectors, it may be that solvent degreasing is a relatively small part of the overall production process, or that substitutes may be more readily available. It may also be that the higher rates of output growth imply larger investments in new capital equipment and the correspondingly greater replacement of older, higher-emitting degreasing machines.

To get a clearer picture of the effect of the reporting threshold on the overall results, several calculations were performed. First, facilities whose last reporting year was 1993, 1994, 1995, 1996, or 1997 were dropped from the sample. Then,

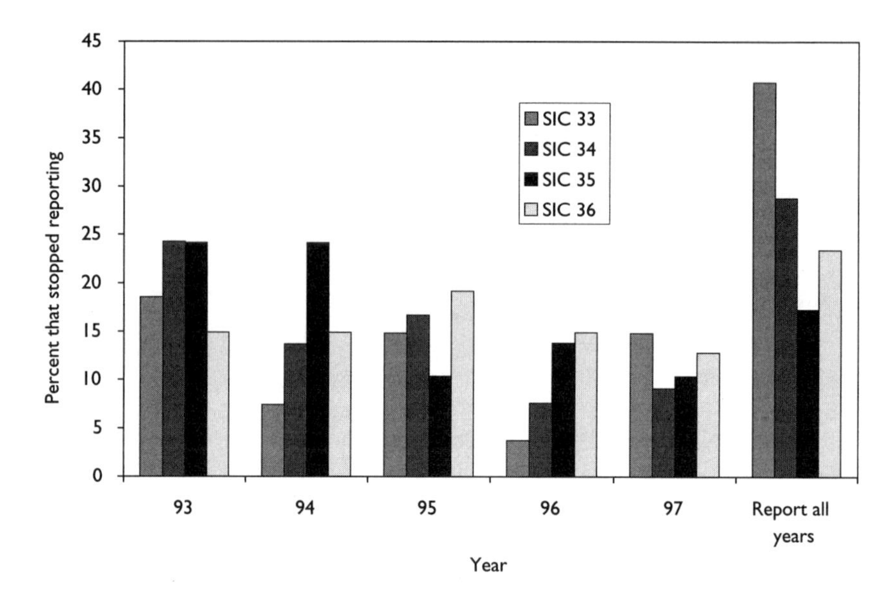

Figure 11-4. *Reporting and Cessation of Reporting of TRI Releases, by Industry, 1993-97*

for each year, the number of facilities that ceased reporting in that given year was divided by the total number of facilities that reported TCE releases.[13] Additionally, for each industrial category the number of facilities that reported every year was divided by the total number of reporting industries to calculate the percentage of facilities for which we have complete data. As shown in Figure 11-4, as many as 25% of the facilities ceased reporting in 1993, although that percentage tended to decline over time. This suggests that a sizable number of facilities reduced their emissions sufficiently to get below the reporting thresholds. SIC 34 (fabricated metals) had the largest number of facilities that ceased reporting in 1993 and 1994. This industrial category also had the largest total number of facilities in the data set. When adjusted for the total number of facilities which reported all years, however, SIC 35 (industrial machinery) showed the greatest percentage reduction in facilities that ceased reporting TCE emissions in 1993 and 1994.

To determine the number of facilities potentially eligible for the early reductions provisions, we adopted fairly conservative criteria. Specifically, we restricted our sample to those facilities that reported TRI releases for all years (1988–1999). Since we do not know the precise amount of TCE released once a facility ceases reporting, we can only assume that they have fallen below the reporting threshold. Thus, it is not possible to determine whether facilities that ceased reporting, especially in the early years, would be able to demonstrate the required 90% reduction needed to qualify for the early reductions provision.

A review of the facility-specific TRI data in the four two-digit industries analyzed indicates that of a total of 46 facilities that reported TCE release data for all years, none qualified for the early reductions program. One facility did demon-

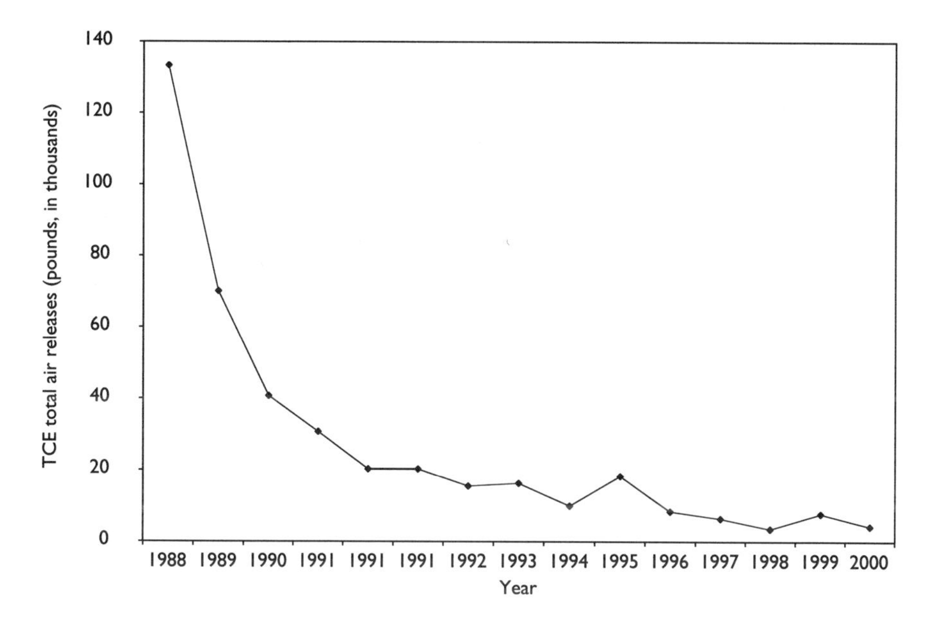

Figure 11-5. *Trends for Facility A1 in SIC 36*

strate a 95% reduction; however, emissions went up again, and therefore it was discarded as an early reductions candidate.

A 1994 General Accounting Office report (GAO 1994) that examined the effectiveness of the early reductions program as applied to a large group of MACT standards (including the halogenated degreaser rule) arrived at quite similar results. Of a total of 8,000 to 13,000 major sources subject to the MACT standards, only 40 applied for the early reductions program. Only 12 of the 40 applications were approved by EPA.

Since we did not find any candidates for early reductions through our TRI analysis, we decided to look at reductions that occurred prior to the compliance date (December 27, 1997). The same criterion for selection was used with this new date: we looked for a 90% reduction in TCE emissions by 1997 compared with 1988 (or 1989). Of the 46 facilities that reported all years, only 4 had a 90% reduction: 2 facilities in SIC 33, and 2 in SIC 36.

Both of the SIC 33 facilities are listed as steel pipe and tube manufacturers. Both are owned by the same parent company. One facility is located in Pennsylvania, the other in Delaware. Trends for them are shown in Figures 11-5 and 11-6. In SIC 36 our calculations indicate that two companies with different product types would have fulfilled our selection criteria. These are shown in Figures 11-7 and 11-8.

Interviews with Firms

Interviews were solicited from the TRI-listed contacts for each of the four facilities (A, B, C, and D) identified as potential candidates for the early reductions

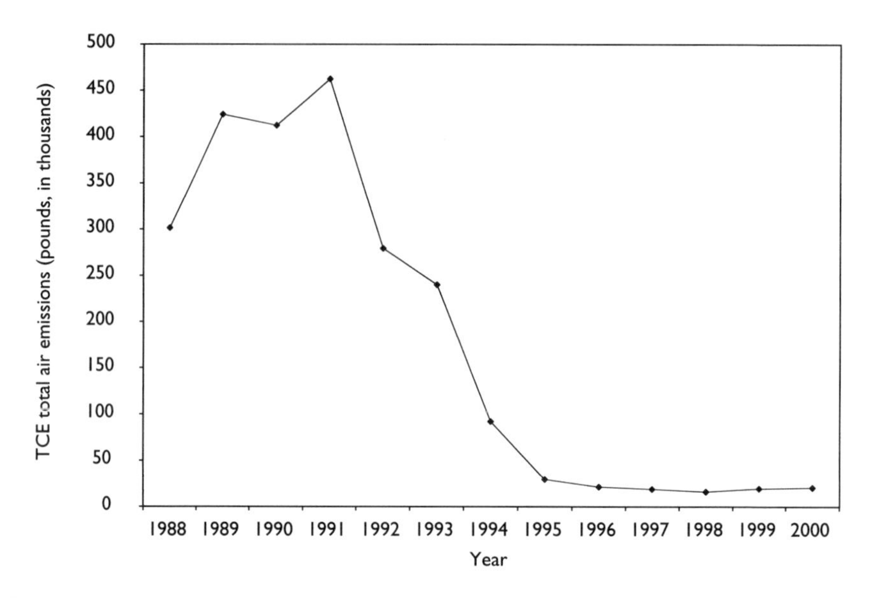

Figure 11-6. *Trends for Facility B, the Delaware SIC 33 Facility*

provisions. One facility (D) refused an interview. The remaining three facilities were owned by two different parent companies. The parent companies were manufacturers of ceramic capacitors (SIC 36) and steel tubing (SIC 33).

Phone interviews with a facility in SIC 36 provided no indication that the operators benefited from the early reductions provision or that they were even aware of it.[14] Neither is there any evidence in the TRI data that its release profile is any different from that of other regulated facilities in the same industry. The interviews also revealed that one of the qualifying facilities (facility A1) had moved part of its operations to another plant (facility A2) in 1990. Before the move, however, a phasedown in TCE releases had already begun. In an interview, the facility operator of A1 attributed this decline in emissions to the firm's purchase of a new, closed-system recycling degreaser in the late 1980s as well as to the installation of emissions controls on other units. In the mid-1990s, the older degreasers were shut down.

Facility A2 opened in 1990. Emissions from this facility did not show as large a decrease as in facility A1. However, emissions did drop steadily for three years after the NESHAP took effect. The facility operator we interviewed noted that in 1998, facility A2 made several changes, including switching to an automated parts-handling system and improved processing so that parts would require less degreasing. An additional control system was also added. After a large initial cost, significant savings in labor and operating costs were achieved. Experiments with other solvents and cleaners were also reported, but the company found that TCE worked the best. Interestingly, the company was recently bought out and will be moving most of its operations abroad. Regulations in this new country will not allow them to use TCE, and the facility operator reported that they will likely use perchloroethylene instead.

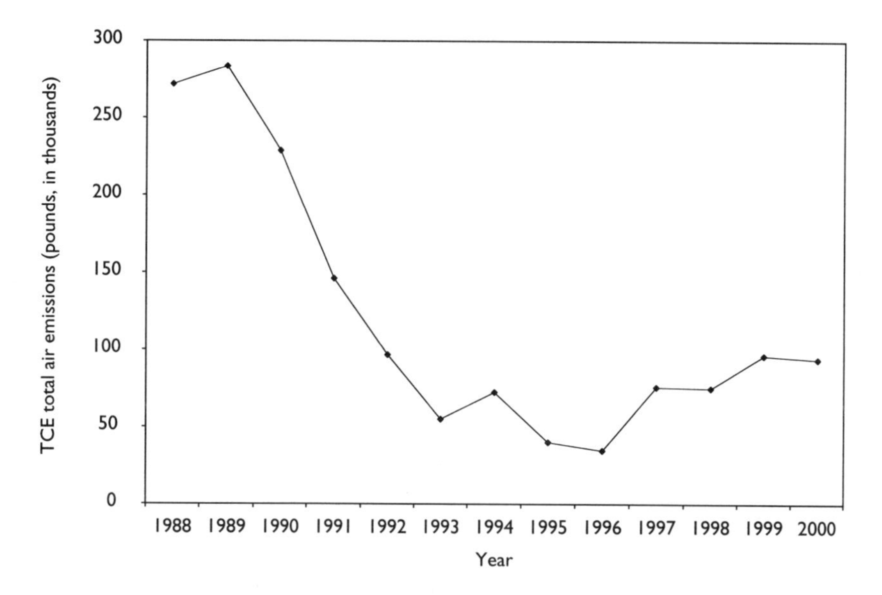

Figure 11-7. *Trends for Facility C, the Pennsylvania SIC 33 Facility*

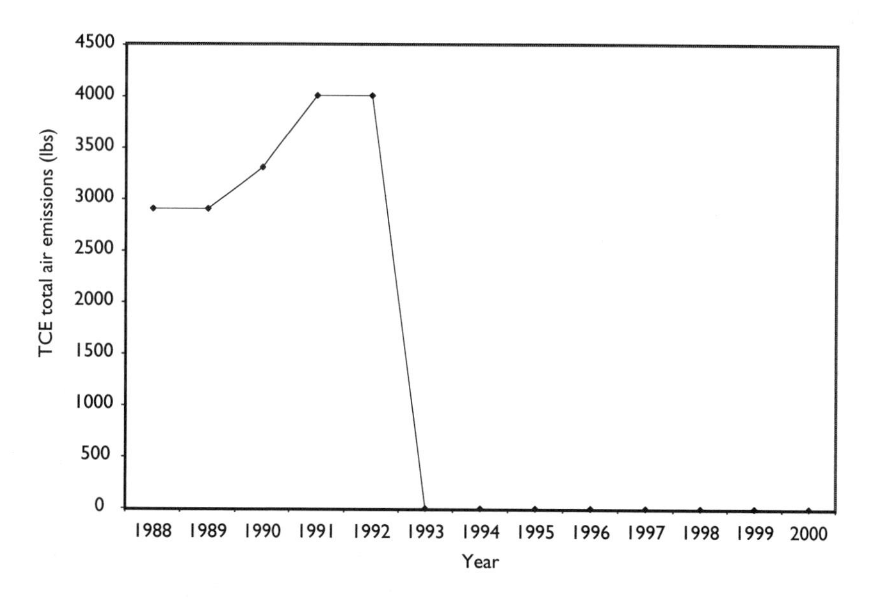

Figure 11-8. *Trends for Facility D, in SIC 36*

Interviews with the steel tubing manufacturer (facility B) revealed that most of the emissions reductions at one of its facilities between 1991 and 1996 were due to a cooperative effort between the facility and the state environmental agency. This company agreed to be a test case for a new NZE degreaser, a low-emissions technology. The principal reason offered by the firm for cooperating with the state agency was to gain experience with the technology in advance of the upcoming regulations. The new degreaser was installed before these regulations came into effect, and the firm reported significant savings on solvent disposal costs. Initial costs of this technology were high—about $0.5 million to $2 million. However, according to the company representative, this investment has paid itself back in significant savings on new solvent and disposal costs. As in the case of the facilities owned by the capacitor manufacturer, there is no evidence that the emissions of this facility in the later years were significantly different than those facilities that did not meet the early reductions criterion.

The steel tube manufacturer also reported another plant (facility C) in a different state, which did not purchase a new degreaser. This plant was smaller and in an older building, which made it difficult for a new degreaser to be installed. Instead, passive controls were installed on their open-top degreaser, a much less efficient technology. These controls were in compliance with the NESHAP. In this case, it would appear that the NESHAP requirements were not stringent enough to induce the company to radically change its degreasing practices.

Discussion

The results presented above paint a fairly clear picture concerning the effectiveness of the regulation but a much cloudier one on the issue of efficiency, especially with respect to the use of economic incentive instruments. The fact that reported TRI releases declined by about two-thirds between 1996 and 1999 is a strong indication of the environmental effectiveness of the NESHAP. Despite the previous decline in emissions in the late 1980s and early 1990s—probably attributable to rules issued under the Resource Conservation and Recovery Act regulation and, possibly, to TRI reporting requirements—there was a three-year period over which emissions stabilized and even increased slightly prior to the imposition of the NESHAP requirements. It was only after the new rule was promulgated that additional emissions reductions were recorded. Although we lack direct *ex post* information on TCE use in small sources, informal discussions with industry sources indicate significant declines in that sector as well.

The evidence on the efficiency of the rule is much less compelling, since we lack any direct *ex post* measure of compliance costs. What we do have is the anecdotal evidence from the steel tubing manufacturer, which reported significant savings on solvent use and disposal costs associated with the installation of a new NZE degreaser. We also have the *ex ante* estimates developed by EPA of relatively modest capital costs (less than $2,000) for small degreasers, plus estimated reductions in solvent use leading to EPA's overall findings of net savings associated with the rule.

Despite the absence of *ex post* cost information, we can draw some inferences about the likely costs or cost savings from the observed pattern of the emissions reductions. The fact that TRI emissions were relatively flat for several years in the mid-1990s prior to the promulgation of the NESHAP suggests that enterprises were not flocking to the new machines before they were obliged to do so. Although this does not preclude the possibility that the firms were uninformed at the time about the potential savings from the new degreasing machines, it does raise serious questions about the claim of net cost savings. If such savings were actually achievable, apparently they were not large enough to motivate the required investments in the early 1990s.

Apart from the cost issues, other factors may have contributed to the unpopularity of the early reductions provision. The General Accounting Office report (1994) mentioned several reasons why the early reductions program was not particularly attractive to firms, including difficulty in developing baseline and historical emissions data. Another reason cited involves the relationship between the federal MACT and state-specific regulations: because it was uncertain whether the MACT would turn out to be at least as strict as the state rules, it was not worthwhile for companies to spend effort applying for early reductions only to find that they needed a higher level of control to comply with the state regulations. Also, if the actual promulgation of the MACT was delayed (as often occurred), then the additional six years to comply might not be needed.

Arguably, the issue is not whether the rule generated net savings but whether the inclusion of a compliance option involving economic incentives served to lower overall compliance costs. On this point we have mixed evidence. We do know that firms took advantage of the emissions bubble, which suggests that one of the economic incentive options proved helpful to them and, presumably, lowered their compliance costs. However, since neither the TRI data nor our follow-up interviews provided any evidence that the early reductions provision was actually used by the affected enterprises, it seems unlikely that this provision provided any benefits to the affected facilities in the form of lower costs or greater flexibility, or to the ambient environment in the form of earlier emissions reductions.

Conclusions

Our case study of the regulation of TCE in degreasers under Section 112 of the U.S. Clean Air Act indicates that the command-and-control elements of the regulation were clearly effective in reducing emissions. In all likelihood, total costs—even if they were not negative—were quite low. However, the economic incentive elements of this hybrid rule were only marginally effective at best in introducing additional flexibility into the system. Although there is some indication that the emissions bubble was helpful to some facilities, we could find no evidence that the early reductions provision was used at all by the affected facilities.

These findings beg the obvious question of why the economic incentives were not sufficiently attractive to the affected firms. Several hypotheses suggest themselves. Perhaps there was enough flexibility in the performance-based rule in the first instance that the potential for further cost reductions through the early reductions provision was minimal. It is also possible that the marginal costs of emissions reductions were sufficiently nonlinear that once a 90% reduction over baseline was achieved, there was little to be gained from delaying future compliance dates. Alternatively, it is possible that the administrative burden or other transaction costs imposed by the permit writers was too great to lure any facilities to the early reductions program. Despite anecdotal information supporting all these hypotheses, we have been unable to locate hard evidence on these issues. The General Accounting Office report, the most specific source of information available, generally lends support to the third of our hypotheses. However, all we can say with confidence is that, for whatever reason, the early reductions provision offered by EPA was not attractive enough for the affected facilities to use it.

Our overall conclusion is that if economic incentive mechanisms are to be effective, firms must have sufficient motivation to use them. In the case of the MACT for TCE, it is hard to argue that the mechanisms served as anything other than window-dressing on a relatively low cost (possibly negative cost) command-and-control regulation. Whether stronger incentives might have yielded a different result is unknown. What is clear is that the voluntary economic incentives incorporated into the rule were not widely used by industry.

Notes

1. EPA has classified most of these solvents as either Class B or Class C carcinogens. Class B indicates probable human carcinogen, which means there are sufficient animal toxicology data but few or no human data. Class C is a possible human carcinogen, indicating that the animal data imply carcinogenicity, but the question remains unresolved. Trichloroethylene and perchloroethylene are between a B and C classification; methylene chloride and carbon tetrachloride are Class B carcinogens; and 111-trichloroethane is Class C. Currently EPA is revising the health risk information for trichloroethylene under its new cancer risk assessment guidelines.

2. In the Clean Air Act, the solvents are regulated as hazardous air pollutants (HAPs), a group of substances believed to have no threshold for cancer and certain other health effects. Under the Safe Drinking Water Act, they are subject to maximum contaminant levels. Under the Clean Water Act they are controlled via effluent guidelines. Spent solvent (the waste solvent) must be disposed under hazardous waste disposal guidelines stipulated in the Resource Conservation and Recovery Act. There are also relevant occupational exposure limits and state laws beyond these federal statutes. For major sources, allowable emissions of halogenated solvents are typically incorporated in source-specific permits.

3. Existing machines are defined as those which were constructed or reconstructed before November 29, 1993, the date the standard was proposed in the Federal Register. Existing machines have until December 1997 to comply; new ones must be in compliance by December 1994.

4. A major source is defined as "any stationary source or group of stationary sources located within a contiguous area and under common control that emits or has the potential to emit, considering controls, in the aggregate, 10 tons per year or more of any hazardous air pollutant or 25 tons per year or more of any combination of hazardous air pollutants." CAA §112(a)(1).

5. Federal Register 64(133), July 13, 1999, Rules and Regulations.

6. There is also a restriction on the inclusion of certain high risk HAPs, such as chlorinated dioxins and furans, in the calculations of the 90% reduction requirement.

7. Small facilities are defined as those with 1 to 19 employees.

8. The uniform results across industries apparently derive from the use of the model plant approach to develop the estimates.

9. The low estimates are from SIC 376 (guided missiles, space vehicles, and parts), and the high estimates are from SIC 359 (industrial machinery, n.e.c.).

10. Note that very small sources are exempt from the regulation altogether.

11. Based on consumption data reported in 1991 from the economic impact analysis and in 1998 from the Halogenated Solvents Industry Association. TRI releases as a percentage of consumption was between 30% and 40%. It appears that over time, the percentage of TCE use has decreased; however, the sources for these two estimates differed and the data may not be strictly comparable.

12. That is, facilities were identified as potential candidates for early reductions if $(Q_i - Q_{1993})/ Q_i \leq 0.9$, where Q_i = the quantity released to air in the baseline year; and Q_{1997} = the quantity released to air in 1997.

13. Numerically, this quantity is N_i / N_t where N_i = the number of facilities that ceased reporting in year i (i = 1993, 1994, 1995, 1996, 1997) and N_t = the total number of facilities reporting TCE to TRI, over time.

14. Given the passage of time since the implementation of the rule, the fact that they said they were unaware of the provision is not necessarily conclusive.

References

Michigan Department of Natural Resources. 1995. *Fact Sheet: How the Clean Air Act Affects Halogenated Solvent Cleaning Operations.* No. 9502, July.

University of Tennessee. 1995. *Clean Air Act Compliance for Solvent Degreasers: Regulatory Strategies for Manufacturers Affected by the Clean Air Act Amendments NESHAP for Halogenated Solvent Cleaners.* Center for Industrial Services and Tennessee Department of Environment and Conservation. April.

U.S. Environmental Protection Agency (U.S. EPA). 1993. *Economic Impact Analysis of the Degreasing NESHAP.* Draft. A-92-39. Research Triangle Park, NC. November.

———. 1994. *National Emission Standards for Hazardous Air Pollutants: Halogenated Solvent Cleaning.* Federal Register 59(241): 61801. December 2, 1994.

U.S. General Accounting Office (GAO) 1994. *Toxic Substances: Status of EPA's Efforts to Reduce Toxic Releases.* Report to the Chairman of the Environment, Energy, and Natural Resources Subcommittee, Committee on Government Operations, House of Representatives. GAO/RCED-24-207.

CHAPTER 12

Lessons from the Case Studies

Winston Harrington, Richard D. Morgenstern, Thomas Sterner, and J. Clarence (Terry) Davies

IN THIS CHAPTER WE consider what the six pairs of case studies have taught us. We are especially interested in looking at the cases from two vantage points. First, they allow us to observe and speculate on similarities and differences between European and American approaches to policymaking. Although we are unable to draw definitive and general conclusions, we do find some fault lines that seem to distinguish policymaking on the two sides of the Atlantic. Second, these cases provide lessons regarding the actual performance of different environmental instruments. Accordingly, we revisit the hypotheses laid out in the Overview regarding the characteristics of economic incentive (EI) and command-and-control (CAC) instruments for pollution abatement.

Transatlantic Similarities

The most important common characteristic in U.S. and European environmental policies is that countries employ a mix of traditional regulatory approaches and economic incentives. This is true not only for a nation's total portfolio of environmental policies but also, more surprisingly, for each nation's approach to individual problems. As Table 12-1 shows, every policy we reviewed except one actually contained a mix of EI and CAC instruments.

The mix of instruments is, in part, due to the inherent characteristics of the instruments themselves. Most traditional regulatory approaches rely on economic penalties as the primary negative incentive for compliance. However, in this study we have not included these penalties in our universe of economic instruments. More importantly, all incentive approaches rely to a greater or lesser extent on regulatory intervention. The traditional starting point for the standard environmental economics texts is that pollution is an externality (i.e., those who benefit from polluting do not bear the costs of the pollution, and vice versa), and thus free markets by themselves are not capable of solving pollution problems.

Incentive instruments such as permit trading make sense only in the context of fixed limits on total emissions from all sources. Even effluent charges, which may be the instruments that come closest to a pure market approach, rely on a non-market regulator to establish the level of the charge. As Jonathan Wiener (2003) states, "Nor is the advocacy of 'market-based instruments' based on the premise that 'the market' can solve all environmental problems; it is rather an effort to correct what are recognized to be market failures by adopting government policies that reconstitute incentives in environmentally desirable directions."

There are historical as well as theoretical reasons for the mix of EI and CAC policies. Incentive approaches have become politically acceptable only in recent years. On both sides of the Atlantic, regulation was the initial approach to most environmental problems. Thus in most cases where incentives have been used, a preexisting set of regulatory policies was in place. However, the mix can evolve in many ways. In the case of U.S. regulation of lead in gasoline, a regulatory policy led to an incentive approach, which then reverted to a ban, another regulatory policy.

Sometimes regulatory policies lead naturally to incentive approaches. This was true of the first U.S. emissions trading policies, which evolved as a way of accommodating some growth in areas with substandard air quality levels. Harrington's case study of the effluent guidelines program captures this type of evolution. A basic regulatory program is beginning to look much like an incentive program as discharges to publicly owned treatment works (POTWs) become more important than direct discharges and as the POTWs increase their charges on pollution sources.

The extent to which our cases found mixed EI and CAC approaches seems greater than can be accounted for by theoretical and historical reasons alone. Based on our cases, it seems that incentive and regulatory instruments have different advantages and drawbacks, they distribute costs and benefits differently, they appeal to different constituencies, and these differences are recognized by both regulators and politicians. This recognition makes it attractive to employ a strategy of mixed approaches. A mix has the potential to maximize both efficiency and effectiveness, to appeal to multiple constituencies, and to avoid some of the pitfalls of both obtuse bureaucratic regulation and unbridled market incentives.

Does the mix of approaches that we found in the real world undermine the hypotheses with which we began the study? We do not think so. The case study authors have largely succeeded in focusing on the incentive or regulatory aspects of the policies they studied. Analytically, the two types of approaches can be separated, the authors have done so, and the overall findings with respect to our original hypotheses constitute the bulk of this chapter.

European and American Regulatory Performance

One versus Many

Our six case study pairs illustrate that American and European regulators have to deal with similar environmental problems, but they often deal with them in different

Table 12-1. *Comparison of Policy Instruments in The Case Studies*

Case	*Incentive elements*	*Regulatory elements*
E.U. acid rain: Germany (Chapter 1)	—	Stringent technology-based standards for utility boilers
U.S. acid rain (Chapter 2)	Marketable permits distributed to existing power plants (1990)	BACT for new power plants (1977); RACT for existing plants (1977); new source review (1977)
U.S. industrial water pollution (Chapter 3)	Direct dischargers (1972): state tradable permit programs in water-quality-limited river basins Indirect dischargers (1972): tradable rights to POTW capacity (New Jersey); sewer surcharge fees on BOD, TSS, various measurements of nitrogen	Direct dischargers: permits based on effluent guidelines Indirect dischargers: federal pretreatment standards for some industries and pollutants; local limits for other industries
E.U. industrial water pollution: the Netherlands (Chapter 4)	Pollutant discharge fees primarily for oxygen-demanding substances	Discharge permits issued by district water boards
E.U. NO_x emissions: France and Sweden (Chapter 5)	NO_x emission fees. $3,000/tonne in Sweden recycled to industry based on output; $40/tonne in France used to subsidize abatement	France and Sweden: emissions permits required for all sources
U.S. NO_x emissions (Chapter 6)	Trading program in ozone transport region (1999); NO_x SIP Call (2003)	Technology-based standards for new (1971) and existing (1996) utility boilers
U.S. and E.U. ozone-depleting substances (Chapter 7)	U.S. response to Montreal Protocol (1987): tradable permits for production and consumption; excise tax on permits. E.U. response to Montreal Protocol (1987): tradable production or import permits; individual country actions (Austria, deposit-refund system for refrigerants; Denmark, tax; Sweden, fee on successful applications for exemptions)	U.S. pre–Montreal Protocol: prohibitions in specific applications (1979–1987); labeling requirements in individual states (1975) U.S. response to Montreal Protocol: prohibition of small-quantity sales; SNAP rules E.U. pre–Montreal Protocol: aerosol bans in Norway and Sweden (1979) E.U. response to Montreal Protocol: comprehensive controls in Austria, Denmark, Finland, Germany, Italy, Netherlands, Sweden No comprehensive legislation in FR, GR, IR, PO, SP, UK

U.S. leaded fuel (Chapter 8)	Supply side: trading and banking of permits through interrefinery averaging (1982–1987)	Supply side: catalysts in new vehicles (1975); refiners required to provide unleaded fuel (1974); lead-content standards (1979–1982) Demand side: prohibition of leaded fuel in vehicles with catalytic converters, enforced by inlet restrictors
E.U. leaded fuel (Chapter 9)	Demand side: differential fuel taxation, making leaded fuel more expensive than unleaded (1985, Sweden and Austria; all E.U. countries by 1990)	Supply side: catalysts in new vehicles (1986); mandated availability of unleaded fuel Prohibition of leaded fuel in vehicles with catalytic converters, enforced by inlet restrictors (1985–1990)
E.U. trichloroethylene (Chapter 10)	Product tax (Norway, Denmark)	Technology-based standards (Denmark); product ban (Sweden)
U. S. trichloroethylene (Chapter 11)	Within-facility emissions bubble; early adoption incentives	MACT standards for hazardous pollutants

Notes: BACT: best available control technology; BOD: biochemical oxygen demand; MACT: maximum achievable control technology; POTW: publicly owned treatment works; RACT: reasonable available control technology; SIP: state implementation plan; TSS: total suspended solids.

ways. The most obvious difference is that the European Union consists of many countries. On the other side of the Atlantic, the United States has allocated principal responsibility for environmental rulemaking to the federal government, even though the states retain primary responsibility for permitting and enforcement. In the United States, this centralization of environmental policymaking is primarily the result of a series of landmark statutes passed between 1969 and 1980.[1] It is not clear that these centralizing moves were part of a grand plan; rather, they appear to have been prompted more by *ad hoc* concerns. First, some environmental problems crossed state lines. More importantly, there was an air of crisis at the time, a concern that environmental problems had to be dealt with right away or there wouldn't be any environment left. Most of the states had, in the minds of many, demonstrated that they could not act quickly enough or forcefully enough to deal with the multitude of environmental problems they faced. The federal government was also sufficiently powerful to stand up to the large corporations that were presumably the primary source of environmental degradation. Federal authority over environmental policy also avoided the much-feared "race to the bottom," in which polluters would shop around for lenient states willing to sacrifice environmental quality for new jobs and economic growth.

In contrast, the environmental tide in Europe reached flood stage at different times in different countries, beginning in the late 1960s in the wealthiest nations of western Europe, especially the Nordic countries, and sweeping south and east to reach the countries of the old Soviet empire by 1990 or so. Each country adopted its own policies according to its own timetable. But more recently, the gathering momentum of economic, social, and political integration in the European Union has also provided a centralizing impulse to all kinds of policies, not least environmental policy.

Our European cases reflect this mix of country-specific and E.U.-wide policy initiatives. For example, the trichloroethylene (TCE) case study concentrates on Sweden's ban but also explores Norway's tax and Germany's quite stringent regulation (just short of a ban). The mix of country-specific and E.U.-wide policies is also seen in the leaded gasoline case. Individual countries had their own policies on introducing vehicles with catalytic converters. But if a country mandated catalytic converters, its citizens' intercountry travel would require *all* countries to introduce unleaded fuel, a measure that was fully implemented in the European Union by 1989. The European Union also implemented regulations in 1981 specifying the maximum and minimum content of lead in leaded fuel, and in 1998 it specified a complete switchover to unleaded fuel by 2005. However, the heterogeneity among European countries on this issue, as on other issues, is large. Individual countries completed the phaseout of lead as early as 1994 (Austria), but as of 2000, leaded fuel made up nearly 20% of the gasoline supply in neighboring Italy.

Selection bias is clearly an issue for the cases included in this volume. This bias occurs in the regulations selected for study and in the particular nations chosen to represent Europe. In most instances the European cases were chosen because an innovative policy had been adopted in a particular jurisdiction. Thus, most of the European regulations selected for this volume represent the exceptional

European action, whereas the U.S. regulations are, by their very nature, national in scope. In that respect the U.S. actions can be thought of as representing the average rather than the exceptional policy.

Further, these regulatory comparisons completely ignore the issues of timing and preexisting environmental conditions. Earlier pollution reductions are more valuable than those undertaken at a later date. Thus, the observation that the U.S. and European nations both phased out chlorofluorocarbons (CFCs) ignores the fact that the United States began the CFC phaseout almost a decade earlier than the Europeans. Similarly, although a few northern European nations beat the U.S. timetable on the phaseout of lead in gasoline, most lagged behind—in some cases by a decade or more. Moreover, neither the preexisting pollution levels nor the difficulties of achieving particular reductions are fully considered in the case studies.

Ex ante *Analysis*

Our case studies also suggest that European and American regulations differed substantially in the amount and nature of *ex ante* analysis. This difference is largely attributable to the long-standing requirement imposed on U.S. government agencies to carry out a Regulatory Impact Analysis (RIA) to project the economic consequences of regulatory proposals and make it possible to compare the predicted direct effects of the regulation with the cost.[2] For each regulatory alternative to control pollution, analysts had to estimate the environmental effects, the abatement costs, any other indirect costs, and often any economic dislocations (such as plant closures or unemployment) likely to result from the regulation.

There was no European counterpart to this requirement. Very likely regulators, regulated companies, and other parties in Europe were just as concerned about the benefits and costs of regulations. But without a formal requirement to produce a public report, there was no paper trail available to researchers. Although our case study authors were in most cases able to reconstruct or infer estimates of the environmental consequences of the regulations in question, in only two cases could they find an *ex ante* estimate of anticipated costs. In three other cases, *ex ante* estimates of environmental effects were easy to produce because the policy in question was a ban on discharge of a particular substance, and the anticipated change equaled the current discharge.

Perhaps coincidentally, it is interesting that both cases in which emissions reductions exceeded expectations involved market-based approaches, and both cases where actual performance fell short of expectations relied on traditional command-and-control measures. This result is discussed in more detail in Harrington et al. (2000).

Regulatory Stringency and Effectiveness

A third potential point of comparison between European and U.S. case studies lies in the stringency and effectiveness of the corresponding regulations. In light of recent transatlantic environmental controversies, such as the U.S. refusal to sign the Kyoto Protocol on global climate change and the genetically modified food fight, it would be easy for the casual observer to conclude that Europeans are much more concerned about environmental quality than Americans are.

It is useful here to keep in mind the relative diversity of E.U. environmental policies compared with the U.S. case, where uniform national policies have enforced a substantial degree of homogeneity. When U.S. and E.U. policies are compared, the European representatives are usually the countries of northern and western Europe, which tend to have the most stringent policies.

The cases do not support the notion that European environmental regulations, even in northern and western Europe, have consistently been more stringent, either in design or in performance. We have a case (emissions of sulfur dioxide, SO_2) in which the European policy was more stringent in design and more effective in performance than its U.S. counterpart; it was also much more expensive per unit of emissions reduction achieved. We have a case (TCE) in which the European environmental objective in one country (Sweden) was more ambitious than the U.S. goal yet the environmental consequences are comparable. We have a case (industrial water pollution) in which the U.S. policy was more ambitious, at least initially, and yet lagged in performance well beyond the statutory deadline. We have still other cases (CFCs and leaded fuel) in which the outcomes in European countries and the United States are almost the same.

The Importance of Litigation

One reason frequently given for the apparent differences in outcome is American litigiousness. It is a rare U.S. environmental regulation that does not end up in court. In contrast, litigation is unusual in Europe. In the United States, proposed rules are challenged in regulatory proceedings and, after promulgation, in federal district courts. These district court decisions are frequently appealed to the circuit courts, and even further appeals to the Supreme Court are not uncommon. Every one of the U.S. policies in our sample of cases provoked courtroom challenges, and some, such as the effluent guidelines, provoked hundreds of challenges. In the European TCE case, Sterner reports that the Swedish ban was challenged by the Swedish chemical industry; such legal contests are relatively unusual in that culture.

American litigation has had several effects on regulatory productivity. First, it has delayed promulgation of rules. Numerous lawsuits have prevented implementation of the Clean Water Act's best available technology (BAT) rules and tied up the resources of the Environmental Protection Agency (EPA) in revising the remanded rules. On the other hand, environmental groups have been successful in using lawsuits to sped up rulemaking (a Natural Resources Defense Council suit put EPA on a schedule of regular issuance of BAT rules, for example) and in some cases led to more stringent rules. Generally speaking, though, most litigation, especially during the early years (1970–1985) served to relax and delay the rules. (See Melnick 1983 for the classic account of the gap between legislative ambitions and performance in the Clean Air Act.)

Product Bans

Product bans, the polar case of regulatory stringency, are well represented in our sample of cases on both the European and the American side. Several regulations fell into this category: both leaded fuel cases, both CFC cases, and TCE in Sweden. Some might add industrial water pollution in the United States, since it established a goal of zero discharge. However, this "ban" is much larger in scale and scope than the others, and it is possible that some in Congress regarded the goal as aspirational or rhetorical.

Of these, the leaded fuel and CFC bans were ultimately successful. Swedish TCE has failed to achieve 100% emission reductions, and although it has come quite close, achieving the goal is taking much longer than first imagined. It should be noted that comparable reductions of TCE were achieved in the United States (which did not impose a ban). U.S. industry also failed to achieve zero discharge of effluents, as required by the statute, but it is a special case.

If a ban is successful, then the postregulation pollution level is zero. Thus it is often easier to determine whether a ban has been effectively implemented than to estimate the effects of a policy that does not attempt to eliminate the pollutant entirely. Ordinarily, the technology for implementing a ban is the substitution of a new process or product, which is easier to observe.

Use of Economic Incentives

Overall, we found little difference between the countries of Europe and the United States in the predilection for using economic incentives. However, we did find a dramatic difference in the type of instruments used. Among our cases, nearly all incentive policies in the European Union are emissions taxes, whereas nearly all such policies in the United States are marketable permits. The only exceptions are found in the policies for ozone-depleting substances, where a supplementary emissions tax was used in the United States and a tradable permit system was used in the European Union. Why is this?

One possibility is that American environmental policies are more likely to have ambient objectives supported by specific targets and timetables for emissions reductions. As noted in the Overview, one characteristic of emissions fees is that setting the level of the fee determines the final marginal cost of abatement, not the pollution abatement target. Europeans, perhaps, are more likely to be thinking in terms of making the polluter pay for the pollution, rather than setting some prespecified ambient (or emissions) target. This may be related to subtle transatlantic differences in the perception of society and its rights over nature.

The difference also could be related to different attitudes to taxes. European taxes are generally higher than in the United States and are structured differently, with more reliance on commodity taxes and less on income taxes. Europeans are accustomed to extremely high tax rates on some products, such as motor fuel, where taxes can make up more than 80% of the total price of the product. Europeans are also more comfortable with the notion of using taxes to achieve policy goals other than raising revenue.

The preference for tradable permit systems in the United States dates back to 1976, when it became clear that some cities would not meet the 1977 deadline for attainment of EPA's national ambient air quality standards. Under the Clean Air Act, this meant no new air pollution sources or expansion of existing sources. However, the agency allowed new or expanding sources to proceed with their plans as long as they could find offsetting emissions reductions among existing sources. Thus were the bubble and offset policies born.

An additional element pushing EPA toward marketable permits was the fact that the agency has no authority to levy new taxes. That power belongs to—and is jealously guarded by—congressional tax-writing committees, and the members of those committees (supported by the Treasury Department in the executive branch) are generally uninterested in using the tax system for anything except revenue collection. (The complexity of the U.S. tax system reflects intense interest in from whom and in what way the money is collected.) Environmental taxes also have been vehemently opposed by the industries that would be subject to the tax.

Equity and Political Influence

In recent years questions about the fairness or equity of environmental regulation have arisen on both sides of the Atlantic. The most common question in public discussions is whether different income or racial groups have been disproportionately affected by either uncontrolled risks or the costs associated with regulations. Accordingly, we asked the case study authors to pay particular attention to this issue. The results are somewhat surprising.

When the question is posed as a potential impact on the general population, only the U.S. case of lead in gasoline seemed to involve significant public debate on equity issues.[3] Interestingly, what was identified in several cases as a significant "equity" issue was the potential burdens of the regulation on segments of industry—mostly on small business.[4] Five of the cases identified small-business impacts as significant (U.S. TCE, French NO_x, U.S. water, U.S. lead, Swedish lead). In each of these instances the governments explicitly addressed the equity issue in the design of the regulation. The most common regulatory response was to exempt at least some categories of small business from the regulation. In the case of U.S. TCE, for example, most small gasoline stations and auto repair shops were exempted.

An additional response to equity concerns can be seen in the different schedules by which particular nations adopted new regulatory requirements. This is particularly true in Europe, where E.U. directives generally permit a good deal of flexibility at the country level. The European case on lead phaseout highlights this issue. Specifically, the authors argue that despite the relatively high exposure to airborne lead among low-income groups—and thus the obvious gains to this group from rapid adoption of the lead phaseout program—several European nations, particularly low-income nations, have been slow to phase out lead. Although part of this response may be explained by the differences in perceived benefits, the authors suggest that this differential timing is based, in part, on cross-national equity concerns.[5] The most common U.S. response of this kind is the distinction made between new and existing facilities. Many U.S. pollution control policies exempt older facilities from the standards that must be met by new facilities

Comparison of EI and CAC Approaches

The selection of individual case studies was governed in large part by a desire to compare performance of different environmental instruments in at least a partially controlled manner. The "control" is the common environmental problem addressed in two or more jurisdictions. We now return to the hypotheses, discussed in the Overview, on the relative performance of economic incentive and command-and-control instruments, asking how those hypotheses are informed by our case studies. Fixing the environmental problem to be addressed makes meaningful some comparisons that would not be clear if we compared different instruments in the abstract. We can make useful comparisons about stringency, abatement cost, technologies employed, introduction of new technology, and speed of implementation. It is these comparisons that yield the most insights about the hypotheses.

However, our approach has some important limits. First, we suffer from the common problem of all case study research—a small number of observations, not randomly selected. Second, these cases differ in many aspects besides the policy instrument chosen, including political institutions, history, and preexisting environmental quality. These differences can affect the outcomes we observe as much as the difference in instrument. The complexity of the regulatory history or the varying structure of regulation in most cases precluded straight-up comparisons of a U.S. case with its European counterpart across all the hypotheses. Accordingly, we make tentative judgments about some hypotheses by examining only one case of a pair. Sometimes such comparisons are facilitated by the application of both types of instruments at different times, such as U.S. leaded fuel.

Third, as discussed at the beginning of this chapter, our instrument comparisons are rarely as clean as we would like. For each hypothesis, what we would like is to measure the quantitative difference in a response between the EI instrument and the CAC instrument. For example, the first hypothesis on efficiency would be supported if we could conclude that the unit cost of abatement is less for the incentive instrument than for the corresponding regulatory instrument. But typically those differences cannot be observed quantitatively, because one case study is unable to report anything of relevance on the variable in question, or differences in another variable hinder our ability to make comparisons, or the different regulatory approaches cannot be analytically separated to a sufficient degree. To continue with the efficiency example, a difference in unit costs between the two policies could be due to a difference in overall stringency, which would have no particular implications for the efficiency hypothesis.

On the other hand, the richness of the case studies means that we can often arrive at a tentative judgment about a hypothesis even if we have results from only one of a pair of studies. For example, the finding of high administrative costs in the U.S. water pollution case is based on the extremely large number of regulations that had to be written and on the extensive delays. This observation required no close comparison with the Dutch opposite number, which incurred no similar delays. However, the observation that the administrative costs of the German SO_2 policy (also regulatory) were relatively low compared with the U.S. acid rain policy raises the question of whether it is the instrument that is driving administrative costs, or American litigiousness.

1. Static efficiency. Incentive instruments are more efficient than regulatory instruments.

The notion that economic incentive mechanisms can deliver emissions reductions at lower social cost than regulatory instruments is, perhaps, the principal textbook advantage of such approaches. Although not all the cases made explicit comparisons between the two instruments, those that did were uniform in their findings of efficiency advantages for economic incentive measures. Some of the larger gains from such measures are found in the U.S. SO_2 and lead phasedown programs. In the SO_2 case, Burtraw and Palmer argue, after accounting for exogenous changes in fuel markets in a consistent manner, that realized costs are almost one-half the levels indicated in information available to legislators in 1990, and perhaps one-half again (i.e., one-quarter) the cost of a hypothetical CAC technology standard.

For the lead case, Newell and Rogers report that the banking program itself, by allowing for a more cost-effective allocation of new investment within the industry, saved more than $225 million over an inflexible command-and-control approach. The whole program, including trading, is estimated to have saved considerably more.

In the Dutch water case, Bressers and Lulofs note that the degree of pollution reduction in the early years corresponded well to the production cost incentives created by the fees, which indicates that discharges were reduced fairly efficiently. In contrast, in the United States, the *ex ante* estimates of marginal abatement costs varied by a factor of 30 across a sample of industry categories. Thus, even though no additional information was available *ex post*, Harrington suggests it was not very likely that cost-effective reductions were achieved.

In the German SO_2 case, Wätzold notes that there can be limits to the static efficiency advantages of economic incentives. In cases where the regulations are extremely stringent and everyone has to do all that is technologically feasible to meet the emissions goal—as in the case of the German SO_2 regulations—the gains from trade are very limited.[6] In that case, economic incentives are not likely to yield significant savings without jeopardizing achievement of the stringent quantity objectives.

By contrast, there is solid evidence that economic incentives achieved substantial cost savings when used to orchestrate the elimination of CFCs and lead in gasoline. Presumably they worked in these cases of maximal stringency because there were cost heterogeneities that could be exploited during the phaseout period. In the German SO_2 case, the similarity of the power plants and the abatement technologies available suggest relative uniformity of costs, thereby reducing the potential advantage for economic incentives. In the Swedish TCE case, many of the problems encountered were due to the fact that incentive-compatible mechanisms were not used. It was the small number of high-cost abaters who caused much of the commotion.

Overall, it appears that the cases in this volume do lend support to the textbook proposition that economic incentives are more cost-effective than command-and-control approaches to pollution control.

2. Information requirements. Generally, EI instruments require less information than CAC instruments to achieve emissions reductions cost-effectively.

The claim of smaller information requirements for market instruments refers to the burdens placed on the regulator as opposed to the regulated entity. It applies particularly to those cases in which abatement costs vary considerably among plants. The critical question is, how much information about costs does the regulator need to establish cost-effective standards? For CAC approaches, it is evident that the regulator needs a great deal of cost information before the standards can be established. In contrast, for EI instruments, the regulator does not require the same detailed cost data. Rather, much of it can remain with the regulated entities, who in turn will have obvious incentives to develop accurate, facility-specific information that they might not otherwise be motivated to collect or share with regulators. (See the discussion on the cost revelation hypothesis, below.)

In addition, in the case of EI instruments, the regulated entities also have strong incentives to act so that the resulting emissions reductions are cost-effective, especially when questions arise about the technological or economic feasibility of particular regulations. In a regulatory system, it is often difficult for regulators to judge claims of infeasibility of any kind, since feasibility can depend heavily on site- or firm-specific factors. In an incentive-based system, what is feasible is revealed in the firm's response.

However, several other kinds of information are needed to make successful environmental policy, including the enforceability and environmental effects of proposed regulations. In these areas, incentive instruments have no obvious information-economizing advantages over regulatory instruments. Because we treat these types of information specifically below (when we consider the hotspot and monitoring hypotheses), we focus here only on the cost information.

In the cases in which incentive instruments (taxes) were used in Europe, it does not seem that the authorities had to acquire a great deal of information about plant-level costs. For incentive instruments used in the United States, it is difficult to determine whether the choice of instrument affected information requirements. First, because of the emphasis on *ex ante* studies in the United States, some information about costs for leaded gasoline, flue gas desulfurization, and NO_x control had to be collected prior to implementation (or had already been collected to implement earlier regulation). Second, the novelty of economic incentives, together with the need to meet emissions reduction targets specified in legislation, required careful analysis of regulatory proposals prior to issuance to raise the confidence of all parties that the new regime was going to work as anticipated. Still, it appears clear that the information requirements at the plant level were lower than they would have been for a traditional permitting program.

When we turn to traditional regulation, in the United States we find substantial information requirements even when no attempt is made to write cost-effective regulations. For technology-based standards, it was originally thought that the main task of information collection was to identify a standard and a technology that would meet that standard, and that that task could be accomplished by considering broad classes of industries, with no need for more-detailed data collection. However, Harrington reports that for EPA's effluent guidelines, the sheer heterogeneity of industry and industrial processes was staggering. The information requirements of technology-based standards turned out to be formidable because of the multiplicity of industries and technological processes and often the need to set several standards for each process. Although the same abatement technologies could be specified for several different industries, the agency nonetheless had to provide evidence that the designated technology was a feasible choice for that industry. Furthermore, equity considerations—both substantive and procedural—placed additional information collection burdens on EPA. Typically, the agency has been besieged with claims that great harm to petitioning firms will follow from the regulations, often supported by extensive documentation, which in turn places additional burdens on the agency to respond. Under the terms of the Administrative Procedures Act, EPA is required to take these claims seriously, often by conducting further analysis and sometimes by collecting new data. In Europe there is less evidence of this particular problem, although it is partly evident in the Swedish prohibition on TCE.

3. Dynamic efficiency. The real advantages of EI instruments are realized only over time, since they provide a continual incentive to reduce emissions and permit maximum flexibility in the means of achieving those reductions, thus encouraging more efficient production and abatement technology.

Both prescriptive regulatory policies and incentive mechanisms like cap-and-trade programs can provide incentives to find low-cost ways to achieve reductions. Unlike the prescriptive policies, however, the cap-and-trade program provides an incentive to harvest low-cost emissions reductions at specific facilities even after the performance standard is achieved because those reductions can avoid investments at other facilities.[7]

The evidence from the case studies on dynamic efficiency lends general, although not universal, support for the textbook view that market-based instruments provide greater incentives than regulation for continuing innovation over time. The U.S. lead and SO_2 cases, as well as the Swedish NO_x case, offer strong support for this hypothesis.

In the case of the U.S. lead phaseout program, the authors note that the pattern of technology adoption was consistent with an economic response to market incentives and plant characteristics. Specifically, they found a significant divergence in the pattern of technology adoption among refineries with low versus high compliance costs: low-cost refineries (i.e., expected permit sellers) significantly increased their likelihood of adoption relative to the high-cost facilities (expected permit buyers) under market-based lead regulation compared with individually binding performance standards. Interestingly, in the case of the

U.S. SO_2 program, the authors note that innovation occurred principally through changes in organizational technology (i.e., the organization of markets), and through experimentation at individual boilers rather than through more traditional measures, such as patentable discoveries. They argue that under a prescriptive regulatory approach, the incentives would not have existed for some of these discoveries, such as fuel blending and scrubber performance improvements.

Sterner and Millock report a somewhat similar finding. The Swedish NO_x charge created strong incentives for fuel switching, modifications to combustion engineering, and the installation of specific abatement equipment, such as catalytic converters and selective noncatalytic reduction. Equally important, the fee created incentives to *use* the equipment to fine-tune combustion and other processes in such a way as to minimize emissions. The Swedish experience suggests a strong connection between the monitoring requirements and the observed emissions reductions via fine-tuning. The monitoring, in turn, became a reality only because of the high charges, which had to be based on accurate emissions figures.

In the Dutch water pollution case, although much of the initial response represented little more than good housekeeping measures, subsequently more advanced, so-called process-integrated measures were also taken. Furthermore, the Netherlands became a world leader in the development of new water purification technologies, including nitrate bacteria and membranes. In addition, engineering consulting firms sprang up in the country in response to the new policies. Overall, it is estimated that on average the unit abatement costs for organic pollutants dropped by half between 1986 and 1995.

Regulatory approaches can also create incentives for innovation. For example, Wätzold finds that the German SO_2 ordinance was truly technology-forcing. The required stringency put the regulation at or beyond the technological frontier for flue gas desulfurization, and utility officials were very concerned whether compliance was even possible. However, the pollution abatement industry rose to this challenge. Indeed, Wätzold observes that the "regulatory ratchet"—the incentive for firms to avoid innovation if it simply means that they will be subject to more stringent regulation—does not apply to vendors of pollution abatement equipment, who have strong incentives to demonstrate advanced technology, regardless of policy instrument.

Similarly, Burtraw and Evans report that some experimentation with innovative postcombustion controls occurred for compliance with the U.S. reasonable available control technology standards of the ozone transport region (basically, the Northeast), even though the abatement policy was relatively inflexible. They also note that those plants that engaged in experimentation in the region were often treated differently by regulators and some even received subsidies from the Department of Energy. Thus, innovation incentives in regulatory regimes often come about through administrative procedures or exceptions that are supplemental to the regulation.

A similar finding of possible innovation under command-and-control is also found in the U.S. water pollution case. As Harrington notes, if the adoption of process changes instead of end-of-pipe treatment is taken as the measure of innovation, then one can clearly observe a significant increase in the use of innovative technologies during the period of the CAC regulations. Although it is impossible to say what process change would have been employed in an economics incentives regime, the results do suggest that command-and-control is not without effect in encouraging out-of-the-box thinking in abatement. On the other hand, EPA conducted *ex post* studies of the response to the effluent guidelines in a few industry subcategories and found much less use of process changes than anticipated.

Overall, the evidence suggests innovation occurs under both regulatory and economic incentive regimes. The dynamics and pattern of innovation may differ between the two types of policy. It may also vary significantly for other reasons between various countries. Those that are centers of chemical engineering, such as Germany and the United States, are more likely to expect, and get, technological improvements than other countries.

4. *Effectiveness.* Regulatory policies achieve their objectives faster and with greater certainty than incentive policies.

The evidence on the comparative effectiveness of the different instruments is mixed. In both TCE case studies, which were dominated by command-and-control approaches, there is evidence that substantial emissions reductions were achieved in a short time. The dramatic character of the prohibition in Sweden appears to have speeded up research into alternatives and likely benefited users (and regulators) even outside Sweden. However, the Swedish ban could be considered unsuccessful if judged strictly as a ban, since use has not stopped entirely, even after 10 years. In a direct comparison with the high tax rate on chlorinated solvents in neighboring Norway, Sterner found the economic incentive policy more cost-effective and at least as effective. Sterner also suggests that a lower but broader tax on many chlorinated solvents, as was used successfully in Denmark, might also have reduced toxic exposure, not quite as effectively as the existing Swedish ban but at much lower cost. In the United States, Loh and Morgenstern report that the observed emissions reductions—roughly the same percentage as achieved in Sweden—were based almost entirely on regulatory mechanisms. Apparently, the early reduction program developed by EPA was not sufficiently attractive to industry to encourage widespread participation.

In the case of leaded gasoline in Europe, the authors argue that a tax differential alone—without also the required use of catalytic converters and a limit on the lead content of fuels—would have slowed the phaseout significantly. Similarly, the authors of the German SO_2 case highlight the rapid reductions achieved under the command-and-control system. Specifically, they note that large emissions reductions were required within five years for plants installing scrubbers and within two years for plants that switched fuels—clearly a faster pace than in the market-based U.S. system. They also note that at least in the early years of the program, there was a good deal of overcompliance, as firms tended to operate with significant safety margins in order to avoid both the penalties and the adverse publicity associated with violations.[8] Although there has been considerable overcompliance in the U.S. system as well, the fact that the excess reductions could be used at a later date has different environmental implications. In Germany the excess reductions could never be used, whereas in the United States they could be added to the pollution load at a later time.

Interestingly, at least two cases point to significant environmental gains from both approaches, albeit with some undesirable side effects over the longer term. The prescriptive approaches adopted to reduce U.S. NO_x emissions led to emissions reductions of about 14% from coal-fired power plants, if measured under somewhat fictional assumptions that they did not affect other operational and investment decisions. However, it is widely believed that the new source performance standards provided an incentive to extend the life of existing plants and thereby avoid costs associated with pollution control at new plants. This perverse incentive is likely to have undermined the accomplishments of the program to some degree. At the same time, the authors argue that in the Title IV NO_x program, the absence of an aggregate cap may be responsible for the less-than-anticipated reductions in emissions from coal-fired boilers during the 1990s. In the U.S. water pollution case, the authors point out the difficulty of making highly prescriptive environmental regulations as stringent in practice as they appeared at the time they were debated in Congress. Thus, they note that for many industries covered by the effluent guidelines program, the command-and-control regulations regarding the best available technology (BAT) were mired in so many details that they were delayed for more than a decade past their statutory deadlines.

Finally, in the Swedish TCE case study, Sterner suggests that unsuccessful regulatory policies may have broader implications for the credibility of environmental institutions. In response to the regulation, Swedish users of TCE were able to act in a concerted manner to persuade the public and the environmental authorities that complete implementation of the ban would cause undue harm. The authorities allowed numerous waivers and exceptions, which may have undermined the authority of the Swedish environmental agency, emboldened firms to oppose other regulations, and demoralized those firms that did comply with the regulation, perhaps giving them reason to think twice about cooperating with the agency in the future. It is doubtful whether a similar problem would arise with incentive instruments. After all, the firm's cost is capped by an emissions fee or the price of an emissions permit. The fee or permit price, moreover, is known to the regulator, which pretty much eliminates the possibility of bluffing by the firm. Even if it decides not to abate, it can simply pay the fee rather than challenge the authorities. This option may, however, create trust and credibility problems between the authorities and the environmental community.

5. Regulatee burden. Regulated firms are more likely to oppose EI regulations than CAC instruments because they fear that they will face higher costs, despite the greater efficiency of EI instruments.

Experience on both sides of the Atlantic suggests that no government ever put this hypothesis to the test, which in a way is strong support for it. Although recent legislative proposals in the United States and the European Union have called for partial auctioning of allowances, historically permits have been allocated *gratis.* However, with cost-of-service regulation, such as that used in regulating electricity rates, the regulatory burden is generally less under *gratis* allocation.

In Europe, regulatory burdens were reduced in the French and Swedish NO_x cases by returning the collected emissions fees to the industries from which they had been collected. In Sweden the fees were returned on the basis of energy produced. In France the revenues were used to subsidize abatement investments by the firms that paid the fees. In France the burden was low in any case because the tax rate itself was so low—its primary purpose was to give firms an additional incentive to comply with emissions limits.

Hammar and Löfgren argue that the use of a tax differential—with a lower tax rate on unleaded gasoline—tended to reduce the overall burden of fuel taxation on the refinery sector. Using a different logic, the authors of the U.S. lead phaseout case also conclude that the regulatory burden of the regulation was reduced by the use of a rate-based program *cum* banking. On both sides of the Atlantic, the refiners were able to pass most of the additional costs forward to consumers.

In the first phase of Title IV of the U.S. SO_2 program, the case study authors argue, the regulatory burden associated with permit costs was not an important factor for either producers or consumers. This is because most of the U.S. electricity sector was regulated according to cost of service, and prices reflected allowance costs at the original cost to the firm. Since allowances were distributed for free, they typically were not reflected in electricity prices. This approach is politically appealing but, as the authors note, may create costs in the form of a misallocation of resources in the general economy.

In the Dutch surface water pollution case, the fees were used to support the construction of treatment facilities; the incentive effects were unanticipated. They were hardly a burden since the contribution to collective treatment replaced much of the firms' private abatement costs. Rather, the fee system served as a device to distribute treatment costs on the basis of the polluter-pays principle. Against this yardstick, no extra direct costs are imposed on industry as a result of the tax.

In most cases in which EI measures were used, explicit efforts were made to recycle tax revenues[9] or otherwise limit the burden on existing sources—for example, by grandfathering allowances. Especially if the firms were able to pass the costs forward to consumers, it is likely that the regulatory burden was significantly reduced, and possibly even completely offset. In contrast, under a CAC system, which generates no revenues, there is no obvious way for government to offset regulatory costs—nor is there the same need for so doing.

6. Administrative burden. Regulatory policies have higher administrative costs.

Although it is clear that implementation of the CAC-oriented effluent guidelines program—one of the major elements of the U.S. water pollution case—imposed high administrative costs on EPA, several other cases carry more mixed messages. In the U.S. lead phaseout, the authors argue, the complexity and flexibility of the program increased the likelihood of both intentional and unintentional violations, especially by smaller refiners and inexperienced fuel blenders. This, in turn, increased the monitoring and enforcement costs of EPA. The authors believe that much of this problem is attributable to the program's reliance on a ratio of lead use to total output, rather than an overall cap on lead usage.[10]

The U.S. SO_2 cap-and-trade program has gained a reputation for low administrative costs, a feature that has made it popular with both EPA and industry. At the same time, the authors of the German SO_2 reduction program argue that there is no evidence that the administrative costs of designing and implementing their CAC-type program were higher than for a comparable EI-based program.

In the case of the U.S. NO_x program, the authors argue that the Ozone Transport Commission's CAC-based Phase I was probably more difficult to develop than the Phase II trading program, but only slightly so. They note that *gratis* pollution allocations present a regulator with rent-seeking behavior on behalf of market participants, thereby forcing the regulator to establish rules for allocation and, subsequently, for verification of allowance claims. In addition, regulators still bear the burden of demonstrating that incentive-based polices are "feasible." In the case of the NO_x SIP Call program, this required the identification of available abatement controls, their applicability to U.S. facilities and coal types, and electricity market modeling.

In Europe, the costs of administering the incentive-based (tax) measures used to regulate NO_x emissions are thought to be relatively low. For example, in Sweden, the environmental agency estimates that central administrative costs are approximately 0.6% of total yearly tax revenues. Monitoring requirements are an order of magnitude higher. In the French case, however, monitoring relies to a large extent on existing regulatory structures for control of standards-based regulation. A fair amount of flexibility was granted to the individual firms so that they could choose whether to use direct measures or apply emissions coefficients set by the regulatory agency.

The authors of the Dutch water pollution study generally support the view that CAC approaches have higher administrative costs, although they note that the permit-granting and enforcement activities associated with the fee program have been substantial. In the same vein, the authors of the U.S. TCE case note that a report by the U.S. Government Accounting Office identified the complexity and cost of establishing facility-specific baselines—a prerequisite to participating in the early reductions program—as an important barrier to the success of that program.

Overall, the evidence on this hypothesis is mixed. Although there is some indication that administrative burdens associated with regulatory rules are higher than for incentive-based rules, there are also a number of counterexamples. The extent of the burden depends on the context and nature of the policy action.

7. *Hotspots and spikes.* The performance of all pollution abatement instruments is seriously compromised for pollutants with highly differentiated spatial or temporal effects, but more so for incentive than for regulatory instruments.

Incentive approaches offer the clearest advantages for controlling pollutants for which location does not matter (such as stratospheric ozone depleters and greenhouse gases) or where location of emissions cannot easily be affected by any policy. For example, though lead has high spatial differentiation, authors

Newell and Rogers point out that environmental hotspots are not a significant concern in the United States because the pollution is created through gasoline consumption, not production, and there is likely little or no relationship between the location of refineries and automobile exhaust across the country.

In principle, a prescriptive approach to regulation could do a better job than incentive-based measures in targeting specific areas or time periods. In both TCE cases, for example, the pollutant is a potential workplace hazard whose effects are localized. On its face, this would tend to support prescriptive regulation in which regulators have the authority to prescribe more stringent controls where necessary to preserve environmental quality. Although an incentive-based approach could also protect specific areas or time periods, this would require a relatively complex design.

In practice, however, the situation is not so clear: incentive-based measures may work to the detriment or to the benefit of any particular area. Evidence presented about the U.S. SO_2 program suggests that emissions trading has (serendipitously) benefited geographic areas that contain a disproportionate number of sensitive ecosystems and has led to aggregate health benefits in addition to those that would have resulted absent trading. However, this has not prevented state environmental authorities and public utility commissioners from trying to interfere with specific emissions trades that would have the effect of increasing emissions in their state. Recently, moreover, some concerns about environmental justice have arisen with respect to the SO_2 trading program, although the evidence thus far presented suggests that the effects are *de minimis*.

Ironically, in the one market case study in which a hotspot issue arose, the problem was that the emissions fee was deeply discounted, not that it was insufficient to achieve environmental quality. Specifically, in the northern Netherlands, a financially distressed industry (potato starch) was for some time allowed to pay much lower emissions fees than other industries, significantly delaying the achievement of acceptable water quality in the region. But this hardly counts as a mark against incentive instruments, since similar exceptions are routinely granted in regulatory regimes.

Ultimately, if hotspot problems do develop in EI regimes, there are potential remedies. The authors of the U.S. NO_x case note that a hybrid approach may be useful. They observe that although the ozone transport region's trading program confers considerable flexibility in achieving abatement requirements beyond the reasonable available control technology, these standards are still in place during the trading season. In effect, there is a limit to the concentration of pollution that can be released from any source, so the potential for emissions hotspots is reduced.

8. Monitoring requirements. The monitoring requirements of EI policies are more demanding than those of CAC policies because they require credible and quantitative emissions estimates, whereas regulatory policies at most require evidence of excess emissions or the absence of abatement technology.

Although only a limited number of cases report information on monitoring requirements, the results do not generally support the notion that incentive-based approaches are more demanding than command-and-control policies. In the case of SO_2, it appears that both the German and the U.S. programs adopted continuous emissions monitors (CEMs), although the U.S. authors say that other techniques, such as coal sampling and engineering formulas, could have been used to estimate SO_2 emissions at less cost and nearly as accurately. CEMs, they argue, were expensive but necessary to achieve political consensus. Similarly, in the U.S. NO_x case, Title IV required CEMs (at least for major sources), so the monitoring requirements were the same under all the programs after 1990. Previously, the new source performance standards did not require CEMs.

What is more, in some cases it appears that sophisticated monitoring can improve the efficiency of economic incentive instruments. A particularly interesting story emerges in the case of European NO_x controls. Here, the authors argue, the high fees made emissions more visible to both managers and regulators. The perceived importance of accuracy in emissions measurement increased with the monetary payments based on these emissions numbers. In fact, one of the principal "discoveries" of the Swedish program was that emissions were very sensitive to small changes in plant operations. Detailed monitoring was the only way plant engineers could determine the effects of small changes in temperature and other combustion conditions on the overall operation and, particularly, the cost-effectiveness of the facility.

The U.S. lead case generally supports the notion that the difficulties of monitoring are not significantly different under the alternative approaches. Specifically, EPA delegated the responsibilities of data collection and assimilation to the refiners themselves, which then reported their figures to the agency. Figures on lead usage were easily checked against sales figures of additive suppliers. Gasoline volume was less easily monitored, however, and more enforcement cases involved misreported output than misreported lead use. In the view of the authors, though it may be true that the marketable permit program required a greater quantity and variety of information than a command-and-control policy would have, the collection of this information was fairly straightforward and inexpensive.

Overall, based on our limited sample, there is no strong and consistent evidence that incentive-based policies pose more onerous monitoring requirements than prescriptive ones. New programs of both types, operating on both sides of the Atlantic, increasingly require similar, high-tech methods for measuring emissions, assuring compliance, and the like. As noted, in at least one case (Swedish NO_x), the stringent monitoring requirements helped firms achieve certain operational efficiencies by fine-tuning the temperature and other combustion conditions in their boilers.

9. Tax interaction effects. Adverse tax interaction effects exist for both types of instruments but are likely to be larger with EI than with CAC policies achieving the same emissions reductions.

The theoretical literature suggests that interaction of environmental regulations with preexisting regulations or taxes causes the social cost of new regulations to be higher than would be measured in partial equilibrium analysis. One important hidden cost stems from the interaction of the program with the preexisting tax system. Any regulation that raises product prices potentially imposes a hidden cost on the economy by lowering the real wages of workers. This can be viewed as a "virtual tax" magnifying the significance of previous taxes, with losses in productivity as a consequence.

Economic instruments allow for more efficient allocation of emissions reductions among regulated firms than prescriptive approaches. However, particularly if these efficiency savings are not great, economic instruments are likely to impose a greater cost through the tax interaction effect. The reason is that they drive up a firm's marginal production costs not only by the abatement costs but also by the cost of the emissions embodied in another unit of output. The corresponding price increase serves to erode further the real wage. This tax interaction effect can be at least partially offset if abatement costs under the EI mechanism are lower than under the CAC, or if the environmental policy raises revenue that can be used to reduce reliance on distortionary taxes, or at least to mitigate the price impact of the regulation.

None of the EI cases in this volume used the revenues to reduce other tax rates. In the U.S. SO_2 and NO_x cap-and-trade programs, permits were grandfathered, and in the former, Butraw and Palmer argue that the difference in the tax distortion may have made the policy almost as costly as the CAC program. In the Dutch water pollution case, the fee revenues were returned to industry to support new investments. In the Swedish NO_x case, revenues were rebated to firms based on generation output. Theoretical analyses have shown that this tax-rebate mechanism is approximately equivalent to a tradable performance standard (Fischer 2001); both encourage abatement but relieve firms of the additional cost, on average, of the emissions embodied in output. Consequently, one would expect a lesser tax interaction with this mechanism. On the other hand, the weaker price increase also sends less of a signal to encourage conservation as a means of reducing emissions, so these allocation mechanisms are still less efficient than optimal revenue recycling. However, in all these cases the authors argue that it would not have been politically acceptable to use the revenues to offset other (distortionary) taxes.

Goulder et al. (1997) investigated the magnitude of the tax interaction effect in the context of the SO_2 program and found that it added 70% to their estimated compliance costs, under the assumption that electricity prices are set in the market rather than by regulators, which is increasingly the case. However, if prices are set by regulators based on the cost of service, then the regulatory burden is much lower because allowances under Title IV were distributed at zero original cost. If the government were to auction the SO_2 allowances and use the revenue to reduce preexisting distortionary taxes, the additional cost falls to 29% of estimated compliance costs.

10. Effects on altruism. Economic incentives encourage the notion that the environment is "just another commodity" and reduce the willingness of firms and citizens to provide environmental public goods voluntarily. Regulatory policies are consistent with a norm that requires every discharger to "do his best" and thus provide a better basis for a change in social and personal attitudes about one's responsibility to the environment.

Among our case studies, we found several examples of voluntary pollution reductions, but they didn't seem connected in any systematic way to the choice of policy instrument. Instead, a firm's willingness to overcomply seemed to depend on its overall situation. During the 1970s and early 1980s, there was little evidence that firms or trade associations were willing to make voluntary reductions; indeed, they fought regulations by whatever means they had at their disposal. Thus, in the U.S. water pollution case, industries fought fiercely against the establishment of the effluent guidelines, and EPA had to litigate virtually every rule. Gradually, regulators and firms began to look for compromises. As relationships improved, some well-known national firms made it a practice to exceed performance requirements of the effluent guidelines by a substantial margin, so that questions about compliance would never arise. Something similar appears to have happened in the TCE cases, where the stringent regulation first provoked an almost "antigreen" reaction among some companies. But then the Swedish firm SKF, after TCE was banned in its home country, decided to phase out TCE in all its plants around the world.

In Germany, the average SO_2 emissions rate (in mg/l) achieved by 1995 was only 38.5% of the emissions standard, evidence of altruistic behavior by utility boilers during the 1980s. After 1995, however, partial deregulation of the German electric power industry made these voluntary reductions seem like an unaffordable luxury, and they have begun to disappear.

Some signs of voluntary behavior appear in the EI cases as well. In the Swedish NO_x case, some (mainly municipal) firms overcomply. In the Dutch water quality case study, Bressers and Lulofs noted that among employees at many firms there was a genuine desire to reduce pollution, and the fees may have reduced the conflicts within business between benefiting shareholders and doing the right thing for the environment. However, a regulatory policy would very likely have had the same effect. Finally, there have been several cases in which firms have used tradable emissions permits to make charitable contributions. For example, in 1997 Niagara Mohawk Power Co. donated 15,000 tons of SO_2 credits to the Environmental Resources Trust, which retired them.[11]

11. Adaptability. Compared with CAC instruments, EI instruments can be changed more quickly and easily in response to changing environmental or economic conditions.

The main difference in adaptability of an instrument appears to stem from the process required for updating. Tax changes generally have to pass parliament (an exception is if the tax law is written to allow for automatic updating—for instance, by indexing to inflation). This may also apply to certain regulatory changes but not to a prescriptive approach tied to the definition of a technology, such as best available control technology standards, which can be updated through an administrative process. In this case, technology vendors have an incentive to introduce technologies that can form the basis for a new definition of the technology standard. Similarly, regulatory bodies or local authorities may in some cases also have the right to change fees that do not go to the treasury.

Yet a review of the actual cases indicates that EI and CAC approaches can be quite similar in their inability to adapt to new information. For example, a well-known flaw of Title IV of the Clean Air Act was that as new information became available about the relative benefits and costs of SO_2 reductions, the cap could not be changed short of an act of Congress.[12] A more prescriptive approach, such as the NO_x provisions of Title IV, shares this attribute.

The French NO_x tax was notably slow to change, since its levels were fixed (too low) for five years. However, the authors argue, an advantage of the French tax was that it allowed for government and regulatory agencies to collect and improve information on emissions levels and abatement actions undertaken by firms in different industry sectors. In this sense, it yielded a distinct advantage compared with the prior regulation. In the Dutch water pollution case, the provincial authorities, which must approve fee increases by the water boards, tend to be more reluctant to grant increases during recessions than during periods of economic health.

An alternative to a firm cap would be one that adjusted in response to new information. Others have suggested similar trigger mechanisms on emissions caps to provide economic relief if costs are greater than expected (Pizer 2002), but as the SO_2 case study authors argue, such an approach might be more politically acceptable if coupled with a mechanism that provided additional environmental improvement when costs were less than expected. A safety valve that relaxes the cap when allowance prices hit a specified level—or lowers the cap when allowance prices dip below a floor—acts like a tax system in this regard by incorporating new information about costs.

Perhaps the most interesting situation involving adaptability can be found in the U.S. effluent guidelines program, which appears to be changing in ways that no one anticipated when it began. Back in 1972, the focus of the program was on the technology-based standards for direct dischargers. In recent years, direct dischargers, while still important in some industries, have gradually become fewer in number and less important in environmental terms. Furthermore, among indirect dischargers it is likely that waste surcharges are having increasingly larger incentive effects as rates are being raised by local POTWs for revenue purposes. Thus, as Harrington notes, this quintessential regulatory program may be gradually evolving into a hybrid program with important incentive elements.

12. Cost revelation. With incentive instruments, it is easier to observe the cost of environmental regulation. Theory tells us that for a firm subject to an emissions fee, the marginal cost is the same as the fee rate; in a tradable permit regime, the marginal cost is the market price of the permits. With regulatory instruments, a firm must abate to a pre-specified quantity; there are no fees or permit prices to which marginal costs can be equated.

To begin, we are reminded by the authors of the Dutch case study that the equating of marginal costs of abatement to the effluent tax rate or to the price of permits is a theoretical result, not an empirical observation. Based on their research, Bressers and Lulofs argue that firms do not generally know what portion of their costs is driven by abatement concerns. However, they also point out that the firms have to make a calculation of how much to abate, just as they have to calculate the quantities of other inputs they use. The choice of abatement level has to be based on something, and it is almost certainly closer to the point equating price and marginal cost than would obtain in a regulatory regime. Overall, the cases provide strong support for the hypothesis.

Clearly, the economic incentive instruments in our sample elicited considerable information about the cost of abatement, but there were also complicating factors in several cases. Probably the most successful case in this respect was the U.S. acid rain program. The cap-and-trade scheme provided a way to observe marginal costs and infer total costs. However, originally this information was not widely disseminated because allowance prices did not have to be reported to EPA. Independent allowance-trading firms have since developed indices to make such information more readily available. Also, the EPA allowance auction can reveal important information about prices, and the first EPA auction in 1993 was particularly important in this regard. Such information is not available with a prescriptive regulation. At the same time, even actual price information can be misleading and require careful interpretation. During the first few years of the program, the allowance price fell to $100 per ton or less, which according to most observers is far below the long-run marginal cost of abatement. Apparently, some utilities made major investments in flue gas desulfurization, creating a glut that caused prices to crash. Since the mid-1990s, they have recovered to $150–175 per ton.

In hybrid regulatory-incentive systems the information revealed by the economic instrument depends on whether that instrument is the driving force behind control actions taken. At $40 per ton, the French NO_x tax is probably too low to have incentive effects (which are provided by the emissions standard in force), but the Swedish NO_x tax is something else entirely. Its rate of $4,000 per ton is almost certainly the binding constraint, so the level of the tax clearly tells us something about marginal abatement costs.

In the U.S. leaded gasoline case, Hahn and Hester estimate from anecdotal evidence that the price of lead removal was under $.01 per gram prior to banking, and from $.02 to $.05 during the banking phase, when standards were becoming increasingly stringent (Hahn and Hester 1989). However, this system was based on lead concentrations, not on total lead in fuel, which meant that some assumptions were required to get to total lead. Had the program been designed as a cap-and-trade scheme in the spirit of the SO_2 trading program, with clearly specified lead allowances rather than the lead averaging scheme, an even clearer market price would likely have emerged.

The one instance where incentive instruments do not reveal the costs is when limits are imposed on the pure price incentives. This is referred to in the economics literature as a corner solution. Consider, for example, the use by several European countries of a tax differential policy to ensure that the price of leaded fuel remained above the price of unleaded. That is, the countries did not want to force the removal of lead from fuels, at least until about 1995, when valve seat recession in older engines ceased to be an issue. Thus, they were actually *seeking* a corner solution. As Hammar and Löfgren report, the tax differential—together with the fact that it was successful—does reveal an upper bound of the cost of removing lead from gasoline. Not surprisingly, that differential was much larger than the imputed cost of lead removal in the United States, where the tradable permit program did elicit cost information from the refinery industry. However, the U.S. lead permit trading program did not reveal the cost of eliminating lead from gasoline because it had switched back to a regulatory program by the time of the final phaseout.

International Trade Effects. We briefly return now to the international trade hypotheses stated in the Overview. These hypotheses are corollaries or applications of the 12 hypotheses stated above:

(a) The complexity and often opaqueness of construction and operating permit requirements, which typically accompany direct regulations, tend to favor domestic industry over foreign-owned firms.

(b) By favoring end-of-pipe treatment and more stringent requirements for new plants, regulatory instruments provide innovation incentives for domestic abatement technology producers, giving them an advantage in world markets.

(c) By imposing greater regulatory burdens on regulated firms, incentives leave those plants more vulnerable to import competition.

Only a few of the cases had international implications, and of those, the issue of instrument choice came up in only a limited and indirect way. Two case studies—industrial water pollution in the Netherlands and SO_2 in Germany—cited policy-inspired gains to domestic producers of abatement technology, the subject of (b) above, but one was a regulatory instrument and one an incentive instrument. In addition, as reported in the Dutch study, Dutch and Belgian authorities imposed very different regulations on their declining cardboard industries in the early 1980s. The Dutch held their industry to the same standard as other industries in the country, while the Belgians, taking into account the industry's parlous state, imposed regulations that were lenient by comparison. Years later, the environmental difference was great, but the relative economic position had barely changed. That is, differences in abatement effort in the two countries had little effect on the overall fortunes of the industry in the two countries. Since different instruments were used, this case would seem to be a refutation of (c) above: the greater burden of the Dutch policy apparently did not adversely affect the industry's prosperity overall. In addition, to some it could also be seen as support for the notion that stricter regulation can adhere to the advantage rather than the disadvantage of industry—the so-called Porter hypothesis.

Except for those fairly minimal instances, the choice of instrument did not seem to have international trade effects, an outcome that probably shouldn't be considered too surprising. After all, several recent reviews have found only limited support for the notion that environmental regulations have had a significant adverse effect on competitiveness (Dean 1992; Jaffee et al. 1995; Levinson 1996). If differences in regulatory stringency do not have trade effects, it is unlikely that differences in instrument will.

Conclusion

Simple and dramatic conclusions, the staple of newspaper headlines, rarely emerge from collections of detailed studies such as those presented in this volume. Yet, at the risk of oversimplifying, we start with the most basic observation of all: based on a dozen cases drawn from Europe and the United States, it appears that environmental regulation is alive and well. Although this comes as no shock to policy experts, it remains surprisingly common to hear complaints, emanating largely from the business community, that environmental regulations are not very effective in achieving results. Further, they argue, environmental agencies routinely underestimate the costs involved.

The case studies in this volume document significant environmental results. Averaged across all the cases, emissions fell by about two-thirds compared with baseline estimates. Although any comparison with an estimated baseline is, by its very nature, hypothetical, the authors' ability to document the credibility of the baseline assumptions as well as the actual emissions reductions supports the basic observation that regulations can achieve environmental results.

The case studies also show that there are no magical solutions. All forms of regulation have pluses and minuses. One of the important findings of our studies is that the administrative burdens associated with both regulatory and incentive instruments are significant: the accomplishments of regulations do not come easily.

Also interesting is the fact that the case study authors were able to find or re-create *ex ante* estimates of expected emissions reductions in all of the U.S. cases and four of the European cases. Comparison of the *ex ante* with *ex post* observations suggests a reasonable degree of accuracy in the estimates. Not surprisingly, the cases in which emissions reductions were greater than expected involved incentive instruments. The cases in which reductions fell short of expectations involved regulatory approaches. This finding, consistent with other literature, suggests that regulators may be unduly pessimistic about the performance of incentive instruments or unduly optimistic about the performance of regulatory approaches, or perhaps both.[13]

The continued growth in popularity of incentive instruments is due in part to the actual and perceived success of existing examples, of which the ones considered in this book are among the most prominent. This growing interest is consistent with the results of our case studies, which we think generally support the continued use of market-based instruments. This can be seen in Table 12-2, which summarizes the above discussion of the hypotheses. In the table we have sorted the hypotheses so that the ones favoring incentive instruments appear first, followed by the ones favoring prescriptive regulation. In each group, there are six hypotheses, and in each three are supported and three are not. This arrangement makes it appear as though our "competition" between incentive and regulatory instruments ended in a dead heat.

However, we would argue that these hypotheses are not all of equal importance. In our view, the most important are efficiency (both static and dynamic), effectiveness, and regulatory burden. Many of the remaining hypotheses—

Table 12-2. *Summary of Outcomes*

Hypotheses favorable to EI instruments	Supported?	Comments
1. Static efficiency. Incentive instruments are more efficient than regulatory instruments.	Yes	If the emission standard is stringent enough, as in the German SO_2 ordinance, then there is no advantage to incentives.
2. Information requirements. Generally, incentive instruments require less information than regulatory instruments to achieve emission reductions cost-effectively.	No	All policies turned out to require much information, although not necessarily for the purpose of achieving cost-effectiveness.
3. Dynamic efficiency. The real advantages of incentive instruments over regulation are only realized over time, because unlike regulatory policies they provide a continual incentive to reduce emissions, thus promoting new technology, and they permit a maximum of flexibility in the means of achieving emission reductions.	Yes	This often shows up not in patentable innovations but in site-specific changes to equipment and operating practices.
6. Administrative burden. Regulatory policies have higher administrative costs. During the pre-implementation phase, greater information is required to prepare emission standards.	No	
11. Adaptability. Compared to incentive instruments, regulatory instruments can be changed more quickly and easily in response to changing environmental or economic conditions.	No	Many primarily regulatory policies show adaptability by adopting incentive instruments.
12. Cost revelation. With incentive instruments, it is easier to observe the cost of environmental regulation.	Yes	

Hypotheses favorable to regulatory instruments		
4. Effectiveness. Regulatory policies achieve their objectives more quickly and with greater certainty than incentive policies.	No	Does not apply at the aggregate level.
5. Regulatory burden. Regulated sources will tend to prefer regulatory instruments to incentive instruments, because of the strong possibility that they have to pay more under incentive even though the social costs may be less.	Yes	The only major incentive policies that have been adopted have overcome this problem by designing instruments to be revenue-neutral (i.e., grandfathered tradable permit systems or recycling of effluent tax revenues).
7. Hotspots and spikes. The performance of all pollution-abatement instruments is seriously compromised for pollutants with highly differentiated spatial or temporal effects, but more so for incentive than for regulatory instruments.	Yes	Incentives can be made local, however, as is illustrated by congestion fees in some cities.
8. Monitoring requirements. The monitoring requirements of incentive policies are more demanding than those of regulatory policies because they require credible and quantitative emission estimates.	No	Monitoring requirements of both instruments have been exacting.
10. Effects on altruism. Economic incentives encourage the notion that the environment is "just another commodity" and reduce the willingness of firms and citizens to provide environmental public goods voluntarily.	No	

monitoring requirements, hotspots, administrative burden, and so forth—are special cases of the more important hypotheses and are of secondary importance. In addition, questions of effectiveness and efficiency were at the core of the controversy over the initial selection of policy instruments in the 1970s and 1980s. As advertised by their proponents, economic incentive instruments do appear to produce cost savings in pollution abatement, as well as a steady stream of innovations that reduce abatement costs. Also, the main concern of opponents of incentive instruments—that they would not work—is not borne out in these case studies. In the cases presented in this book, they worked quite well.

However, the finding of economic efficiency of incentive instruments is tempered by one other strong finding from these comparisons. As discussed in the preceding section, the regulatory burden hypothesis—the idea that polluters prefer CAC regulation because EI instruments entail a tax payment or purchase of permits in addition to abatement expenditures—received strong support. Indeed, for all but one of the incentive instruments examined, the actual or potential revenue raised by incentive instruments has been reimbursed to users, either by explicit tax distributions (as in the Swedish NO_x tax) or by grandfathering of emissions permits. The only exceptions were sulfur in Sweden and the Dutch effluent fees, which were used to finance wastewater treatment facilities. (In fact, that was their intended use; their incentive properties didn't emerge until later.)

Using revenues in this way, of course, means they cannot be used for other purposes, thus short-circuiting one of the chief advantages of economic incentives—that they generate a source of revenue to (potentially) overcome the problems raised by regulation. For example, they could be used to correct a preexisting tax distortion exacerbated by the instrument, or to overcome a hotspot problem by subsidizing additional abatement, or to correct a perceived or actual inequity in their application. In almost all real-world cases, those opportunities have been foreclosed by the need to gain political support by easing the regulatory burden imposed on polluting firms. This apparent inability, in practice, to use the revenues generated by incentive instruments to address hotspots or other regulatory problems may be particularly important vis-à-vis adverse tax interactions. In fact, as Burtraw and Palmer note, the failure to auction off SO_2 emissions permits almost completely nullified the efficiency advantage of the SO_2 trading program over the most plausible regulatory alternative.

The case studies, we hope, have laid to rest some misconceptions and strengthened support for other hypotheses. They also have shown the complexities of the real world—the importance of history and context and the fact that the details of a policy can make a big difference in its impact. We think the cases also have shown the usefulness of policy analysis. Nations can learn from one another and from past successes and failures, but only if the experience is analyzed in a careful and responsible way. This volume is intended to contribute to such analysis.

Notes

1. Initiatives include the National Environmental Policy Act of 1969, the Clean Air Act of 1970, the Federal Water Pollution Control Act Amendments of 1972, the Toxic Substances Control Act of 1976, the Resource Conservation and Recovery Act of 1976, and the Comprehensive Environmental Response, Compensation, and Liability Act of 1980 (Superfund).

2. Executive Order 12291, reinforced by Executive Order 12866.

3. In the United States, measurements of increased blood levels among urban children—where minorities were disproportionately represented—were widely discussed in the press and in policy circles.

4. It is not clear that issues of distribution affecting small *firms* should necessary be labeled equity issues, since owners of small firms may well be rich *persons*. However, it is quite common to do this, and it may help some interest groups further their agendas.

5. The United States focused on the toxic aspects of lead in gasoline, while the Europeans emphasized the air quality issues.

6. See also Sterner (2003, Chapter 12).

7. Note that there is a special form of interaction between the static and dynamic efficiency concepts. For static efficiency it is crucial that the tax level is set at the "right" level. Dynamic efficiency, however, hinges on the *discovery* of initially unknown abatement opportunities, implying that we cannot really know that "correct" value *ex ante.*

8. For the same reason, they argue that it is also not so important that suppliers of abatement equipment provide such strict guarantees of abatement performance. Interestingly, they note that this reported overcompliance was reduced once the liberalization of the energy market took place.

9. An exception is sulfur taxes in Sweden. These, like most environmental taxes, typically go to the treasury. For most of the energy taxes (but not sulfur), there are special reductions, which imply that industry pays much less than households.

10. This particular type of cap created an incentive to increase output, particularly by fuel blenders.

11. Retiring of Emission Credits to Speed Battle on Acid Rain, *The Buffalo News,* August 27, 1997.

12. A related concern is that tradable permits may instill a property right that would be difficult to change. This was forestalled in the design of Title IV by an explicit statement that allowances did not constitute a property right.

13. In six cases, the authors judged the estimates to be reasonably accurate predictions of actual outcomes. In two cases the authors found that actual emissions reductions were larger than predicted *ex ante* (U.S. NO_x and Sweden lead phasedown). In the other two cases they found that actual reductions fell short of the predicted levels (Sweden TCE, U.S. water). See Harrington et al. (2000) for other *ex ante* and *ex post* findings.

References

Dean, J.M. 1992. Trade and the Environment: A Survey of the Literature. In *International Trade and the Environment,* edited by P. Low. Discussion Paper 159. Washington, DC: World Bank, 15-28.

Fischer, C. 2001. Rebating Environmental Policy Revenues: Output-Based Allocations and Tradable Performance Standards. Discussion Paper 01-22. Washington, DC: Resources for the Future.

Goulder, L.H., I.W.H. Parry, and D. Burtraw. 1997. Revenue-Raising vs. Other Approaches to Environmental Protection: The Critical Significance of Pre-Existing Tax Distortions. *RAND Journal of Economics* 28(4): 708–31.

Hahn, R., and G. Hester. 1989. Where Did All the Markets Go? An Analysis of EPA's Emissions Trading Program. *Yale Journal on Regulation* 6(1): 109–53.

Harrington, W., R. Morgenstern, and P. Nelson. 2000. On the Accuracy of Regulatory Cost Estimates. *Journal of Policy Analysis and Management* 19(2): 297–322.

Jaffee, A.B., S.R. Peterson, P.R. Portney, and R.N. Stavins. 1995. Environmental Regulation and the Competitiveness of US Manufacturing: What Does the Evidence Tell Us? *Journal of Economic Literature* XXXIII: 13263.

Levinson, A. 1996. Environmental Regulations and Industry Location: International and Domestic Evidence. In *Fair Trade and Harmonization: Prerequisites for Free Trade?* Vol. I, edited by J. Bhagwati and R. Hudec. Cambridge, MA, and London: MIT Press, 429–58.

Melnick, R.S. 1983. *Regulation and the Courts: The Case of the Clean Air Act.* Washington, DC: Brookings Institution.

Pizer, W.A. 2002. Combining Price and Quantity Controls to Mitigate Global Climate Change. *Journal of Public Economics* 85(3): 409–34.

Sterner, T. 2003. *Policy Instruments for Environmental and Natural Resource Management.* Washington DC: Resources for the Future Press.

Wiener, J.B. 2003. Whose Precaution After All? *Duke Journal of Comparative and International Law* 13: 207–62.

Index

Selected Books from RFF Press

- *Climate Change Economics and Policy: An RFF Anthology*
  MICHAEL A. TOMAN, EDITOR
  Paper, ISBN 1-891853-04-X

- *The Economics of Waste*
  RICHARD C. PORTER
  Cloth, ISBN 1-891853-42-2 Paper, ISBN 1-891853-43-0

- *Improving Regulation: Cases in Environment, Health, and Safety*
  PAUL S. FISCHBECK AND R. SCOTT FARROW, EDITORS
  Cloth, ISBN 1-891853-10-4 Paper, ISBN 1-891853-11-2

- *India and Global Climate Change: Perspectives on Economics and Policy from a Developing Country*
  MICHAEL A. TOMAN, UJJAYANT CHAKRAVORTY, AND SHREEKANT GUPTA, EDITORS
  Cloth, ISBN 1-891853-61-9

- *The Measurement of Environmental and Resource Values: Theory and Methods, Second Edition*
  A. MYRICK FREEMAN III
  Cloth, ISBN 1-891853-63-5 Paper, ISBN 1-891853-62-7

- *Painting the White House Green: Rationalizing Environmental Policy Inside the Executive Office of the President*
  RANDALL LUTTER AND JASON F. SHOGREN, EDITORS
  Cloth, ISBN 1-891853-73-2 Paper, ISBN 1-891853-72-4

- *Policy Instruments for Environmental and Natural Resource Management*
  THOMAS STERNER
  Cloth, ISBN 1-891853-13-9 Paper, ISBN 1-891853-12-0

- *Private Rights in Public Resources: Equity and Property Allocation in Market-Based Environmental Policy*
  LEIGH RAYMOND
  Cloth, ISBN 1-891853-69-4 Paper, ISBN 1-891853-68-6

- *Public Policies for Environmental Protection, Second Edition*
  PAUL R. PORTNEY AND ROBERT N. STAVINS, EDITORS
  Paper, ISBN 1-891853-03-1

- *Regulating from the Inside: Can Environmental Management Systems Achieve Policy Goals?*
  CARY COGLIANESE AND JENNIFER NASH, EDITORS
  Cloth, ISBN 1-891853-40-6 Paper, ISBN 1-891853-41-4

- *The RFF Reader in Environmental and Resource Management*
  WALLACE E. OATES, EDITOR
  Paper, ISBN 0-915707-96-9

- *Technological Change and the Environment*
  ARNULF GRÜBLER, NEBOJSA NAKICENOVIC, AND WILLIAM D. NORDHAUS, EDITORS
  Cloth, ISBN 1-891853-46-5

- *True Warnings and False Alarms: Evaluating Fears about the Health Risks of Technology, 1948–1971*
  ALLAN MAZUR
  Cloth, ISBN 1-891853-55-4 Paper, ISBN 1-891853-56-2

For more information, visit www.rffpress.org